# EATING CLEAN
## FOOD SAFETY
## AND THE CHEMICAL HARVEST

### Selected Readings

Introduction
by
Ralph Nader

Prepared with the Assistance of
Michael Fortun

Center for Study of Responsive Law
1982

ISBN#0-936758-05-8

# Table of Contents

Introduction *by Ralph Nader* .............................................. I
Filthy Food .................................................................. 1
Additive Avalanche .......................................................... 51
The Pollution Stew .......................................................... 135
Defending and Strengthening the Food Safety Laws ..................... 177
Taking Steps Toward Safer Food ....................................... 237

# INTRODUCTION

### For the Safety of the Food Supply "You're the Difference"
*by*
*Ralph Nader*

Everybody eats — about one thousand hours worth a year — but only a few stand guard publicly and full-time over the safety of America's food supply. They number several thousand civil servants and their task is enormous. They are supposed to make the various food safety laws work on your behalf by inspecting, testing, analyzing, detecting, prosecuting, seizing, recalling and otherwise implementing the standards that affect your meat, poultry, fish, fruits, vegetables and dairy products.

Periodically, I hear from these conscientious "guardians of the food supply," people such as meat and poultry inspectors, analytical chemists and health specialists. Never have I known them to be more worried than they are at the present time. On the one side, they witness the Reagan Administration cutting their modest budgets, staff and enforcement resources. They see food industry lobbyists weaving a net of influence over Congress to weaken further the already weak and under-used food safety laws. On the other side, they are overwhelmed by the floodtide of pollutants, animal drug residues, chemical additives and unsanitary conditions which contaminate the lengthy food chain from the fields and ranches to the marketplace and, finally, to your dinner table.

These guardians need your help. And, you deserve their greater effectiveness. The situation calls for a distinctive kind of patriotism — the coming together of American consumers who care enough to learn they can make a difference in removing toxic and filthy materials from food long before it reaches the kitchen. What is at stake is people's health. The mission is clear — to end an epidemic of silent, cumulative violence. It is a violence that can afflict people with a wide range of sickness from recurring stomach disorders to malignant cancers, even to damaging the genetic inheritance of children, grandchildren, and future generations.

But you may say: "What can *I* do? How can one person be heard?" This resource book of selected readings responds to those who wish to know more about the hazards in our food and what can be done about them. It brings a hopeful message. People can organize to support government and community initiatives to insure their food is safer.

Currently, America's food supply is abundant. But, its quality is jeopardized. However, almost no one feels the pinch of these contemporary perils entering the human body. And, it is easy to think that if something doesn't pinch, then it doesn't hurt. Wrong! The chemicals and wastes of an industrial society soiling its own nest are silent marauders that, by and large, do not provoke thresholds of immediate pain. The pain and agony come later. This means that we have to develop an informed and anticipatory strategy to build an adequately protective base of science, technology and law.

There have been quite enough tragedies, crises and studies to warrant serious concern that the emerging evidence of harm may be the tip of the iceberg.

**Item.** After Allied Chemical dumped Kepone, a toxic pesticide, into a tributary of the James River in Virginia, it took some years before the state government discovered the presence of this poison in the river's fish and oysters which were commercially harvested. Kepone-contaminated fish were turning up in fish markets as far north as New York City and Philadelphia. For about five years, ending in late 1980, the state closed off the James River to most types of fishing. But, for years, many people had been ingesting traces of Kepone, unknowingly.

**Item.** In June 1979, a spare electrical transformer at the Pierce Packing Company in Billings, Montana — a hog slaughter and packing plant — was damaged, causing approximately 200 gallons of a coolant containing polychlorinated biphenyls (PCBs) to contaminate meat and bone meal that was then sold to feed manufacturers and farmers. The contaminated feed spread through at least ten states — polluting poultry, eggs, pork products and a variety of processed foods, eventually found in seventeen states.

Several weeks elapsed before the Food and Drug Administration (FDA) tracked the poison to its source. In the meantime, great quantities of animals (800,000 chickens, 4,000 hogs) and food had to be destroyed. An official of the U.S. Department of Agriculture (USDA) testified that "it is entirely possible that an occurrence such as the recent PCB incident could go undetected by our monitoring system, for a long period of time. It is also possible that a single incident of this size could go entirely undetected... we were lucky in picking up this single incident relatively quickly."

**Item.** In 1973, a fire retardant, polybrominated biphenyls (PBB), was accidentally mixed with a feed additive in Michigan and fed to dairy cows. Tragedy piled upon tragedy. Whole farms were contaminated in the soil, the silos, and the equipment. Tens of thousands of cows and other farm animals were sacrificed. Over 1,000 farmers and their family members who consumed meat and dairy products from contaminated animals showed damage to their immunological systems, livers and neurological functions. Tests indicated that about 90% of Michigan residents have PBB in their bodies. No one knows the long-term effects of PBB. It has turned up in embryos and mother's milk and is spreading throughout the environmental cycle.

**Item.** Environmental Protection Agency studies show that in most human mother's milk tested since 1976, there are PCBs.

**Item.** For many years, until 1980, DES, a synthetic estrogen, was administered to cattle to make them put on weight faster. DES is a long known human and animal carcinogen. Some of that DES, along with other animal drug residues, found its way into human beings. The meat industry resisted government attempts to ban DES use most strenuously, which accounted for the years of delay.

**Item.** Infants and young children are more sensitive to contaminants such as PCBs, lead and mercury due to less developed immunological and detoxification systems, different diets and daily food intakes relative to body weight. Still, EPA tolerance levels for pesticides are set for adults, which may not provide comparable protection for the very young.

**Item.** On February 16, 1978, a General Accounting Office official told a Congressional committee that "with few exceptions, neither the USDA nor FDA can locate and remove from the market raw meat and poultry found to contain illegal residues. Most was sold to the public." At the same hearing a USDA official admitted that "there are serious weaknesses in our program." Between the years 1974 and 1976, USDA randomly sampled for animal drugs, pesticides, and general environmental contaminants such as mercury and lead. These sample tests, the latest publicly available, showed worrisome rates of violation: cattle (14.96%); calves (8.84%); swine (15.83%); chickens (5.08%); turkeys (8.46%). To make matters worse, the USDA has no legal authority to quarantine suspected animals.

What a mess! And, in addition, there are food-borne diseases, from which millions of Americans suffer annually. These diseases are rarely traced to filthy, unwholesome and spoiled products at meat, poultry and fish processing plants, although such conditions still prevail and escape the law. Culprits still try and too many of them succeed in marketing dead, dying or diseased animals. There are no mandatory inspection standards at all for fish processing plants. A bill to cover fish, as laws cover poultry and meat, has failed to pass Congress.

The FDA regularly lists its seizure cases (for rodent filth etc.), and it knows that such seizures only scratch the surface of the problem. In 1979, beleaguered USDA inspectors, sometimes having to resist pressure from company managements and their own superiors in the Department, condemned as unfit for human eating over 55 million chickens, more than 100,000 cattle, some 40,000 calves, 26,000 sheep, 243,000 pigs and over 37 million pounds of processed meat. Jim Murphy, a leader of government food inspectors, cited USDA data showing that inspection prevents as many as 6,500 cases of tuberculosis and hundreds of thousands of cases of tapeworm and salmonellosis every year.

Murphy is so upset at what the current USDA leadership is doing to reduce inspection standards and budgets that he has launched a publication called "The Inspector." In an article titled "How Safe is America's Food?", Murphy recounts proposals of Agriculture Secretary John R. Block to alter the system of full-time government inspection at meat plants; weaken the application of sanctions; speed up chicken production lines so that 4,200 chickens per hour can whiz past inspectors; and permit companies to operate under self-inspection systems, euphemistically called "voluntary total quality control."

Murphy asks two good questions: "How many slaughter houses, packing and processing plants can be expected to discard diseased or contaminated meat when the loss means a downturn in profits? Can consumers be assured that the meat and poultry products they buy are safe, wholesome and free of disease under production line speedups and self-inspection systems?"

More insidious and less subject to safety surveillance are intentional additives to food and drink. The chronic pollutants that rest invisibly in the family food basket are even more poorly monitored. Food company officials, unmindful that the burden of proof for safety is on them, make light of these hazards and challenge critics to prove direct cause and effect connections. However, these chemicals do not signal their impacts like the crunch of vehicles colliding with one another. Their signals are more subtle, with effects appearing usually much later. Fortunately, modern science has developed ways to assess which chemicals cause cancer over time or whether they attack the liver, the kidneys, the lungs, or the genes.

Such evidence of mass food contamination warrants neither mass indifference nor mass hysteria. Rather, what is required for a nation of nearly 230 million people are more investigative health programs, not less; more inspections, not less; and adequate resources to enforce the law, not mindless cutting of tiny budgets.

As the history of public health measures in this country reminds us, an ounce of prevention is worth a pound of cure.

In 1905, *The Jungle,* a novel by Upton Sinclair, depicted the filthy, nauseous conditions in Chicago's meat-packing plants. A public uproar resulted in the first federal meat inspection law the following year, In 1967 similar disclosures of United States Department of Agriculture reports on revolting conditions in meat plants all over the country led to another public uproar and the passage of a stronger meat safety law. Then a mandatory poultry inspection law was enacted in 1968. People do react against rodent remains in their pork sausage or frankfurters. But, somehow, it is more difficult for people to react as forcefully to reports of pesticides, animal drugs and industrial wastes in their hamburger, chicken, fish or bowl of fruit salad. A more sensitive psychological understanding of silent, cumulative chemical violence is needed to curtail uninvited contaminants.

Before his retirement from office in 1978, one of this country's greatest Congressmen, John Moss of California, conducted a series of hearings on cancer-causing chemicals in food. His Subcommittee on Oversight and Investigations (of the House Committee on Interstate and Foreign Commerce) found that "pesticides were a ubiquitous source of chemical residues in food. Two hundred seventy-one pesticides were registered for agricultural use in the United States. Forty-one cause cancer, genetic mutations, birth defects and/or reproductive disorders. Another 92 lack safety studies for these seriously injurious effects. Yet, they continue to be sold and widely used. . ."

Representative Moss concluded: "American consumers cannot be sure that the meat, poultry, fruits, and vegetables they buy are not tainted with potentially dangerous pesticide residues. We all have to eat. Because of the nature of chemical contaminants, we are forced to rely on the Federal government to protect us against potentially dangerous chemicals we cannot see, smell, or taste. Our examination leads us to believe that we cannot rely on the Federal government to protect us." (For more detail, see the conclusions of this report excerpted in Section Five.)

Today, the federal government provides even less protection. In fact, its primary activity is to huddle with food company officials to find out how they want to weaken both the regulatory process in the governmental departments and the laws themselves before Congress. This sad state of affairs need not continue. Both corporate executives and politicians know that the situation can change dramatically if the American people bestir themselves. As long as these power-brokers of business and government fear an aroused people, there is hope. And, they most definitely do. That is why they go to such extremes to avoid informing you about the deplorable situation.

When was the last time you heard anyone from the United States Department of Agriculture, the Environmental Protection Agency, or the Food and Drug Administration make a public statement on the poor state of food safety? All they need to do is reach into their own agency's files and ask some questions as Representative Moss and his staff did. But, they have chosen not to do so. Nor do the large food processors and soft drink companies enlist your assistance to fight against contamination, even when it comes from external sources and is not originally their fault. A notable example is polluted drinking water which they use in their food processing activities.

Rather than inform you and ask for your help, these conglomerates devote their vast chemical and advertising resources to lowering their customers' expectations to four appeals: making food "melt in your mouth," artificially tasty, colorful with dye, and easy to prepare. None of these appeals assures nutritious, clean, safe and fairly priced food. None of these appeals reflects the health interests of consumers and their children. On the contrary, these superficial temptations are mostly ways to manipulate buyers into paying more money for less food value with no questions asked. However, people start wanting to ask questions when they learn that a 12-ounce can of Coke or Pepsi contains 9 teaspoons of sugar or that the average person eats more than five pounds of chemical ingredients a year. Some 1,800 chemicals are routinely added to highly processed food for color, flavor, texture and long shelf life.

This chemical barrage has overwhelmed natural tastes for natural flavors. One food advertising executive boasted, "There has been a definite shift in food preference to the taste of processed foods. If you give someone strawberry ice cream made with fresh strawberries, you'd have a totally unacceptable product. People would say, 'I wouldn't eat that artificial stuff.'" What he is saying is that his corporate clients have chemically programmed the tastebuds of millions of people to crave their food substitutes.

Even the alleged convenience of these "convenience foods" is overrated. In her recent book, *Keep It Simple,* Marian Burros was re-emphasizing old knowledge when she presented several dozen menus which take no more than thirty minutes sat home to prepare a meal from scratch which is cheaper, safer, quicker to fix, better and more tasty. The old knowledge became a best-seller in 1981.

Other old knowledge may be even in greater demand. As people find out that overly refined, processed food, some of which is called "junk food," contributes to diseases such as cancer, heart ailments, diabetes and stroke, they can become downright old-fashioned, returning to unprocessed foods. For greater consumption of sugary, fatty, salty foods and refined grains imposes too heavy a penalty on both health and pocketbook. Moreover, junk food diets can aggravate obesity, dental caries, gum disease and constipation.

There are, of course natural substances in food (for example, certain fungal toxins that can contaminate peanuts and grains) which can be harmful. Some persons are particularly allergic to some natural foods. But such realities cannot be used by industry officials either to excuse their unconscionable proliferation of dangerous, risky and useless substances that are present in the food chain or to avoid their reasonable duty to inspect for them. It is not only food companies such as Campbell Soup, Coca-Cola, Frito-Lay, General Foods, General Mills, ITT Continental Baking, Kellogg, Nestle, and Ralston Purina which bear responsibility. Also involved are companies such as Allied Chemical, DuPont, General Motors, U.S. Steel, International Paper and many other firms where pollution of the air, water and soil adversely affect the environment in which food grows.

Powerful as these corporations and chemical manipulators are, they are no match for the sovereignty of an active citizenry. The tools are available but they need to be put in place. Modern scientific instruments and techniques can detect the presence of hazardous chemicals in food. Our country surely can devote the resources to accomplish the necessary tasks. The laws can be strengthened and fairly implemented by a government made to represent its people.

The readings in this book are meant to support your determination that the food chain can be cleaned up significantly. You start with public opinion on your side. Polls regularly report that the American people want stronger food safety regulation. A "can-do" attitude and an enthusiasm for working with other citizens in a common quest for "eating clean" are prescriptions for success. Just being part of the discovery and change process will sharpen your consumer skills (as you buy food for the family) and build your sense of civic strategy. That, at least, has been the experience of people who have become involved. Most important, organized citizen activism for the public health can set an enduring example for children.

Together with other Americans you can defeat attempts to dilute the food safety laws and programs. Sections Four and Five provide connective leads which you can enlarge upon. Often it takes just a few like-minded friends in your community to make your voice heard. The consumer groups, whose addresses are listed in Section Five, are ready to help you, especially the Center for Science in the Public Interest and the Community Nutrition Institute, in Washington, D.C.

Once you decide to make a difference, you'll start a momentum that will bring results — the kind that will be good for consumers. ■

# FILTHY FOOD

## Section One

**FILTHY FOOD: The Never Ending Struggle for Sanitation**

Without non-stop care, meat and poultry plants tend to attract numerous uninvited guests — certain bacteria, insects, rodents and their remains, paint flaking into the processing areas, disease, dirty water and just plain accumulations of filth. Periodically, United States Department of Agriculture reports find their way into public print and the awful conditions described in many plants turn the stomachs of many Americans. Less frequently, a Congressional committee will look into the "dirty meat problem" and demand better inspections, stronger enforcement and improved standards. Mostly, however, the battle for sanitation is decided daily by government inspectors, many of whom are conscientious, but unfairly pressured, and some of whom are indifferent or on the take.

Section One of this collection of readings is designed to introduce you to the politics of meat and the situation surrounding the inspection of meat plants. Two books are excerpted. *Sowing the Wind,* by Harrison Wellford, describes the efforts of the consumer movement to strengthen the meat inspection law in the late Sixties. *Prime Rip,* by Wayne Swanson and George Schultz, reports the powerplays of the meat lobby against their perceived nemesis, Assistant Secretary of Agriculture Carol Foreman, during the Carter Administration. Both selections make for engrossing reading about the continuing, even nagging persistence of these health hazards to an unalerted public. A digest from a recent General Accounting Office (an investigative arm of Congress) report illustrates this point. Namely, that the job of assuring sanitary conditions and wholesome meat and poultry remains an urgent one. Note the recommendations which GAO made to the Department of Agriculture.

Next comes an article from a government monthly publication, *FDA Consumer,* entitled "Gumshoeing the Food, Drug and Cosmetic Act at the Grass Roots Level." Here is a picture of how the Food and Drug Administration's [FDA] inspectors do their work. Another article describes what happens when the FDA finds food that is unfit for human eating, and includes a list, for your reference, of 21 Food & Drug Agency offices and 128 resident FDA inspection posts around the country. You may wish to note the office address nearest you so you can make ready use of their services. They work for you, after all.

Lastly, there is included a concise FDA consumer memo called "Food Poisoning: the 'Infamous Four'," with useful advice about how you can reduce the risk of food-borne illness by following careful kitchen practices during the preparation and cooking of meals. The memo also describes the origins and symptoms of the four most common sources of food poisoning.

# I

# The Politics of Meat

There is no more visceral consumer issue than contaminated meat. The consumer movement, to a significant degree, was founded and later resurrected on complaints about rotting and diseased meat. Americans are emotional about meat, perhaps because they consume so much of it (nearly 116 pounds of beef and veal and fifty pounds of poultry per capita each year), and because they feel peculiarly defenseless against its adulteration and contamination. At least since the U.S. Army served rotten canned meat to American soldiers during the Spanish-American War, the threat of unwholesome meat, whether real or fancied, has provoked politically potent outpourings of grass-roots protest. In 1970, for example, when word leaked out that the Department of Agriculture might permit the sale of chickens with cancerous tumors,* the President's consumer advisor, Mrs. Virginia Knauer, was deluged with angry letters which, for pure vitriol, surpassed comments received on any other consumer issue. (Messages from the White House to USDA, we are told, were only slightly more polite.) Because of such incidents, the head of federal meat and poultry inspection, the nation's chief guardian of meat purity, sits in one of the hottest and least secure seats in government.

With the exception of federal controls on banks and railroads, meat inspection is the nation's oldest regulatory system. As early as 1865, a federal law was passed prohibiting the importation of diseased cattle and swine. At its best, meat inspection is a classic regulatory

* For a full discussion of the chicken cancer affair, see Chapter 4.

function, protecting the ethical businessmen from being undersold by less scrupulous ones and protecting the citizen where he cannot protect himself.

The ingenuity of food chemistry and processing technology long ago overwhelmed the consumer's natural detection devices—seeing, smelling, tasting—which might have protected him from bad meat. Seasoning agents, preservatives, and coloring agents can now serve effectively as cosmetics to mask the true condition of meat products.[1] The use of cheap fillers and additives such as water, cereal, and fat, unless carefully controlled, gives the consumer less and less protein for his dollar. A case in point is the hotdog, the fat content of which increased from 19 percent to 33 percent between 1937 and 1969.

The total effect of unwholesome meat on human sickness can only be estimated. National health surveys estimate that five to ten million cases of acute intestinal illness, many of them meat related, occur annually in the United States, but most go unreported in official records.[2] Meat animals harbor a number of diseases potentially harmful to man. Trichinosis and hog cholera in pork, brucellosis in beef, and staphylococcus and salmonella bacteria in processed and fresh meats may be directly harmful, while animal cancer—103,000 cattle carcasses were held back by federal inspectors in 1969 for removal of carcinomas and cancer eye—poses potential long-term risks according to some scientists.

Poultry is an even more fertile breeding ground for disease organisms affecting man. Twenty-six diseases are known to occur in both man and fowl—the most serious of which are salmonellosis, psittacosis, and Newcastle Disease.[3] Meat inspection is also necessary to protect the consumer from hidden chemical contaminants such as mercury, PCB (polychlorinated biphenyls), stilbestrol, and other synthetic hormones, antibiotics, and chlorinated hydrocarbons such as dieldrin and DDT. With the widespread and often careless use of these chemicals, it has become increasingly difficult to keep them out of the food supply.

The ease with which meat can become adulterated and contaminated, coupled with the extreme competitive pressures in the industry, makes it a risky candidate

From the book, Sowing the Wind, by Harrison Wellford, copyright 1971 by Center for Study of Responsive Law, published by Bantam Books, Inc., 666 Fifth Ave., New York, NY 10019.

for self-regulation. Profits are among the lowest of any major industry.* As a result, the industry tends to be subject to Gresham's law, with bad meat driving the good out of circulation in the marketplace. While the consumer's need for something more than corporate ethics to protect him from unsafe meat now seems obvious, it was not always so. In the nineteenth century, inspection was limited to meat slaughtered for export. The meat packers themselves initiated the legislation to protect the image of American meat in world markets,[4] but they strenuously opposed extending the same protection to the American people. Albert Beveridge, the Republican Senator from Indiana, who was an early consumer champion, gave this reaction to the primitive state of meat inspection at the turn of the century:

I looked at our own meat inspection laws, and was horrified to find that, while our laws demanded careful inspection of meats intended for foreign consumption (simply because other countries would not otherwise take them), there was practically none at all for meat sold our own people.[5]

The first law requiring mandatory inspection of domestic meat came about, as did the Wholesome Meat Act sixty years later, through the efforts of a few crusaders who skillfully converted information to action while capitalizing on a national mood for reform. In 1906, Beveridge became an active supporter of meat reform after reading Upton Sinclair's sensational novel, *The Jungle*. While Sinclair conceived his book as an attack on capitalism per se, its larger message was lost in the public indignation provoked by his lurid description of filthy conditions in the Chicago stockyards:

There would be meat stored in great piles in rooms, and the water from leaky roofs would drip over it, and thousands of rats would race about on it. It was too dark in these storage places to see well, but a man could run his hand over these piles of meat and sweep off handfuls of the dried dung of rats. These rats were nuisances, and the packers would put poisoned bread out for them. They would die, and then rats, bread, and meat would go into the hoppers together. This is no fairy story and no joke. The meat would be shoveled into carts, and the man who did the shoveling would not trouble to lift out a rat even when he saw one. There were things that went into the sausage in comparison with which a poisoned rat was a tidbit.

* * * *

There were open vats near the level of the floor; their peculiar trouble was that they [men] fell into the vats; and when they were fished out, there was never enough of them left to be worth exhibiting; sometimes they would be overlooked for days till all but the bones of them had gone out to the world as . . . "pure leaf lard."

Reading these passages, it comes as no surprise that meat consumption in the United States fell by 25 percent in the year following publication of *The Jungle*.[6]

Public opinion, aroused by Upton Sinclair and other muckraking journalists, overcame the opposition of the House Agriculture Committee, which tried to bottle up the meat reform bill. The crucial factor was the active support of President Teddy Roosevelt who had read Sinclair's book in manuscript and secretly commissioned a study of the Chicago meat packers which confirmed the scenes in *The Jungle*. In June, 1906, Roosevelt signed into law the Meat Inspection Act which, he assured the public, would "insure wholesomeness from the hoof to the can."

The meat act, which remained virtually unchanged for the next sixty years, is regarded by historians as a critical achievement of the reformist spirit, which, leavened with suspicion of big business and hostility to corruption became known as Progressivism. With the Pure Food and Drug Act passed the same year, the meat act was the first consumer protection measure instigated by grass-roots consumer protest. Unfortunately the consumers tended to view the act as a symbol that bad meat was no longer a problem and ceased their vigilance after the act was passed. In the ensuing decade, the public continued to be victimized by rotting, diseased, and contaminated meat.

The failures of the original meat act were due in part to weak enforcement and to inspectors who conspired

---

* In 1969, the meat packers had a 1 percent return on sales and a 9.6 percent return on net worth. By comparison, the drug industry had a 9.6 percent return on sales and 19.9 percent on net worth.

with large meat packers to evade its standards, but more important, there was a major loophole in the law. The early legislation applied only to meat sold in interstate commerce. Meat packers and processors who sold their meat within the confines of a single state were exempt from federal inspection. As late as 1967, nearly 15 percent of the meat slaughtered in the United States and 25 percent of the processed meat were not inspected according to federal standards. In many cases it was not inspected at all.

The states were loath to assume responsibility for this meat. In 1967, twenty-two states did not require mandatory inspection of livestock before and after slaughter; and eight states had no meat inspection at all.[7] The danger to the consumer was especially great in fourteen states, where nonfederally inspected meat accounted for over 40 percent of all meat slaughtered. A 1963 USDA survey of state plants, kept secret until 1967, revealed deplorable conditions which would have shocked even Upton Sinclair. In Delaware, the survey records:

In addition to the very grave and urgent problem posed by the distribution of food derived from diseased animals, the attached report details extremely bad and revolting dirty food handling methods without any regard for rudimentary sanitation. Rodents and insects, in fact any vermin, had free access to stored meats and meat products ingredients. Hand washing lavatories were absent or inadequate. Dirty meats contaminated by animal hair, the contents of the animal's digestive tract, sawdust, flies, rodents and filthy hands, and the tools and clothing of food handlers were finely ground and mixed with seasonings and preservatives. These mixtures are distributed as ground meat products, frankfurters, sausages and bologna. Due to the comminuting process and seasoning of these products, most of the adulteration could not be detected by the consumer.[8]

A second investigation in 1967 confirmed this abysmal picture. A USDA investigator in Norfolk, Virginia, reported: "I found abscessed beef and livers, abscessed pork liver, parasitic livers mixed with the edible product. The owner was with me at the time. On being questioned on the mixture, he said, "These will be rechecked later."[9] At a North Carolina state inspected packinghouse, an observer reported instances of "snuff spit on the floor, sausage meat fallen on the same floor which was then picked up and shoved into the stuffer."[10] These conditions, described by veteran USDA inspectors, were repeated with revolting consistency in many of the intrastate plants surveyed by USDA.

The surveys clearly showed that Sinclair's work remained unfinished. In 1967, a tiny coalition of reformers dedicated themselves to cleaning up the state inspected plants. In doing so, they revitalized the consumer movement, and gave it a broader base than ever before.

The pooling of forces by labor, public interest lawyers, consumer oriented Congressmen and journalists, their imaginative use of the media to take the message of reform directly to the people, their readiness to use proper names in describing the abuses of Swift, Armour, and other national companies, became a model for public interest advocacy in the sixties. And yet their success, which expanded the sense of the possible for a new generation of "muckrakers," was far more difficult to preserve than it was to win. The campaign for passage of the Wholesome Meat Act in 1967 and its subsequent fate in the hands of the meat inspection bureaucracy in the Department of Agriculture, give revealing insights into the strengths and weaknesses of the consumer movement.

*The First Skirmishes*

The USDA state surveys were the catalyst for meat reform in 1967. The meat packers learned that USDA inspectors were reporting filthy and tainted meat from state plants as early as 1963, and they began to prepare their defenses. Their strategy was to disperse the forces of consumer reaction by creating state meat packing associations which would lobby aggressively in state legislatures for state inspection programs and thus escape tougher federal scrutiny. Pressured by the various trade groups of the meat industry and by the State Commissioners of Agriculture, USDA kept its surveys secret for four years.

## The Politics of Meat

Whenever questions were raised about the adequacy of inspection in state plants, the trade groups scrupulously avoided any description of actual conditions. But there were occasional glimpses of the public's defenselessness. Aled P. Davies, Vice President of the American Meat Institute and the unofficial leader of the meat lobbyists in Washington, inadvertently dramatized the consumer's plight in a 1967 House committee hearing. Under forceful questioning by Representative Tom Foley (D.-Wash.), he suggested that consumers worried about uninspected meat should carry their hamburger with them on trips across state lines:

FOLEY: "Can you tell me a good way to determine what is interstate and what is intrastate in a hamburger?"
DAVIES: "Ask to see the package from where it came."
FOLEY: "But you are a great deal more skilled in knowing what you are doing than most citizens."
DAVIES: "I can smell."[11]

Mr. Davies in fact was underestimating the skill of his constituents in the meat industry. Intrastate packers using antibiotics, preservatives and seasoning agents, and other additives banned under federal law had more than enough chemical ingenuity to mask the odor of decaying meat.

The execrable conditions in state inspected plants undermined the standards of the entire meat industry. These plants provided an outlet for the sale of what the industry called "4-D meat" (for dead, dying, diseased, and disabled) which could not pass federal inspection. It was not immediately clear why large meat packers opposed upgrading the state plants, for most of their plants were already federally inspected.[12] The answer was relatively simple: the competitive advantage which 4-D meat enjoyed over more wholesome products attracted not only marginal operators, but also some of the largest meat packing firms in the country. The Department of Agriculture surveys included reports on Armour, Swift, and Wilson and Company, which operated nonfederally inspected intrastate processing plants to compete in local markets where, as Senator Mondale reported, "out of reach of the federal government" they could reap profits "by passing off sick meat to consumers and by using additives not permitted under Federal regulations."[13] In their defense, the large packers claimed they were forced into this practice because their federal plants faced serious competition from unregulated establishments. The fact that major packers whose names were household words were implicated in the deliberate sale of unwholesome products undercut the traditional defenses of the meat lobby. These practices could no longer be dismissed as the work of small fly-by-night operators.

The conditions in the state plants were still secret when Congressman Neal Smith (D.-Iowa) first began to call attention to Upton Sinclair's unfinished business in 1960. As a farmer who used to attend cattle sales as a young man, he discovered that the same people always seemed to step forward to buy diseased cattle when they appeared on the block. He investigated and found that they were buyers for packing plants which did not come under the federal meat inspection law. Concerned, he introduced legislation to eliminate these practices in 1960—and reintroduced it each year for seven years afterwards. The House Agriculture Committee just as persistently refused to hold hearings on the bill.

Smith's dedication was shackled by his preference for working with the "system" of the House. As one observer put it:

Neal kept doing favors for the boys in the club [the House Agriculture Committee], and hoped they would give him hearings in return. But they were just stringing him along.[14]

His campaign was stymied until 1967, when Congressman Harold Cooley (D.-N.C.), a rigid foe of consumer legislation, stepped down as chairman of the full committee after losing his seat in the mid-term elections. Robert Poage (D.-Tex.), a slightly more flexible man with some Populist leanings, took Cooley's place, but the key change was in the chairmanship of the Subcommittee on Livestock and Grains, which had jurisdiction over meat inspection. This post went to

*The Politics of Meat*

Graham Purcell (D.-Tex.), a responsible legislator who had both Poage's and Neal Smith's respect.

Smith now saw his chance to break the logjam on meat reform and he took it. He was ably assisted by Ed Mezvinsky, a young lawyer from Iowa, who joined Smith as his legislative assistant in 1967. Alert and quietly aggressive, Mezvinsky used the information Smith had gathered and went outside the committee system to publicize the need for meat reform. One of the men he informed was his occasional handball partner, Nick Kotz, a reporter for the *Des Moines Register*. Another was Ralph Nader, fresh from his automobile safety campaign, who had been looking into poultry inspection at the time.

On February 16, 1967, President Johnson gave a mild blessing for a meat bill in his consumer message to Congress. Consumer protection was a good consensus issue, and Johnson was looking for new approaches. After a blitz of major social legislation, the Great Society had bogged down. Consumer issues carried the promise of new constituencies. However, according to Joseph Califano, Johnson's top domestic advisor, the White House staff was not really persuaded that the time was right for strong consumer legislation. In fact, meat reform had such low priority in the White House that Secretary of Agriculture Orville Freeman, whose Department drafted the Administration's bill, did not even have to check it out with the White House staff.[15]

The political stakes shifted in the summer of 1967 when Ed Mezvinsky's lobbying began to pay off. At a luncheon on July 16, arranged by Mezvinsky, Nader and Kotz persuaded Representative Thomas Foley (D.-Wash.), a member of the vital House Agriculture Committee, to sponsor Smith's reforms in the committee. Nader and Kotz provided the opening shots for the meat fight with articles in the *New Republic* and the *Des Moines Register* respectively. Checking up on allusions to conditions in state plants which he noticed in speeches by several USDA officials, Nader obtained alarming data on unsanitary practices in state inspected plants, although he had not yet seen the actual reports. His article of July 15 focused attention on the state surveys and compared them to conditions which had led to passage of the Meat Inspection Act in 1906:

It would be misleading to compare such intrastate operations today with those conditions prevailing at the turn of the century: As far as impact on human health is concerned, the likelihood is that the current situation is worse. The foul spectacle of packinghouses in the earlier period has given way to more tolerable working conditions, but the callous misuse of new technology and processes has enabled today's meat handlers to achieve marketing levels beyond the dreams of their predecessors' avarice. It took some doing to cover up meat from tubercular cows, lump-jawed steers, and scabby pigs in the old days. Now the wonders of chemistry and quick-freezing techniques provide the cosmetics of camouflaging the products and deceiving the eyes, nostrils, and taste buds of the consumer. It takes specialists to detect the deception. What is more, these chemicals themselves introduce new and complicated hazards unheard of sixty years ago.

Simultaneously, Nick Kotz, acting on tips from USDA inspectors, was conducting his own investigation. On July 14, someone in USDA leaked to Kotz the Department's confidential reports on intrastate plants in several midwestern states. Kotz and Nader then went to USDA together to demand release of the rest of the state surveys. At first, Rodney Leonard, Administrator of the Consumer and Marketing Service, denied their existence.* An inquiry by Senator Mondale (D.-Minn.) got the same reply. Eventually, Kotz and Nader did receive a copy of the state reports from USDA but the Department still refused to make them public officially.

During this period, a meat reform bill was taking shape in the House. The parent bill, H.R. 12144, initially supported by the Administration, was sponsored by Congressman Graham Purcell and drafted largely by his assistant, John Rainbolt. While an improvement on earlier drafts proposed by USDA, the Purcell bill still did not require any action from the states. Instead, it attempted to cajole the states into reform by promising to pay 50 percent of the costs if

---

* Leonard now acknowledges that political and industry pressures dulled his sensitivity to consumer problems in 1967. After leaving the Department, he became a leading consumer advocate on food issues.

they decided to accept federal standards. Consumer and labor groups backed a much stronger bill, H.R. 12145, known as the Smith-Foley Amendment, which promised to make federal inspection mandatory in state plants grossing $250,000 a year.*

Opposition to the bills was led by the three major meat packing associations: NIMPA (National Independent Meat Packers Association), the AMI (American Meat Institute), and the WMPA (Western Meat Packers Association). At the state level, the State Commissioners of Agriculture represented by NASDA (National Association of State Department of Agriculture) opposed the bills on the grounds that the federal government should not interfere in state affairs.

At the federal level, the Department of Agriculture was very ambivalent about its own stake in the struggle. Although USDA had had damning information on the state plants since 1963, it gave only a summary report to the House Appropriations Committee and withheld details about specific conditions in specific plants. In a letter to the House Committee on Agriculture on July 20, 1967, Nader described the costs of this policy of secrecy:

Unfortunately, the Department lost an important opportunity to give momentum to its findings because it did not release the state-by-state backup studies which would have given specific substance to the general conclusion. Numerous, concerned employees of the Meat Inspection Service surveyed the respective states where they were stationed. By and large, these state reports are the product of sensitive, observant public servants who wanted a sickening situation remedied. Their high superiors in Washington, however, placed other considerations above prompt and full disclosure.

USDA's reluctance to release the reports was in part a result of pressure from Jamie Whitten, chairman of the House Appropriations Subcommittee on Agriculture,

* Many long strategy sessions by the consumer coalition preceded the introduction of this bill. According to Richard Falknor, an assistant to Congressman Foley and a key figure in these discussions, some suggested that FDA, not USDA, be put in charge of the meat program and that failures in the federal system as well as the states be exposed. Ultimately, the decision was made to keep the strategy simple: emphasize the virtues of federal meat inspection and push it as a model for the states. As it turned out, this simplicity of approach was vital to the success of the media campaign.

and Harold Cooley, when he chaired the House Agriculture Committee. Both were afraid of embarrassing state programs. USDA was also motivated by the conflict of functions which keeps a strong consumer orientation from developing in the Department: it is charged with promoting the *sale* of meat as well as conducting its *inspection*. While in fact strong meat inspection promotes the sale of beef in the long run, most meat packers had a shorter-range view and saw the surveys only as a potentially costly embarrassment. Their view was effective, for in 1967 as today, the meat packers and the cattlemen are much more formidable adversaries for USDA officials than the unorganized consumer.

For the most part, the Department played the classic bureaucratic role of the "man in the middle," reeling from pressure on all sides. From a narrow organizational viewpoint, the federal meat inspectors might have been expected to see the meat bill as an opportunity to do a little empire building. This did not prove to be the case. Jurisdiction over intrastate plants would add to their bureaucratic turf, to be sure, but this consideration was outweighed by others. First, even if the consumer coalition managed to pass a strong bill over the opposition of the farm committees, the purse strings for the new program would be securely held in the committees' unsympathetic hands. The inspectors anticipated a nightmare where vast, new responsibilities over thousands of small plants would be thrust upon them with no guarantee of adequate funds to do the job. And it was the meat bureaucrats, not the consumer champions, who would have to defend the program year after year behind the closed doors on Capitol Hill.

There was also a factor of convenience: meat inspection under the best conditions is a rough and dirty job, with the inspector under constant pressure from the plant managers. Few wanted assignments to the smaller and dirtier state plants, where managers were unaccustomed to the discipline of federal standards and where at least in some states political manipulation was a way of life.

They were uneasy for another reason as well. No

one was more aware than the federal inspectors themselves that filthy, adulterated meat, while more common in state plants, was also present in some federal establishments. If they assumed responsibility for the state plants, they could expect that the consumer investigators would soon begin to look at inadequacies in the federal plants as well. The inspectors ran the same risk, however, if the debate on Capitol Hill was prolonged. Public exposure of conditions in the state plants therefore made them very nervous. As one retired Consumer and Marketing Service (C&MS) official said: "We read Nick's [Kotz's] columns with a grin and a prayer." In retrospect, Ralph Nader described the attitude of USDA's meat division toward the Wholesome Meat Act as "a ball of putty pulled this way and that by each new pressure."

Secretary of Agriculture Orville Freeman was less ambivalent. In his ceaseless effort to give his department a low profile with vested interests and thereby spare his President controversy, he was not about to tackle the whole agribusiness establishment, from the meat packers through the State Commissioners of Agriculture to the farm committees, for the sake of a consumer bill. He actively opposed mandatory meat inspection, even after the White House was prepared to yield. Moreover, according to one observer on the White House staff Freeman—and many of the career bureaucrats for that matter—felt that Congress was being steamrollered by the hysteria of a handful of militants: "Orville just didn't think there was a problem,"* said one former aide.

Freeman's main ally was the National Association of State Departments of Agriculture. NASDA had strong organizational and political reasons for opposing mandatory inspection. Aled Davies, of the American Meat Institute, testified before the House Agricultural Subcommittee on July 17, 1966, that "representatives of the State Departments of Agriculture are in a better position than we to comment on the adequacy of their meat inspection programs," but that was just the problem. Many State Commissioners of Agriculture ran their inspection programs like private duchies where political patronage, not professional qualifications, determined an inspector's chances of advancement.* In many states, the commissioners are statewide elected officials supported by an agribusiness constituency which makes them independent of their governors. According to Congressman Neal Smith and other observers, the slaughterers and processors who had a financial interest in weak meat inspection at the state level frequently made campaign contributions to the commissioners in some states. In Georgia, in particular, meat inspection reform became a hostage of local politics.

One of the most vocal opponents of federal inspection was Phil Campbell, the State Commissioner of Agriculture for Georgia. (President Nixon later appointed Campbell Under Secretary of Agriculture in 1969 and he is now, according to many observers, the dominant influence on meat policy in USDA.) For more than a decade, Campbell had been a powerful force in Georgia politics and was widely regarded as a future gubernatorial candidate. Campbell objected to federal interference and charged USDA with failing to help state inspectors improve their own systems. He vigorously defended his state inspectors against charges of the USDA state surveys. In 1967, he charged that USDA had threatened him with retaliation if he testified in defense of the state inspectors before the Senate Subcommittee holding hearings on meat. In a telegram to Secretary of Agriculture Orville Freeman after the hearings, he stated:

Many times prior to my testifying before the Senate Subcommittee, I and many other Commissioners of Agriculture of the United States received long distance telephone calls from personnel in USDA threatening us with a smear campaign and stating that things would get very nasty if we testified. Within two days after my testimony on November

---

* Freeman played much the same passive role in the hunger controversy one year later. He was incredulous that severe malnutrition could really exist in the Mississippi Delta, until the hungry poor camped on the doorstep of his Department during the Poor People's March in 1968.

* California, New York, and several other states with strong meat laws and high professional standards for their inspectors were exceptions which did not fit this picture.

15, Federal personnel of the USDA Meat Inspection Division in Atlanta received orders to find something wrong in nine federally inspected Georgia plants. I presume they are now making an effort to follow these orders. Mr. Secretary, you know me well enough to know that regardless of the consequences, I cannot be blackmailed.

A skillful politician, Campbell knew his opposition to the federal government was good politics in Georgia but his posturing had little effect on the outcome of meat reform in Congress. Privately, Campbell told USDA officials he was not really opposed to the Wholesome Meat Act, but was only trying to "protect my boys."

Campbell's performance illustrates the fact that opposition to mandatory federal inspection of intrastate plants was firmly rooted in local politics throughout the nation. Nearly every district had a small packinghouse operating antiquated equipment whose manager felt threatened by federal meat standards. The House of Representatives is notoriously receptive to this kind of pressure, especially on matters falling within the jurisdiction of the agricultural committees. It was business as usual, therefore, when the House passed the weaker Purcell Bill on October 31 with a 403–1 vote, but defeated the Smith-Foley Amendment 140–98. This outcome was heavily influenced on the eve of the vote when the State Commissioners of Agriculture, often with their governors on the line, deluged their Congressional delegations with calls demanding that they vote against the stronger bill.

Nevertheless, in some quarters, the vote was a surprise. While it was predictable that the House Agriculture Committee would strongly oppose mandatory extension of federal inspection, few observers thought that a majority of Congressmen would dare to vote against the consumer on the floor of the House. They were saved embarrassment because the meat bill went before the House on a teller vote—not a roll call.* When tellers are used, House members are counted as they file quickly down the middle aisle. Their names are not recorded and no one except observers on the scene can be sure how they voted. Before the teller vote, meat lobbyists led by Aled Davies and representatives from NASDA buttonholed Congressmen in the corridors and cloakrooms and then watched from the gallery to make sure they voted against mandatory federal inspection. Only 238 of the 435 members of the House voted. After the vote, Neal Smith noted:

Some members told me they wanted to help by at least staying off the House floor on the crucial teller votes. But they said they couldn't because the meat lobbyists were in the House gallery watching them.[16]

The consumer coalition was defeated in the House because they could not counter the local pressure on Congressmen organized by the State Commissioners and the meat lobbies. If they were to resurrect the concept of mandatory inspection in the Senate, they would have to personalize the need for reform at the local level. Here the USDA surveys, which revealed revolting conditions in hundreds of state inspected meat plants around the nation, were the key. By an imaginative use of the media to publicize the surveys, the reformers set out to make the struggle for mandatory federal inspection an intimate local concern. This campaign went into high gear in November after defeat of the Smith-Foley Amendment. It was led by a coalition of journalists, Congressmen, and consumer advocates. In addition to Nick Kotz and Ralph Nader, Arnold Mayer of the Amalgamated Meat Cutters Union, Congressmen Neal Smith and Tom Foley (and their legislative assistants, Ed Mezvinsky and Richard Falknor) and Senator Walter Mondale (D.-Minn.) were key participants.*

Kotz pressed his investigations and alone wrote more than fifty pieces on the subject of meat inspection. More than any other single journalist, he insured passage of the Wholesome Meat Act (he was rewarded with a Pulitzer Prize). Ralph Nader conceived the plan of sending detailed reports on individ-

---

* The teller vote, one of the many anachronisms in the House rules which shield Congressmen from public accountability, was abolished in the House rules reforms of 1971.

* Senator Joseph Montoya (D.-N. Mex.) was also prominent but considerably less militant than this group. He initially opposed any mandatory federal inspection but later amended his position.

ual plants verified by the USDA surveys directly to their hometown newspapers. Many newspapers picked up the story and began to investigate local plants, generating a whole new wave of exposés.*

Nader and Kotz constantly pressed USDA for more information. In November, 1967, Kotz uncovered USDA lab tests of meat products from intrastate plants collected in nationally known supermarkets. Of 162 samples, 123 had not met federal standards. This injected a new public concern. The housewife could no longer assume she was protected merely by buying from prestige national chains. Bad, nonfederally inspected samples were found at Safeway, A&P, Kroger, and other national stores. By constantly hammering home these reports, they began to build momentum nationwide for new meat legislation. Most significantly, they made meat reform good politics at the county courthouse level.

At first, the national media were slow to pick up the story. The break came on November 12 when Public Broadcasting Laboratory devoted the first of three programs to the need for nationwide meat inspection. Other programs followed on November 17 and 26. It was a graphic series which showed what public television can do when it focuses on a flagrant public abuse. Before these programs, with the exception of Cowles Publications (for which Kotz wrote), the *New Republic*, and *The Nation*, the printed media were largely silent. *Life*, *Look*, and *The Saturday Evening Post*, the big circulation magazines, gave no sign of recognizing that a vital consumer issue was at stake. The women's magazines stayed true to the timorous tradition of avoiding controversy with the food industry and shielded their housewives from the bad news about this basic consumer commodity.** But after the first PBL program, the three major networks felt obliged to follow its lead, and prospects for an effective bill steadily improved.

The meat fight reached its climax in November. The consumer coalition's effort in the House, while it failed, at least insured that meat reform would be a major issue in the Senate. The publicity barrage organized by the reformers after the defeat succeeded in arousing public opinion and the meat industry suddenly found itself under attack in many forums. Senate mail was heavy with demands that all meat up to retail level be inspected. But the question still remained: who was to inspect, the states or the federal government?

The bills before the Senate Subcommittee on Agricultural Research and General Legislation reflected the shift in the national mood. In addition to H.R. 12144, the House-passed bill, there were two other bills: S. 2218, known as the Mondale bill, which would have authorized immediate federal inspection of all state plants which did not meet federal standards; and S. 2147, introduced by Senator Montoya, which embodied the House bill, but allowed for federal take-over if the states could not come up with a mandatory system at least equal to federal standards within three years.

Most observers agree that a key figure at this stage was Aled Davies, Washington representative of the American Meat Institute. Davies is one of the most astute lobbyists representing any industry in Washington. Quick to react to the flow of power on any issue, he shifted his position three times during the meat fight. At first, he was opposed to any bill. Then after the first batch of Kotz's stories, he came out in support of the Purcell bill, which would throw a bone to the consumers by giving lip service to state reform but left compliance strictly voluntary. When the bills reached the Senate, he literally sat in Senator Montoya's office and directed the forces against any immediate requirement of mandatory federal inspection of substandard plants.[17] Davies's shifts were given a forceful prod when Kotz got hold of the American Meat Institute's secret membership list and matched it with the USDA surveys of filthy plants.

---

* The women's page editor of the *St. Louis Globe-Democrat* received 1,379 letters and 229 telephone calls from anxious housewives who had read an account of conditions in state inspected plants in the St. Louis area. She reported this response to the annual meeting of the American Meat Institute.

** This was not always so. In 1906 when Upton Sinclair was leading the first meat crusade, the *Ladies' Home Journal* and *Collier's* were famous muckraking magazines.

Fifty of its members, the cream of the industry, saw their corporate names, many of them household words, linked firmly to rotting and diseased meat.

The key arena at this point was in the White House. The Administration was committed to some bill, but how strong a bill no one knew. Davies's views had a powerful sponsor in the person of Mike Manatos, a White House staffer who handled the Senate for Presidents Kennedy and Johnson between 1961 and 1968. When the Purcell bill passed the House, Manatos, at Davies's bidding, tried unsuccessfully to ram it through the Senate by getting a vote on the Senate floor without referral to Committee.

This unusual procedure was stopped by Senator Walter Mondale, who informed the Senate Committee on Agriculture that he would take the Senate floor and filibuster the bill if the Committee went along with Manatos's plan to avoid hearings. In the White House Manatos was opposed by Sherwin Markham and Larry Levinson, two young lawyers on Joseph Califano's staff. They decided early in the game that the White House had nothing to lose and much to gain from a strong consumer bill.

The public response to the meat exposés by Nader and Kotz and especially the fact that activist Senators like Walter Mondale had joined the outcry made meat reform seem far more promising politically in November than it had at the time of the President's consumer message in January. The President needed a new domestic issue, and he needed a Congressional victory. The Great Society's poverty and civil rights programs were now on the rocky road to implementation, but Congress was increasingly reluctant to act on new initiatives from the White House.* And the Vietnam conflict, now a major war, with escalating inflation as well as violence, was rending what remained of Johnson's "consensus." Meat reform had the double advantage of being a new consensus issue (nobody was really for bad meat, at least in public, and public opinion was far more tolerant to-

---
* According to the *Congressional Quarterly*, the success box score of Presidential legislative initiatives had declined from nearly 70 percent in 1965 to 47 percent in 1967.

ward extensions of federal power in 1967 than it would be two years later) and being relatively cheap, at least compared to the domestic reforms of 1965 and 1966.

These considerations may help explain why Califano, in October and November, became increasingly opposed to USDA's position against mandatory inspection. When Tom Hughes, one of Freeman's aides at USDA, sent a memo to the White House asking for opposition to the Smith-Foley bill, Markham stopped it cold. There was, as one observer noted, "one hell of a fight in the White House."

Betty Furness, the President's consumer advisor, eventually became the public spokesman for the group nicknamed (only half facetiously) by Freeman's assistants as Califano's White House Mafia. When appointed in March, 1967, she seemed a most unlikely candidate for this role. A former television personality, she had replaced the doughty Esther Peterson who, as the first consumer advisor to the White House, had become a sort of consumer's Jeanne d'Arc through her vigorous attacks on corporate fraud and malfeasance. In the political climate of 1966 and early 1967 she came to be regarded as a political liability to the President, particularly after she used her White House forum to support a national boycott of supermarkets. Out of loyalty to Esther Peterson, most consumer activists dismissed Miss Furness as a sham and tended to ignore her.

Ralph Nader, however, saw Furness's tenuous position as an opportunity to put pressure on the White House. The occasion for turning the screw was apparently fortuitous. Shortly before the House vote on the meat reform bills, Arnold Mayer, lobbyist for the Amalgamated Meat Cutters and Butchers Workers, convened a quiet strategy session of labor and women's groups. When the press unexpectedly showed up, Nader seized the occasion to throw down the gauntlet to Miss Furness and demand that she live up to her role as consumer advocate by supporting a strong bill. Whether this sudden splash of limelight influenced her is not known, but in any case on October 30, the day before the House vote, Miss Furness surprised everyone by contradicting the Department of Agriculture and announcing her personal support for mandatory inspection. No one knew, at that time, where the President stood.

When Betty Furness was called to testify before the Senate Agricultural Subcommittee on November 15, the White House was suddenly faced with a deadline for commitment it could not escape. Califano's staff wrote Betty Furness's statement, and she gave the administration's endorsement to S. 2218, which called for immediate federal take-over of all plants operating with inspection below federal standards. Only six days before, Rodney Leonard, presumably speaking for the Administration, had endorsed the weaker S. 2147, the Montoya bill. Queried as to the reason for this sudden shift, Miss Furness won applause and laughter with her reply: "I don't think we should have to be looking askance at our hamburgers and frankfurters for another two years." Nick Kotz later said of Furness's performance: "She really proved herself before that Committee. She was great in the testimony, and handled the questions like a pro." From that point on, the President's Special Assistant for Consumer Affairs was a staunch ally of the consumer coalition and gained stature for herself and her office.

Her statement naturally left Freeman frustrated and angry. Several days before, acting on the assurances of Freeman and Manatos, Senator Montoya had claimed before the same Committee that the White House favored his bill. Moreover, Califano had failed to screen Furness's statement with Freeman in advance. Freeman, with good reason, felt he was being undercut by the White House staff.* Freeman was so angry that he sent a telegram to the heads of the state meat inspection programs reiterating his support of the weaker Montoya bill. Once again, the question became who speaks for the President. According to Califano, he took the issue to President Johnson, who decided to go with

* This was not the first time Freeman felt betrayed by the White House staff. In early 1966, against his better instincts he was ordered by the White House to call a press conference and express his pleasure at a recent decline in food prices. This statement helped the President, but Freeman was harshly attacked by every agriculture interest group in the country, and never really got over it.

the tougher bill, S. 2218. Califano then told reporters that Betty Furness spoke for the White House.

At this point, the conflict between White House staff and the Secretary of Agriculture became bitter indeed. After consultation with Senator Walter Mondale, Califano decided that the Administration should send a letter to Public Broadcasting Laboratory endorsing the Mondale bill (S. 2218) on the next program in its series on meat inspection. He thought the letter would have greater impact if it came from the Department of Agriculture. But when Califano asked Rodney Leonard, the Administrator of the Consumer and Marketing Service, to send the letter, he refused. Freeman, Leonard warned, was absolutely adamant. Califano then sent a member of his staff over to USDA to get a sample of official Departmental stationery. Califano's aides, Markham and Levinson, drafted a letter to PBL, then called Leonard over to the White House late one evening and ordered him to sign it. Freeman, by all accounts, was livid. A few days later at Freeman's instigation, Senator Spessard Holland, a patriarch of the agriculture establishment in the Senate, personally attacked Califano's staff in the Congressional Record, charging that a little group of northern liberals in the White House was leading the Administration astray by pushing for a strong meat program which they knew nothing about.

These shifts in the White House's position and the legions of incriminating facts, marshaled against the meat industry by Nader, Kotz, Public Broadcasting Laboratory, and finally, the major networks and mass circulation magazines, broke down the resistance of the large packers. Consumer confidence in meat was shaken and the packers began to fear that the continuing struggle on Capitol Hill might affect sales.

One turning point occurred on November 2, when Kotz revealed a startling blunder by Blaine Liljenquist, head of the Western States Meat Packers Association. On the eve of the House vote, Liljenquist solicited meat packing companies to contribute to a fund for Congressmen who were working "to preserve our free enterprise system," a code phrase for Congressmen who opposed mandatory inspection of intrastate plants. Robert Poage, Chairman of the House Agriculture Committee, and Aled Davies correctly sensed that Liljenquist's clumsy act might threaten the integrity of their effort. They first tried to cover up for him, then expressed outrage. Davies repudiated his colleague's letter "as a most unfortunate, ill-timed and utterly stupid activity," and the meat lobby's solid front began to dissolve in internal feuds.

By mid-November, Aled Davies decided to cut his losses. He lectured the other trade associations that if they did not "move forward firmly and honestly, we have only ourselves to blame." He criticized the House bill as containing "lots of carrots but not much stick." When he persuaded Swift and Armour to support a strong bill in the Senate, opposition in the rest of the industry folded.

Here was the inside power of the lobbyists put to work for a public good. Davies personally delivered the key votes in the House and Senate Agriculture Committees. In rapid succession, he persuaded them to be for no bill, then a weak bill, and finally a relatively strong bill as he dealt with changing levels of public concern.

In the end, a compromise was reached.* Mondale's plan for immediate take-over of all intrastate plants which did not have federal standards was scrapped, but the principle of mandatory inspection for all meat was established. The states would be given two years to develop an inspection program at least equal to federal standards. The federal government would pay 50 percent of the costs to do so. If the states were well on the way to developing a meat program on December 15, 1969, they would be given an additional year to comply. If not enough progress had been made, the federal government would take over. The Secretary was empowered to require immediate federal inspection of *any* plant which he found producing meat dangerous to public health. This was a great improvement over the House bill which lacked any mandatory provisions for federal take-over of intrastate plants. The compromise bill, however, gave the Secretary of Agriculture great discretion (with its possibilities of delay) in determining how states measured up to federal standards.

On December 15, 1967, President Johnson signed the Wholesome Meat Act into law. Upton Sinclair, in a wheelchair, was at his side. A coalition of consumer and labor groups and crusading journalists had shown that even without initial Administration backing, a major piece of consumer legislation could be pushed to enactment in a six month period, once attention focused on the issue. Most impressive of all, they did it over the opposition of the committees on agriculture in the House and initially the Senate.

This victory, especially the techniques used to win it, became a watershed for public interest advocacy and the consumer movement. The reformers had applied the classic strategy of divide and conquer. They divided the executive branch by using the White House to neutralize the Department of Agriculture; they divided the meat industry and the farm committees into warring camps. Second, they freed food protection of its abstractions and made meat reform good politics at the local level and a brand-name embarrassment at the corporate level. Finally, by holding back some information and pacing their exposés, they kept their momentum up and their opponents off balance. The lesson of the meat fight was that disclosure of facts to the public, especially the technical detail of issues usually reserved for industry and government insiders only, could generate an overwhelming political leverage for reform.

The success of this approach was entirely unexpected in 1967. It is fashionable among political scientists, historians (such as Richard Hofstadter), and many political analysts to stress the irrationality of politics, to downgrade the value of getting facts to the people. They point to many elegantly conceived proposals for reform, from taxes to campaign financing, which bring no response from the public. Real power, they say, resides in the "insiders" world of closed committee sessions, special interest lobbies, and the special

---

* During the House-Senate conference on the bill, efforts to weaken it were intense, Mondale and Foley asked Kotz to ask Nader for his advice on the proposed compromises. The message came back: "Do what you think is best." In leaving them on their own, Nader was putting his former allies on notice that, as far as the consumer movement was concerned, there would be no permanent entangling alliances.

relationships between the career bureaucrats and subcommittee chairmen. Focus on this arena naturally breeds a contempt for public opinion. In 1967, this view was the source of the conventional wisdom among Washington political pros which held that consumer protection could never be good politics on Capitol Hill. The meat reformers in 1967 went over the heads of the insiders. They gambled on the ability of the public to respond if the need for reform were personalized at the local level.

The Wholesome Meat Act was a *bona fide* accomplishment for the consumer movement. But in the continuing effort to assure the purity of meat and poultry, the victory was only a skirmish. Reminiscing about the meat fight three years later, Ralph Nader put the struggle for the Wholesome Meat Act in perspective when he said of Aled Davies: "The meat packers never realized how much he saved them. At first we had concentrated on standards of sanitation. The other issues—chemical adulteration of meat, microbiological contamination, misuse of hormones and antibiotics, pesticide residues, ingredient standards—we were just getting into that when the meat lobby decided it was time to put the lid back on the meat industry's Pandora's box. Filth became a shield against even more serious revelations. Congress relieved itself with the Act and forgot about the rest of the mess."[18]

The depressing sequel to any victory like this is that consumer groups do not follow up. The massive publicity which attends one of their successes masks the fact that their advocacy resources are very slim. The constituency which supported the cause of pure meat in 1967 was, as the saying goes, "a mile wide and an inch deep." With a few exceptions, it was (and is) a public *unorganized* constituency whose only form of political power was public opinion. It lacked a Washington association and paid lobbyists to represent its interests in the agencies, to make personal contacts with influential political leaders, or to offer or withhold campaign contributions for Congressmen or Presidents. It lacked newsletters or trade journals to keep its members informed about the fate of its legislation or to mobilize its power when a threat arises.

Thus while public opinion without organization may be effective in achieving passage of a law, it is often futile in influencing the administration of the law after it passed. Consumers lack the manpower to ride shotgun for the Act as it passes through the appropriations committees and the bureaucracies in the following months. On these fronts, there is a perpetual advocacy gap between the representatives of consumer interests and of private industry. As one Civil War buff on the House Agriculture Committee stated (for some reason "off the record"):

You guys are like Lee's Army of Northern Virginia. You can gather your forces and pull off a big one, but on all the other fronts, our boys keep coming on.

As we shall see, the subsequent history of the Wholesome Meat Act gives these words the ring of prophecy.

# Notes

CHAPTER 1

1. Statement of Ralph Nader before Subcommittee on Agricultural Research and General Legislation of the Committee on Agriculture and Forestry, U.S. Senate, 90th Congress, 1st Session, November 1, 1967, p. 142.
2. *Government Rejected Consumer Items*, Hearings before a Subcommittee of the Committee on Government Operations, House of Representatives, 90th Congress, 2nd Session, April 2, 3, 1968, p. 88.
3. Dr. W. L. Ingalls, paper presented to the 87th annual meeting of American Veterinary Medical Association, August 21–24, 1950. Hearings before Subcommittee on Livestock and Grains of the Committee on Agriculture, House of Representatives, 90th Congress, 2nd Session, February, 1968.
4. Gabriel Kolko, *The Triumph of Conservatism* (New York, 1963), pp. 98f.
5. Claude G. Bowers, *Beveridge and the Progressive Era* (Cambridge, 1932), p. 227.
6. *Ibid.*, p. 230.
7. House of Representatives Report No. 653 on Federal Meat Inspection Act (1967), p. 4.
8. Ralph Nader, "Watch That Hamburger," *New Republic*, August 19, 1967, p. 15.
9. *Meat Inspection*, Hearings before the Subcommittee on Agricultural Research and General Legislation of the Committee on Agriculture and Forestry, U.S. Senate, 90th Congress, 1st Session, November 1967, p. 92.
10. *Ibid.*, p. 182. Statement of Leslie Orear, Director of Publications, United Packinghouse Food and Allied Workers, AFL-CIO, who helped expose unsanitary meat conditions in Chicago.
11. Ralph Nader, "Watch That Hamburger," *loc. cit.*, p. 15.
12. *Meat Inspection*, p. 168.
13. *Ibid.*
14. Interview with Nick Kotz, January, 1971.
15. Interview with John Califano, January, 1971.
16. *Des Moines Register*, November 2, 1967.
17. Interview with Nick Kotz, January, 1971.
18. Interview with Ralph Nader, December, 1970.

When the heavyweights of the meat industry came to Washington during the Carter administration, they invariably spent some time in the USDA headquarters building just a few blocks from the Washington Monument doing battle with an assistant secretary of agriculture. They would climb to the second floor and walk down the hall to the first door past the secretary of agriculture's office. Here they were ushered into a richly paneled office furnished in greens—bright-green carpet, lime-green and green-plaid furniture, leafy green plants all around. They were met by a delicate five-foot-two-inch woman with curly red hair, a pleasant smile, and a disarming Arkansas drawl. To these macho meatmen, she must have looked like a pushover. But they soon enough learned otherwise. The first indication was her viselike handshake. Then her straightforward greeting that made it clear she had no time for unnecessary pleasantries. This woman meant business. Carol Tucker Foreman, assistant secretary of agriculture for food and consumer services, was every bit as tough as any of the meatmen.

And the meatmen didn't like it. They didn't like it that this woman could boss them around. Some of them didn't like it that she was a woman, period. And they all didn't like it that she was asking pointed questions and making difficult demands from USDA headquarters.

The meatmen were used to thinking of the USDA's top bureaucrats as their friends. An assistant secretary of agriculture was usually one of the boys, a man sympathetic to the needs of meat producers, slaughterers, and processors. Certainly the assistant secretary was not a woman. Few industries are still as male-dominated as the meat industry, so taking orders from a woman is not something meatmen accept easily. Their attitude is neatly summarized by one packer, "The Old Timer," who writes a column for *Meat Industry* magazine: "For one thing, I've yet to see a woman who could pack a front quarter of bull . . . or side of beef . . . or bump legs. It has just never been a woman's place around a packing house. Of course, some of the sidelines we've developed are well suited to their size, strength and aptitudes. Portion control comes rapidly to mind along with most weighing, packaging and, of course, clerical work. There have been some pretty fair truck drivers as well. But, by and large, it's been a man's world in the meat business as long as I can remember." It was bad enough that the meatmen had to deal with a woman, but Foreman was something even worse as well: a consumer advocate. The meatmen who in the past could count on the USDA to take their side in the fights with those pesky consumers were now faced with an

---

From the book, <u>Prime Rip</u>, by Wayne Swanson and George Schultz, copyright 1982, published by Prentice-Hall, Inc., Englewood Cliffs, NJ 07632.

administrator who had been quoted saying things like "Consumers have not been just ignored by this department, they've been abused."

Carol Foreman was clearly not the kind of administrator the meat industry and its supporters were used to. And they made sure she knew it. Former secretary of agriculture Earl Butz called her appointment the "ultimate insult to farmers." A Kentucky congressman said, "If Carol Foreman and her cronies have their way, the nation will exist on wheat germ and organic bean sprouts." A Nebraska congresswoman charged, "Carol Foreman, one of agriculture's biggest enemies, is at work right now discrediting the meat industry and causing the public to lose confidence in American farm products." The *National Provisioner*, trade organ for the meat industry, snidely referred to her as "Chatty Carol" and other derogatory names while attacking her every move. To meatmen she was a "dictator," and analogies linking her to leading tyrants became favorites throughout the industry: her press releases came from the "Idi Amin School of Diplomacy"; she was, simply, the "Ayatollah of the USDA."

What did Foreman do to deserve this vilification? Basically, she took the branch of the USDA whose actions most directly affect consumers, and she tried to make it stand behind consumers. That is what she was supposed to do, but it shocked the hell out of the meat industry. Foreman was a voice for consumers and a tough administrator who tried to make the USDA more responsive to the needs of consumers. She cleaned up some of the corruption in meat grading and inspection, and she tried to make grading and inspection more useful tools for consumers. Meatmen groused that all she was really doing was giving the industry a bad name and piling on unnecessary regulations. But there was something that riled them even more about Foreman: she had the audacity to ask basic questions about the healthfulness of meat products. She wanted to know the effects of all the chemical additives poured into livestock and into meat. She wanted to know if processing and labeling regulations really helped consumers figure out what they were buying. And worst of all, she wanted to know if it was really healthy for Americans to eat all the meat we do.

The meat industry doesn't want anyone talking about these things. Increasingly, talk about chemicals leads to links with cancer and other disorders. Talk about labeling leads to evidence that a lot of what we buy as meat is actually fillers and by-products. And talk about health and nutrition leads to evidence that meat may not deserve its spot at the absolute center of the American diet. The meat industry despises the consumer advocates who harp on these points, and it certainly doesn't want a top government regulator bringing them up. In Carol Foreman, however, it had both.

When Jimmy Carter took office in 1977, Foreman was one of the many "outsiders" brought in to run his administration. At the time she was known as one of the most aggressive consumer-interest lobbyists in Washington, heading the Consumer Federation of America, a coalition of 240 consumer groups nationwide. She was quite familiar with the USDA because it had been one of the Consumer Federation's favorite targets. Foreman and the Consumer Federation sued the USDA twice, first for giving in to the meat industry by relaxing meat-grading standards, and then for allowing the sale of processed meats that contained bits of ground bone resulting from a mechanical deboning process. Foreman also attacked the USDA for listening only to food processors and big farmers while ignoring consumers. Once, while Earl Butz was secretary of agriculture, she and a group of consumer advocates marched into a meeting of a USDA advisory panel wearing gags to draw attention to the fact that they were silenced by the USDA decision makers. In that same room about a year later, Foreman was sworn in as assistant secretary of agriculture.

Foreman was chosen for the post because she was known for more than just symbolic acts. "People expect a consumer advocate to be some hysterical lady out of a supermarket," Foreman told one interviewer shortly after her appointment. "But one thing all Carter's 'consumer types' have in common: We know an awful lot about how government operates and how to work within it. When it's screwing you all the time, you learn." Foreman was a savvy, pragmatic strategist who understood politics and the inner workings of government. She had been born into a political family: her father had been Arkansas state treasurer and her younger brother became Arkansas attorney general and, later, a U.S. representative. She was married to an international vice-president of the Retail Clerks Union (which briefly snagged her appointment as assistant secretary while the Senate considered charges that she was, literally, "in bed" with organized labor). On her own she developed a wide range of governmental experience; she worked as a congressman's executive assistant, as a congressional liaison aide for the Department of Housing and Urban Development, and as chief of information and congressional liaison for Planned Parenthood. If there was any question about how tough and gutsy Foreman was, she dispelled it when she lobbied aggressively on Capitol Hill for Planned Parenthood while in a visibly advanced stage of pregnancy.

As assistant secretary of agriculture for food and consumer services, Foreman was in charge of a staff of thirteen thousand and a yearly budget of $9 billion. She was responsible for programs ranging from meat grading and inspection to food stamps, school lunches, and nutrition research and education. She charged that the USDA catered solely to the interests of food producers, and no matter how much it upset them, she was going to make the USDA the "people's department" Abraham Lincoln had envisioned when he set it up.

"I want to kick some ass, I just don't know where to kick it," she told one visitor shortly after taking the job. It wasn't long, however, before she started booting, and the backside of the meat industry smarted for the next four years. Foreman was not afraid to take stands she knew would infuriate the meat industry. And she was not afraid to tell the industry about her stands in languge it understood. Foreman may be physically small and delicate, but she could come on like a Marine drill instructor when necessary. She was known for one of the foulest mouths in government, an attribute that endeared her to some of her male audiences and at least got her the attention of others. In addition, she surrounded herself with a band of young, idealistic assistants who shared her zeal. These were not the usual bureaucrats who easily accepted whatever the meat industry told them. And they were not the kind of bureaucrats that the meat industry felt comfortable dealing with. Meatmen could not use their old-buddy routines on such assistants as Tom Grumbly, an earnest young man still in his twenties who was not afraid to stand up before meat industry groups and say the industry was corrupt; Sydney Butler, who came to the office in blue jeans and shoulder-length hair until a promotion forced him to clean up his act; and Jody Levin Epstein and other women assistants whose insistence on using three names was an implication of women's liberation that some meatmen found offensive.

For the next four years this small band of idealists took on the meat industry and the USDA bureaucracy. They talked continually about restoring the "integrity" of USDA officials, such as graders and inspectors, whose reputations had been tainted by scandal. They proposed reforms and new programs that for once showed concern for the needs of consumers, not just the meat industry. In the process, they made a lot of enemies.

To the meat industry, the sins of Carol Foreman were many, and they were succinctly inventoried early in 1980 at a gathering of West Coast meat-packers. The setting was the annual Western States Meat Packers convention at the Disneyland Hotel. Here was a spectacle that was in its own way as bizarre as anything to be seen at the famed amusement park just a short monorail ride away. There was a convention hall filled with the latest in meat industry fashions, from space-age sausage makers and burger grillers to the Anyl-Ray Fat Analyzer and the Inject-Jet Bone-In Pickle Injector. There were intriguing signs: "Are you EXTRACTING SHROUDS so they can take up brine?" There were short, bull-necked, leisure-suited packers, tall, weatherbeaten, heavy-on-the-rawhide-look packers, and dapper, citified packers. There were speeches on such important topics as electrical stimulation of meat and world meat research. And Richard Lyng was on hand to accept the Floyd Forbes Award, a bronze steer trophy that is the packers' highest award for service to the meat industry.

Lyng had once been an assistant secretary of agriculture, just like Foreman. He had recently retired as president of the American Meat Institute, the dominant national trade organization, for whom he had feuded with Foreman continually. Now this husky, aging man stood before the assembled packers to accept an award given, among other reasons, for keeping the pressure on Foreman. In his deep, sure voice he highlighted for his Disneyland audience the issues that drove the meat industry crazy about Carol Foreman, and, as it turned out, the attitudes that no longer would be welcome in the USDA under Ronald Reagan.

He warmed up with some general statements about the problems facing the industry, such as inflation, energy, and government regulation. He complained about Washington officials who were making matters worse for the meat industry instead of better: "The fuzzy-wuzzies, in too many instances, are in charge," he griped as his audience nodded agreement. He attacked the Consumer Federation of America as an organization that is "antibusiness" and "antimeat." Then he turned to his favorite target.

"You know that the Consumer Federation was the training ground of the assistant secretary of agriculture, who has given this industry, in my opinion, a dreadful time during the past three years. Most of the industry's regulatory problems during that time were the brainchild of what I would call the Ayatollah of the USDA, Carol Tucker Foreman." Lyng gave a rundown of her many "Ayatollah-like" actions. First of all, he said Foreman "went into the USDA with an antimeat attitude." As evidence, he pointed to the fact that in 1977 she helped sponsor a meatless dinner at the White House. "We objected to that," Lyng said, "but even the secretary of agriculture went to the dinner." Nothing is more sickening to meatmen than the idea of vegetarianism.

Foreman's next sin was to bring in Dr. Robert Angelotti as her top assistant. Angelotti was the abrasive bureaucrat who took seriously the USDA's regulatory powers and who vowed that he would apply them to the meat industry for the first time in decades. Meatmen quickly came to despise him as much as they despised Foreman—if not more. Then, when this reformer was charged with filing false expense vouchers, the meat industry pounced on the issue and forced him to resign. Lyng sarcastically told his Disneyland audience that Angelotti's term had been a "great experience for us." About the only thing great about his leaving was that the meatmen could now focus all their wrath on Foreman.

Lyng moved on to the two controversies that had turned meatmen absolutely livid about Foreman: her stands on the use of the chemical additive nitrite and on a mechanical meat-deboning process.

Sodium nitrite and sodium nitrate are time-honored chemicals in the meat industry, important to the production of processed meats such as bacon, hot dogs, lunch meats, and pet foods. They add color, making hot dogs a pleasant pink rather than an unappetizing gray; they add flavor; and they inhibit spoilage and prevent deadly forms of food poisoning like botulism. But they also may cause cancer. For twenty years, the link between nitrites and cancer has been studied, and enough evidence has been discovered to prompt some countries to limit or ban nitrites. But the definitive evidence of the link has not yet been found, and the American meat industry has resisted all efforts to tamper with its use of nitrites. It's an understandable position, since nitrites are used in an estimated $12.5 billion worth of food—7 per cent of the food supply.

The charge against nitrites is that they can combine with other substances to form nitrosamines, which are known to be potent carcinogens. Nitrosamines have been found in a variety of products—in 1979 there was a major scare when large concentrations were found in some brands of beer—but bacon is the product that has raised the greatest fears. There are indications that nitrosamines can be formed during the process of frying nitrite-cured bacon, and also that they can be formed simply by the interaction between nitrites and other substances in the human stomach. Then, in 1978, a government-commissioned study by an MIT professor found supposedly more compelling evidence. The study concluded that there was an increased rate of cancer of the lymph system in laboratory rats fed nitrites. Based on the results, the USDA and the Food and Drug Administration quickly called for a total ban on nitrites. The meat industry mobilized to fight the initiative against one of its most precious additives, and the battle raged until 1980. That was when an independent review of the MIT study revealed serious flaws, indicating that the rate of cancer in the laboratory rats was actually much lower than originally reported. The USDA and the FDA reluctantly dropped their attempts to ban nitrites, although Foreman warned a House committee reviewing the controversy that "it would be a grave mistake for us to be lulled into a false sense of security." She reminded everyone that there was still serious concern in the scientific community about nitrites, and she promised that the USDA would continue to study nitrites and vigorously push for development of alternative chemicals to replace them.

But Lyng didn't buy Foreman's concern. To him, the assault on nitrites was nothing more than "scientific McCarthyism." "Nitrite has been charged with a lot of things that have not been proven; and more and more, it looks like they will never be proven—nitrite is innocent of these charges. I cannot quantify the tremendous harm that this has done, not only to the meat industry, but also to the consumers of the world."

Even the meat industry's wrath over nitrites, however, paled in comparison to one more issue: the controversy over mechanically deboned meat. It may sound like a rather technical kind of dispute, but it was one that pointed out most graphically how strongly the meat industry will fight being forced to tell consumers what they are buying. For Lyng, it was Foreman's "most irresponsible and most Ayatollah-like action."

The meat industry has developed machines that mechanically separate meat from bone, so ideally packinghouse workers should be spared the time-consuming and inefficient task of scraping out by hand as much meat as possible from inaccessible bones and joints. With mechanical deboning machinery, an estimated five to fifteen pounds of additional meat can be extracted from a single carcass—quite a significant amount. But the machinery also extracts bits of bone and bone marrow. The bits are ground into a fine powder by the process, so there is little danger of unsuspecting consumers biting into large chunks of bone in their wienies. But the bits are certainly not meat, even though they were included in the makings for hot dogs, sausages, luncheon meats, and other processed meats as if they were. Even without the powdered bone, some meatmen know enough about what goes into processed meats to avoid eating them. And the recipes themselves are hardly enticing. Take, for example, the ingredients in one recipe for liver sausage: 160 pounds of pork livers, 100 pounds defatted pork stomachs, 160 pounds pork snout trimmings, 30 pounds pork cheek meat, 5 pounds ring liver seasoning, 8 pounds salt, 40 pounds water. Consumer groups were not about to accept the addition of powdered bone to the list without a fight, so in 1976 they sued the USDA. They wanted to halt the use of mechanical deboning machinery until the USDA made sure manufacturers spelled out on the product labels just what was going into processed meats. The consumers also wanted the USDA to determine for sure whether it was healthful for consumers to eat this powdered bone. Carol Foreman, then heading the Consumer Federation, led the fight, and she won. Mechanically deboned meat was banned, although the following year the industry pushed the USDA to reconsider its decision. This time, the official the meatmen had to deal with was all too familiar to them: Carol Foreman was now assistant secretary of agriculture. Foreman did reconsider the USDA's decision and she approved the use of the machinery. But there was a catch: all meat products containing meat processed by the machinery had to include a statement on their label warning that they included "tissue from ground bone." The meat industry argued that the statement sounded awful, and consumers would never buy products that carried that warning. So the haggling continued. In 1978, the USDA issued a revised regulation. This one decreed that mechanically deboned meat had to be labeled, in lettering at least one half the size of the product name, "Mechanically Processed Beef Product [or Pork, etc.]." And in lettering at least one quarter the size of the product name, "Contains up to $x$% powdered bone." The industry once again complained the statements were repugnant, and consumers would still conjure up images of chomping down on bone bits. Now, few packers use the process while an estimated $20 million worth of mechanical deboning equipment stands idle.

"This is a ridiculous situation in which this product—because of the unreasonable labeling requirements for mechanically deboned meat—is being denied the public, while at the same time, the identical product coming from poultry is not being discriminated against," Lyng charged. And since Foreman originally led the fight against the machinery when she was with the Consumer Federation, he contended she was biased and should have removed herself from deliberations on the issue.

Foreman was equally firm in her stand that the regulations were fair and informative: consumers should be told what they are buying, and consumers shouldn't be paying meat prices for bone and marrow. In addition, since it was admittedly inconsistent that poultry products using the machinery were not labeled, she said she would look into requiring the warning for poultry products as well. Her adamant position only served to further infuriate meatmen. "It's about as easy to negotiate with her on this issue as it is for the United States to negotiate with the people of Iran," Lyng charged at Disneyland, to the delight of the assembled packers.

Now he had gotten all the bitter charges out, and he decided to back off just a bit. He conceded "it may not be fair, and it certainly is an overstatement, to compare Carol Foreman with the Ayatollah, but it brings home the concept of unreasonable and probably inexperienced leadership." Lyng had made his point, however, and the Ayatollah nickname stuck. It quickly spread through the industry, becoming an epithet used with glee against Foreman.

And it hurt. Foreman and her staff were used to abuse, but they considered the comparisons to the Ayatollah a low blow. They didn't expect the meatmen to agree with everything they were doing, but they did expect some respect for going about their jobs in a responsible manner. Instead, meatmen such as Lyng made them out to be irrational zealots simply because they were concerned about considering more than just the parochial interests of the meat industry. Foreman was ostracized for hiring assistants such as Angelotti who did not blindly accept whatever the meat industry said. She was attacked on the nitrite issue because not enough rats had died in one test to convince meatmen of a health hazard well established in other tests. And on mechanical deboning of meat, it wasn't even a question of how many rats died; it was merely a matter of asking meatmen to label products so consumers know what's inside.

All of these attitudes are incomprehensible to Lyng and the Disneyland packers. They still want the good old days when the USDA was "their" department, and nobody was questioning what was going into their products. They hang on to attitudes and practices out of Frontierland, and by not recognizing the changing conditions in their industry and the changing desires of consumers, they are living in a fantasyland. This time, Lyng and the Disneyland packers won out. Less than a year after Lyng gave his speech, Carol Foreman was out of a job. And Richard Lyng was deputy secretary of agriculture, second in command, in Reagan's new Department of Agriculture.

When the Carter administration came to an end, she had yet to come to terms with the virulent criticism from the meat industry, the impatience of her former colleagues in the consumer movement, and the indifference of the bureaucracy. During her four years, Foreman was able to earn a grudging respect from some quarters of the meat industry as a woman who stood up for what she believed in. Producer groups in particular came to respect her as a person, if they didn't always agree with her policies. They were impressed with her gutsiness and her down-to-earth personality, and some even considered her "one of the guys." But many other meatmen never came to an understanding with Foreman; they attacked her viciously and relentlessly until the end. Equally frustrating to Foreman was the fact that while she was fending off charges from the meat industry that she was going too far, she had to contend with charges from the consumer movement that she wasn't going far enough. Even Ralph Nader, who had enthusiastically praised Foreman at the start, later complained that she "sold out" to the food producers. Regardless of what Foreman had set out to accomplish, the fact remained that she and her staff were isolated against the rest of what is one of the slowest, dullest, most corrupt, and most entrenched bureaucracies in all of government. The career bureaucrats who were there before Foreman came and who stayed after she left were not very interested in implementing her reforms, so many of them were stalled.

Possibly she was in a no-win situation, considering all the conflicting desires and interests of producers, consumers, and bureaucrats she had to please. In any event, she ran out of time. Carol Foreman's reign turned out to be just a brief aberration in the history of the industry-oriented USDA. When

Ronald Reagan was elected, it was clear the USDA would turn back to its old ways. Reagan's supporters and advisers were the people who despised Foreman, and now they would have the opportunity to undo her attempts at reform.

If Carol Foreman's tenure with the USDA was important for one reason, it was for focusing attention on the new jungle facing the American meat industry. The concerns today about the American food supply and the American diet are potentially as serious as anything revealed in Upton Sinclair's 1905 classic. Back then, the public was outraged by the gruesome tales of what went into our meat. After all, who could not be revolted by descriptions like this: "There would be meat that had tumbled out on the floor, in the dirt and sawdust, where the workers had tramped and spit uncounted billions of consumption germs. There would be meat stored in great piles in rooms; and the water from leaky roofs would drip over it, and thousands of rats would race about on it. It was too dark in these storage places to see well, but a man could run his hand over these piles of meat and sweep off handfuls of the dried dung of rats. These rats were nuisances, and the packers would put poisoned bread out for them, they would die, and then rats, bread, and meat would go into the hoppers together. This is no fairy story and no joke...." No joke indeed, and the government inspectors were not doing anything about these and other problems. Sinclair writes in another section of *The Jungle*, "If you were a sociable person, he [the inspector] was quite willing to enter into conversation with you, and to explain to you the deadly nature of the ptomaines which are found in tubercular pork; and while he was talking with you, you could hardly be so ungrateful as to notice that a dozen carcasses were passing him untouched." *The Jungle* caused such an outcry that immediate steps were taken to overhaul the U.S. inspection system. The reforms that resulted and the improvements made over the years since then have established the American food inspection system as possibly the best in the world at what it does. But there are serious new problems.

One afternoon during the summer before Carol Foreman left the USDA, she was leaning back in the lime-green armchair in the corner of her USDA office. Her sandaled feet were propped up on the coffee table, and her fingers were on her temples as she tried to explain about the new jungle—a jungle that does not lend itself as easily to graphic horror stories like the ones told by Sinclair. "Keeping dirt and hair and bones out of meat is something the inspection system was set up to do many years ago, and we've conquered those problems and we do them very well," Foreman said. Today's problems are more subtle than rats in the sausage meat, tubercular steers, or workers falling into the vats of Durham's Pure Leaf Lard. "The massive use of chemicals in agricultural production is a phenomenon that has begun since World War II, over the last thirty-five years. And it's only in the last few years that we've begun to discover that the chemicals which have made it possible to increase the food supply very substantially also have some unintended and sometimes very unfortunate consequences. That is, for example, pesticides used on corn crops contaminate the corn and then the contaminated corn is eaten by hogs and the contamination ends up in the hog meat at a health risk to consumers."

This chain reaction gives the USDA hundreds of potential contaminants to worry about. There are growth promoters like DES, there are sulfa drugs fed to swine to prevent disease, there are nitrites and pesticides, and there are PCBs and other toxic chemicals that can contaminate the ground on which animals graze without anyone knowing about it until it's too late. A 1979 study by the U.S. General Accounting Office indicates that 14 per cent of dressed meat and poultry sold in supermarkets may contain illegal residues of drugs, pesticides, and other contaminants. Of 143 drugs and pesticides likely to leave residues in raw meat and poultry, the report says 42 are suspected of causing cancer, 20 of causing birth defects, and 6 of causing mutations. The federal Environmental Protection Agency has limited or abolished the use of some of these chemicals and now the inspection service tests for residues so contaminated meat is not passed on to consumers. "However, that turns out to be a problem that requires a great deal more sophistication than we have," Foreman said. "We were late in recognizing it, not by malfeasance but just because we weren't wise enough as a country to question what the side effects might be to the use of those chemicals when we first decided to use them."

The problem of detecting chemical contamination starts with one very frustrating fact: "You can't see it. Unless an animal has so much of it that he's staggering, you can't see it. It's not like a disease or a broken bone or gross physical contamination. Furthermore, the tests we have to do to find those residues are extremely expensive, technologically very sophisticated, and require a long time to perform. You can't do them on every single animal. We inspect every single chicken for disease, but we can't inspect every single chicken for every single chemical that might contaminate it. By the time we took every sample, there wouldn't be any chicken meat left to sell. Plus the fact that the cost of doing all those tests on the chicken would far outweigh the cost of the chicken itself." That means the USDA must rely on statistical sampling, choosing a small group of chickens or hogs or cattle at random on which to conduct the tests. Then it must hope it sampled a sufficient number of animals to catch any harmful residues. Even so, the process is time-consuming and inefficient, and there is no way the USDA can test for every single chemical that could contaminate meat. "The meat inspection laws never contemplated that there would be a problem that you couldn't recognize immediately," Foreman said. "They say that we shall not stamp as wholesome any product that may be contaminated, but we stamp all of these products that we later call contaminated because we have no proof at the time that they're contaminated. We get it two weeks later. Then we have to go back and try to recall the product. It's very expensive, it's damaging to the reputation of the company, and it's inefficient."

Even more disturbing is the fact that the link between residues and health hazards is seldom direct, so the actual danger presented by residues is open to debate. Particularly frustrating are the potentially carcinogenic chemicals. Cancer may wait ten years or more to strike, and when it does, it leaves no clue to its identity, so who is to say some meat eaten a decade ago was the culprit? In addition, the research on chemical residues is highly contradictory: for each study claiming a chemical is hazardous, there is another showing that it is not. And as the list of chemicals suspected of causing cancer and other maladies grows, so does public cynicism about cancer claims. After all, is it always necessary to change time-honored procedures and eliminate chemicals that have been accepted for years just because a few mice died?

There are enough issues in the controversy over chemicals in meat for an entire book, but the dilemma can be stated briefly. The meat industry says the risks of chemicals must be weighed against the benefits. Chemicals combat disease in livestock and contamination of meat products. They help producers raise healthy animals quickly, lowering the cost of meat to consumers. Therefore, to the cattleman struggling to keep his livestock healthy and his operation economical, or to the meat processor striving to combat botulism and other potent poisons, the benefits of chemicals could not be more clear. But the fear of health hazards from this ever-increasing list of chemicals is very real, and to the consumer, the risk could not be more serious.

The concern about chemicals in meat is part of a larger concern in the new jungle: is it safe for Americans to eat as much meat as we do? In recent years researchers have discovered unsettling evidence that Americans should not only be concerned about the additives that go into meat; they should also be concerned about the meat itself.

The United States has become a country where eating too much, not too little, is a major problem. In the twentieth century, the American diet has become rich in meat, dairy products, alcohol, and processed foods, all laden with fats, cholesterol, sugar, salt, and calories. In the past few decades, public health researchers have linked this diet to many of the major killing and crippling diseases, including heart and blood-vessel disease, stroke, high blood pressure, diabetes, cirrhosis of the liver, and cancer of the colon and breast. Meat, with its abundance of fat, has become a major source of concern. There is research indicating that diets high in saturated fats such as those found in most meats lead to increased risk of colon cancer in men and breast cancer in women. There is concern that even the ways meat is cooked can cause cancer: when fat is combined with the chemicals in charcoal, a reaction takes place that is potentially harmful; when meat is heated at high temperature on a metal surface, a feared carcinogen called benzopyrene can be formed.

In light of these concerns, Foreman pushed a campaign to study the American diet and even to recommend changes in it. One result was that in 1980 the USDA and the Department of Health, Education, and Welfare issued the report "Dietary Guidelines for Americans," a first step toward a national nutrition policy that could reshape the American diet. The guidelines were hardly revolutionary, recommending in summary that Americans should "eat a variety of foods; maintain ideal weight; avoid too much fat, saturated fat, and cholesterol; eat foods with adequate starch and fiber; avoid too much sugar; avoid too much sodium; if you drink alcohol, do so in moderation." There were no direct statements that any one product was bad for you, but there were implications. Particularly, there were implications that we should eat less meat—especially the succulent prime and choice meats most abundant in fat.

These subtle implications were enough to set off the meat industry. "By repeating statements about saturated fats and cholesterol, the Government is perpetuating a theory that is unproven," the National Cattlemen's Association complained when the report was released. The Cattlemen and other industry groups said it was premature for anyone to recommend dietary changes, and they pointed to research (much of it funded by the meat industry) claiming that present dietary habits pose no confirmable health hazards. Most of all, they complained that when the government starts telling people what to eat, it is getting into an area where it has no business. But Foreman had planned to push further into this area. She had expected to expand on the "Dietary Guidelines" and to continue pursuing all the serious questions of the new jungle. Now that she is gone, the meat industry can expect an easier time. There will be less talk about chemicals and the American diet from within the USDA. But Foreman was hardly alone in raising questions about the healthfulness of meat, and the pressures from outside the USDA will only increase.

This new jungle has the meat industry extremely touchy these days. Health fears, combined with high prices, have resulted in a decline in meat consumption that is expected to continue. Beef producers are particularly concerned, because through the first half of the nineteen-seventies they had happily watched consumption rise to the highest levels ever: in 1976 every American ate an average of nearly 130 pounds of beef. Now that figure has dropped to just over 100 pounds, back to the level of the mid-sixties. Meatmen are understandably wary of the economic implications of this trend. But they are equally upset that now they also must contend with attacks on the healthfulness of meat. Until recently, meat went unquestioned as one of the most important sources of protein and other nutrients. It was one of the pillars of the Basic Four Food Groups that every American was taught to worship. Now a growing number of critics are chipping away at meat's reputation, and they keep finding more cause for concern. The meat industry knows it must aggressively defend its products.

The first strong indication to meatmen that they were in for some trouble came in 1970 when Frances Moore Lappé's book *Diet for a Small Planet* was released. This book complained about the "incredible level of protein waste built into the American meat-centered diet" and raised questions about the tie between America's rich diet and hunger elsewhere in the world. The meat industry quickly dismissed the book as the ravings of a food faddist. Nevertheless, it sold more than a million copies, marking the beginning of a new awareness of diet and nutrition that was to grow during the 1970's.

The next major blow came in 1976, when Senator George McGovern sensed enough concern about the American diet to call for a study by his Senate Select Committee on Human Needs. The committee held hearings and in January 1977 issued a report, "Dietary Goals for the United States." The report was intended to be simply a guide of nutritional information, so neither McGovern nor the staff director in charge of the project expected it to be particularly controversial. But they had overlooked the implications of one brief recommendation. It read: "Decrease consumption of meat."

The meat industry was irate when it found out the government was telling the public not to eat its products. This time, the industry counterattacked. Such groups as the National Cattlemen's Association and the National Livestock and Meat Board pressured McGovern to withdraw the report. McGovern, as the senator from a heavily agricultural state, was vulnerable to the wrath of meat producers, so in the spring of 1977 he called additional hearings. The meat producers presented their attacks, and McGovern's staff members and consumer groups defended the report. "We wrote the report with simple guidelines for meals, and one of the things we wanted to say was eat less meat," recalls Nick Mottern, who was on the McGovern committee staff at the time. "We got away with a lot because nobody knew we were doing it; so the first edition was not touched by politics." Mottern, now an earnest, aging activist with a slight paunch and a scraggly beard, doesn't believe any of the testimony presented at the spring hearings refuting the report's recommendations. But when a revised report was issued that fall, the simple four-word recommendation on meat had been changed. It now read: "Decrease consumption of animal fat and choose meats, poultry and fish which will reduce saturated fat intake." That may not seem like a major difference, but Mottern points out that the idea behind the report was to present information that would be easy for consumers to understand and use. By changing the recommendation, he charges that the staff deliberately obscured the importance of reducing meat consumption and misled the public. He argued vociferously against the change, and when he lost he resigned from the committee staff. "Every time you attempt anything in terms of public information that hurts an interest in the food industry, there's concerted action to cut it off," Mottern says now. "Most people don't realize how information is being manipulated."

"Manipulation" is a strong word, maybe too strong. But the nutritional value of meat is a sensitive topic, and the McGovern incident taught the meat industry to gear up to fight for its products. The industry is now well prepared to marshal its strong lobbying forces both in Washington and in the home districts; legislators from agricultural regions know they will pay a heavy price if they go against the industry position. In addition, the industry has always operated strong educational and public-relations programs to keep Americans thinking positive thoughts about meat. Trade groups are a major provider of educational materials on nutrition (obviously stressing the importance of meat) and they are always looking for new ways to hype their products. In 1980 the California Beef Council asked its members for $4.6 million to fund an advertising and marketing campaign to sell "sensual" beef. "Sex sells everything else," said a spokesman for the council, "so why not beef?"

The industry is also heavily involved in research aimed at reassuring the public that meat is a healthful food. Its experts can point to studies that rebut all the charges of potential health hazards associated with meat. Unfortunately, it is here that the charge of manipulation may not be too strong. Ideally, scientific research should be the objective middle ground where health concerns could be calmly and rationally investigated. But science has fallen prey to politics, and the motives, as well as the results, of research conducted by groups that have chosen up sides on the debate over the healthfulness of meat have become suspect.

There are many subtle ways to bias research or to structure studies so they will support preconceived positions. An industry like the meat industry is in a particularly good position to support research that will tell it what it wants to hear. Scientific research of all types relies on grants from the industries that will be affected, so it should be no surprise that sometimes research can be bent to serve the needs of industry. There can be a tendency for researchers to concentrate on how new technology and new chemicals can serve industry rather than to question whether these new applications of science and technology might have detrimental consequences. In addition, industry can woo researchers and professors with consulting fees and appointments to boards of directors, so it can become difficult for researchers to bite the hand that feeds them. More than a few researchers and professors have found grants and research opportunities hard to come by after their research has gone against industry's position.

One of the most distressing controversies over research surfaced in 1980 when a report, "Toward Healthy Diets," was released by the Food and Nutrition Board, an advisory group of the National Academy of Science. The academy is a respected governmental advisory body originally established by Abraham Lincoln to give the government unbiased advice, but this report badly damaged its reputation. The report concluded that the dangers of cholesterol had been exaggerated: for healthy Americans there is no reason to limit cholesterol and fat intake in an attempt to avoid heart disease. It also questioned the link between diet and health that was at the base of McGovern's "Dietary Goals" and Foreman's "Dietary Guidelines." This was the first high-level study to reject the theory that cholesterol is closely linked to heart disease, and a shocking rejection of the conclusions of most recent research.

Cholesterol is a fatty substance produced by the liver that is essential to the structure of cell membranes and to the production of certain hormones. But it is also carried in the bloodstream, and it has been found to be an ingredient of the waxy atherosclerotic deposits that clog the coronary arteries and cause heart attacks. Foods that are high in cholesterol, such as eggs, beef, pork, cheese, and butter, have therefore become the targets of researchers and nutritionists concerned with limiting the intake of cholesterol. And the meat, dairy, and egg industries, of course, have been anxious to find evidence refuting the cholesterol-heart disease link. When the Food and Nutrition Board report was released, the industry groups were ecstatic. "We are pleased that finally sound science has come back into the arena, rather than politics we've felt we've been in for some time," said Peyton Davis of the National Livestock and Meat Board.

But to others, politics is precisely what the Food and Nutrition Board report reflected. It turns out that several members of the research team that prepared the report were paid consultants to food producers affected by the controversy over cholesterol. The scientist who presented the report to the media was a paid consultant to the egg industry; two more scientists had conducted studies funded by the egg, meat, and dairy industries; and the chairman of the board was a consultant for food companies including Pillsbury and Kraft. The report was immediately attacked by groups ranging from the Consumer Federation to the American Heart Association. The Food and Nutrition Board's own Consumer Liaison Panel disassociated itself from the report. The consumer panel, comprising representatives of consumer and nutrition research groups and government, complained of the conflicts of interest among the researchers and charged that the researchers ignored evidence against cholesterol. "We can only conclude that the board is dominated by a group of change-resistant nutrition scientists who share a rather isolated view about diet and disease," reads a statement from the consumer panel.

Other scientific research organizations have also been attacked for their slavish defenses of industry positions. The most notorious is the Council for Agricultural Science and Technology (CAST), an organization for farmers, agricultural scientists, and industrialists that is based at Iowa State University. It supposedly gathers the best minds from universities across the country to study agricultural issues. But it has been called by detractors the "agro-industry truth squad." Virtually all of its reports back the industry line, and some have sparked divisive internal controversies. For example, in 1978 several academic scientists resigned from a CAST task force studying the use of antibiotics in animal feeds. Six of the scientists signed a sharp letter of protest accusing CAST of omitting from a draft of the final report evidence on the risks associated with drug use. Instead, the report stressed favorable evidence on the benefits of drugs. In a review of the controversy, *Science* magazine noted dryly, "CAST devotes much of its time to showing the federal government why chemicals used on the farm are less dangerous than someone has claimed them to be."

That attitude continues. In 1980, CAST issued "Foods from Animals: Quantity, Quality, and Safety," a report that runs down the laundry list of charges against meat. The report purports to be an objective discussion of the pros and cons, but it is hardly an even-handed look at the serious controversies facing the meat industry. Instead, the report is obviously structured to refute criticisms of meat. To read the report is to learn that all the fears about meat are groundless; it blithely dismisses concerns on such topics as the banned growth stimulant DES ("the cancer risk factor from DES residues in edible beef tissues is essentially zero") and cholesterol ("the research of a quarter century has not clearly confirmed the hypothesis" of a link between cholesterol and heart disease), while hardly acknowledging the substantial body of unfavorable research on these topics.

But none of this should be particularly startling. Meat is undergoing an unprecedented amount of scrutiny, so meatmen must find means to defend their products and their way of life. At stake are the livelihoods of the men and women responsible for the largest portion of the food dollar—from the farmers growing the grain used to feed the livestock, to the ranchers raising the livestock, to the packers, processors, and supermarkets preparing and selling meat products.

So of course they find research to support their positions and of course they attack those people who question their products. They will fight the Carol Foremans, who are indeed a threat to the way they have traditionally done business. Even Foreman realized her rules and regulations would not be accepted easily by the meat industry. "Most of them force the food producers to change some of their long-cherished practices," Foreman said. "They resist them in every way they can, and some are sure that either they or the consumer will be devastated by them."

Early in 1980, Foreman looked back on the uproar she had caused in the meat industry. "Three years ago some of the farm and food industry was in a total panic about a consumer advocate coming into the department," she said. "I can't see that the food industry isn't better off now than before. And I'm sure the public is." Less than a year later, Foreman was gone, but the concerns she raised about the integrity of the USDA and the healthfulness of meat are very serious, and they won't go away.

COMPTROLLER GENERAL'S
REPORT TO THE CONGRESS

IMPROVING SANITATION AND FEDERAL
INSPECTION AT SLAUGHTER PLANTS:

HOW TO GET BETTER RESULTS
FOR THE INSPECTION DOLLAR

Inspectors from the U.S. Department of Agriculture's Food Safety and Inspection Service monitor meat and poultry slaughter plants doing business in interstate commerce to make sure that plant operations are sanitary and that products are wholesome, unadulterated, and properly marked.

GAO made this review to evaluate how well the inspectors were carrying out their responsibilities, whether plant managers were complying with inspection program requirements, and how efficiently the Department was using its inspection resources.

Assisted by Service supervisors, GAO made unannounced visits to 62 randomly selected meat and poultry slaughter plants in six States to evaluate plant and inspection staff compliance with inspection program requirements. Sixteen, or 26 percent, of the plants--27 percent of the meat plants and 24 percent of the poultry plants--were not in compliance with one or more of the six basic inspection program requirements.

Eleven plants were unacceptable in sanitation, 7 in pest control, 4 in controls over condemned and inedible materials, 2 in ante mortem and post mortem inspection, and 1 in water supply potability. All plants were acceptable in sewage and waste disposal.

Based on the results at the 62 plants, GAO estimates that 44, or 18.4 percent, of the 238 plants sampled in the six States--181 meat plants in four States and 57 poultry plants in two States--do not acceptably comply with one or more inspection program requirements. Further, 24.9 percent of the meat and poultry slaughtered in those 238 plants comes from plants unacceptable in one or more requirements.

CED-81-118
JULY 30, 1981

The high incidence of unacceptable ratings and the large number of deficiencies found at plants not bad enough to warrant unacceptable ratings show that both plant managers and inspection program staff are not fully meeting their responsibilities.

## SANITATION PROBLEMS

Of the 62 randomly selected plants, 18 percent received unacceptable sanitation ratings. The majority of the remaining plants had numerous sanitation deficiencies that the reviewers did not consider serious enough to warrant unacceptable ratings but which indicated that inspectors and plant managers were not fully carrying out their responsibilities. In some cases plant managers appeared to rely extensively on inspectors to identify sanitation problems rather than having their own controls over sanitation.

Some of the sanitation deficiencies noted were

--condensation dripping on, and thereby contaminating, carcasses;

--very dirty overhead structures and equipment;

--dead flies on work surfaces that meat contacts; and

--meat dragging through dirty drip trays.

In some cases, the deficiencies were due to inadequate cleanup from the prior day's work. In other cases, the deficiencies were more long term.

## PEST CONTROL

The seven plants with unacceptable pest control programs had rodent, insect, or insecticide problems. Rodent problems existed in storage rooms and buildings and in maintenance areas. Some plants had fly or other insect problems inside and outside. One plant had inadequate controls to assure that insecticides were used properly.

## WATER SYSTEM PROBLEMS

Only one plant received an unacceptable water supply rating. However, 39, or 63 percent, of the 62 random sample plants had water system deficiencies that could result in contamination of the plants' potable water supplies. These deficiencies included improper cross-connections between potable and nonpotable waterlines and inadequate back-siphonage protection. (See pp. 21 and 22.)

ACCEPTANCE TESTING PROGRAMS
OF QUESTIONABLE VALUE

The Service's acceptance testing programs for cattle and poultry carcasses are ineffective because they are not being conducted as designed. The programs were designed as statistically valid random sampling programs, whereby sample results would be indicative of slaughter dressing defects (such as grease, hair, or bruises) of the universe sampled. However, inspection and plant personnel carrying out the programs invalidated them by substantially deviating from the prescribed sampling plans and methods. (See pp. 24 to 29.)

ANTE MORTEM AND POST MORTEM
INSPECTIONS AND CONTROLS OVER
CONDEMNED AND INEDIBLE PRODUCTS

Ante mortem and post mortem inspections and controls to assure that condemned and inedible products are not sold as edible products were generally adequate. However, some deficiencies existed in plant facilities and equipment and in inspection procedures. (See pp. 31 to 40.)

Service inspectors devote a significant portion of post mortem inspection time to examining meat carcasses for dressing defects (the presence of contamination or unwholesome or inedible parts) that plants failed to remove. The plants, not the inspectors, should be responsible for checking for dressing defects. (See pp. 37 and 38.)

INSPECTION STAFF SHORTAGES

During 1980 from 6 percent to 10 percent of the Service's slaughter inspection positions were unfilled. As of February 21, 1981, the Service had a shortage of about 7 percent among its authorized 5,995 slaughter plant inspectors. The shortages were due to hiring and budget restrictions.

Because of these shortages, certain inspection responsibilities had been neglected, including supervising line inspectors, performing acceptance tests, monitoring plant conditions and operations, and inspecting processing departments. (See pp. 42 to 45.)

BETTER MONITORING NEEDED

The deficiencies at the randomly selected slaughter plants show that Service supervisors need to better monitor plant and inspection staff compliance with program requirements. One problem is the lack of adequate guidance to supervisors as to what constitutes an acceptable level of compliance. (See pp. 47 to 50.)

RECOMMENDATIONS TO THE SECRETARY OF AGRICULTURE

To better assure that meat and poultry plants produce only wholesome and unadulterated products, the Secretary should direct the Service, among other things, to:

--Strengthen its enforcement of sanitation requirements.

--Require plant managers to fulfill their responsibilities for operating and maintaining plants in a sanitary manner through a system of financial disincentives. Imposing some disincentives, such as levying fines, would require the Secretary to obtain legislative authority.

--Initiate a special one-time effort to identify and correct water system deficiencies and take action to prevent recurrences.

--Revise, and require that Service inspection staff follow, procedures governing the quality acceptance testing program.

--Make improvements governing inspection of edible and inedible meat and poultry products.

--Take actions to assure more effective monitoring of meat and poultry inspection activity.

# Gumshoeing The FD&C Act At The Grass Roots Level

*Foods and drugs are processed in every corner of the country, as are medical devices and radiation-producing products. Likewise, blood banks are to be found in countless communities. To keep up with the work in those many establishments, the Food and Drug Administration has 128 resident inspection posts located in all 50 States and Puerto Rico. This article tells how the investigator from one of those posts, the one at Salisbury, Md., does his job.*

*by Roger W. Miller*

It was quarter to seven in the morning. The sun was weak, the breeze unborn, clouds few. The sputter of a boat's motor, muffled by the harbor water, and the occasional lament of a gull were the scene's only sounds. The smells were of the sea—aromas of salt and the rot of things that lived in a different environment. Low-slung wooden buildings, their sides bearing testimony to the whims of climate, crowded the water's edge. A moored skipjack lay motionless, as if resting up for the next day.

Mike Ellison drove his frill-less, government-owned station wagon past a pile of oyster shells, the tires of the dark brown vehicle crunching the spillover. He stopped alongside one of the buildings where a simple sign read "office." A second, light-colored station wagon followed and lined up next to his.

Ellison emerged from the vehicle, taking a small black notebook—his "diary"—with him. He waited for the driver of the second car to join him.

"It doesn't look like they're working here," he said to the other driver as he nodded toward the nearly empty parking lot. Together the two headed for the office and disappeared inside.

Thus did the day officially begin for Ellison, an FDA investigator who works out of the Agency's resident inspection post at Salisbury, Md.

Resident inspection post is the official term used to designate the 128 local offices of the Food and Drug Administration. With one to eight employees, these offices are located in such places as Augusta, Maine; Tifton, Ga.; Lubbock, Tex.; Bellingham, Wash.; Mayaguez, P.R., and Anchorage, Alaska. Forty-three of the offices, including the one at Salisbury, are one-person operations.

The resident posts are in addition to FDA's 21 district offices, where other investigators as well as chemists, lab technicians, and support personnel work. The resident posts are, in effect, mini-districts with responsibility for specific geographic areas. Ellison's area of responsibility covers 10 counties, known collectively as the Eastern Shore of Maryland and Virginia because they lie east of the Chesapeake Bay.

Located between the Atlantic Ocean and the bountiful Chesapeake, the 10 counties live off the salt water and its legacy. The legacy includes a flat terrain and soil with sand from another geological period. The summers are long and hot on the Eastern Shore and the area's 40 inches of rain annually are well-spaced throughout the year. But the porous soil allows the rain easy escape, and the area's farmers use irrigation to assure that their fields will produce the intended tomatoes, asparagus, corn, peas, carrots, and cucumbers.

The salty water that surrounds the Eastern Shore is a farm of another kind, producing oysters, clams, bluefish, striped bass, and a delicacy that is favored by stomachs across the land. That delicacy is the Atlantic blue crab, a crustacean that may grow up to 8 inches across and that contains some of the tastiest seafood this side of heaven.

As an aquatic farm, Chesapeake Bay is extraordinary. From its waters come half of the Nation's clam catch and one-fourth of its oysters. The blue crab harvest from the bay accounts for about half the U.S. total, or about 200 million of the creatures each year.

It was crabs that Ellison and his companion were seeking that early morning, but the critters were to prove elusive. The quest was for a crab picking and packing plant so that Ellison could do a routine inspection while his companion, microbiologist Ammon Swartzentruber from FDA headquarters, could collect samples as part of a nationwide microbiological survey of crab plants.

For this day, they had selected a crab plant in Cambridge, Md., a town on a tributary of the Chesapeake Bay just 28 miles from Salisbury. (Actually, Ellison's day had begun at 5:50 in the morning when he met Swartzentruber at the Salisbury office.) The plant was one of 45 that Swartzentruber and his cohorts from the Bureau of Foods were to check in the survey. For Ellison, it would be just another routine inspection.

But after conferring with officials at the plant, the two learned that no crabs would be cooked, picked, or packed there that day. It seems that the price of the creatures had jumped over the weekend, as buyers for restaurants sought to meet public demand and purchased the blue fellows by the bushel when the fishermen's boats reached the docks.

After conferring with Swartzentruber, Ellison decided to switch the four big boxes he carried in his station wagon to Swartzentruber's vehicle, while the microbiologist went back to the Salisbury office to do paper work until Ellison would be free later in the day. The boxes were used to keep the 8 pounds of crab meat collected each day in the survey. Inside the boxes were various commercial cooling packs that keep the samples at temperatures under 4 de-

*September 1981 / FDA Consumer*

39

grees Celsius (40 degrees Fahrenheit). The samples are shipped via commercial courier to FDA's microbiology laboratory in Minneapolis (see "Microbe Sleuthing in Minneapolis," FDA CONSUMER, Dec. '79–Jan. '80). At Minneapolis, the samples undergo five microbiological determinations. Swartzentruber tries to work his surveys in with the routine inspections of investigators so as to minimize the disruption for a firm.

In his inspection, Ellison would look for the presence or evidence of flies, rodents, or birds in the plant; the handling of ice; the strength of hand-sanitizing solutions; and evidence of sores, tape, or other such abnormalities on the hands of employees. Inspecting a crab plant is different than inspecting most food plants because the meat is handled directly by workers and because the product must be kept cool. Crabs are bottom dwellers and are usually cooked while still alive. Eating bad crab meat may not result in a one-way trip in a long limousine, but it may make one wish he or she were on that trip.

Not being able to inspect the crab plant hardly caused Ellison to miss a stride. In no time at all, he had decided to drive to the other side of town and inspect another seafood plant. "We spend half our time on seafood and half on vegetables," Ellison explained. He uses "we" frequently, although he works alone virtually all of the time. Asked about that first person plural, he explained that he means "FDA and me."

While food takes up most of his time, Ellison has other, typical FDA investigator duties. The 10 counties he covers also contain one major medical device plant that makes emergency oxygen units, a small drug plant that produces a bulking agent for laxatives, a couple of research farms, a couple of experimental laboratories, and medicated feed manufacturers who produce chicken fodder for the Eastern Shore's sizeable broiler and fryer industry. In addition, Ellison "and FDA" inspect blood banks and audit State contract inspection programs of food warehouses and feedmills.

In a typical week, Ellison likes to start off with a major inspection of vegetable processing or a seafood firm and work in other projects along the way. For example, the week that started out with elusive crabs was rounded out with more thrusts at crab packing plants and inspections of an oxygen repacker and a commercial ice firm in the resort town of Ocean City, Md.

Guidance for the resident post inspectors comes from the monthly work plan sent down by the district office. In Ellison's case, the plan comes 103 miles from Baltimore. However, the work plans aren't always that easy to follow, what with interruptions by shortages of crabs, recalls, and other emergencies.

Ellison's resident post and his duties are typical of resident posts around the country, and yet his work is different. It's typical in that he has the responsibility for seeing that the Food, Drug and Cosmetic Act and other laws entrusted to the Agency are enforced among the 252,700 people in his area. It's different in that most investigators spend much more of their time on drug matters and other concerns.

Ellison's new concern that day was a seafood plant, and it was not yet eight in the morning when he drove into the parking lot of the plant. Donning his white coveralls and a white hard hat, he armed himself with a flashlight, a thermometer, and a notebook. The notebook is his diary, in which he would record his observations as he tours the plant. The diary also keeps track of the day's other activities.

Inside he waited briefly for a plant official to greet him, and then moved to a conference room off the plant cafeteria where he made out a "standard notice of inspection" that he signed and gave to the official. The notice spells out the purpose of the inspection and the laws that are involved.

The plant manager joined Ellison in the conference room and together the two of them started out to tour the plant. Two assembly lines were working that day. The fish had come from Iceland in frozen blocks, the blocks were sawed up to the desired shape and size and moved along conveyor belts, the pieces were breaded or dipped in batter, baked and refrozen, or simply refrozen, and then packed in individual containers. The frozen-baked-frozen cycle keeps germ contamination possibilities to a minimum. Likewise, water used in batter is kept near freezing so that the batter stays close to 50 degrees to avoid the threat of toxicity.

Ellison first did a quick run through the plant so he would know where to concentrate on return. The run-through included storage areas, loading docks, the freezer room (where it's five below zero), the powerhouse (where electric capacitors are clearly labeled as containing PCB's), and the carton storage area.

Ellison was constantly on the lookout for signs of dirt and insects. In this plant, the search was made easier in the storage areas by white painted stripes about 18 inches wide along the edges of the rooms. The white stripe enabled Ellison, flashlight in hand, to detect evidence of rodents and insects. The stripe also serves as a border for the storage areas, telling employees to keep products away from the walls so that they can be inspected and turned over easier.

Rodents like to run along walls and insects gather in corners or seek dust to settle in. Ellison looked for rodent droppings or insect carcasses. Should he find signs of rodent infestation, he would go back to his vehicle for his black light that enables him to spot urine stains. But in this case, he found none. The plant manager explained that the firm has a pest control contractor who comes in every weekend.

In the production area, Ellison periodically tested the handwashing solution used by the workers. For this, he took a strip of paper out of a plastic vial. The strip was dipped into the solution. Ellison watched the dipped portion of the strip change color and then compared the color with coded colors on the label of the vial. On the first such experiment, the dipped color turned out to be a dark purple, indicating between 100 and 200 parts per million chlorine, an acceptable solution.

As the workers went off for a 10 a.m. break, the FDA man watched while tables where fish were handled were sanitized. He waited for the workers to come back on the job to see if they made use of the hand sanitation solution before resuming their tasks.

Ellison used his thermometer to test the temperature of the batter at several points along the assembly lines. His observations also included checking around equipment to see if any past accumulations indicated that the night cleanup crew at the plant was doing less than its job.

He watched as workers "stripped down" bags of batter mix, bread crumbs, and the like. Stripping down means taking off the outer layer of the bag so that no shipping dust gets in the product when it is dumped. The stripping might also reveal hidden insects.

Ellison's inspection also included a visit to the plant's records room where computers provide a processing and shipping history of each item turned out by the plant. This would be helpful should a product have to be tracked down in a recall.

The investigator declined an offer for lunch in the plant cafeteria and chose instead a nearby highway restaurant that he had learned about in his travels about the area.

After lunch, Ellison made a final run through the plant and then met with the plant manager to go over his findings.

"We didn't find anything out of line and so we don't want to spend any more time here than we have to," he told the plant official. However, he did collect information from the plant manager about his standing in the company, whom he reports to, etc. It was information that could come in handy for the Agency some day should an emergency involving the plant arise. Had Ellison found problems at the plant, he would have written up an FD–483 "observations of insanitary practices," which he would give to and discuss with management.

The inspection over, Ellison drove to a nearby pay phone to contact Swartzentruber back at the Salisbury office. The two arranged to rendezvous later to make a try at another crab packing plant.

By this time the day had darkened, the sun having lost a battle to the clouds. A breeze blew off the bay and a mist began to develop. Somewhere in the Chesapeake Bay a blue crab was about to make a fatal mistake by grabbing for the piece of eel on a waterman's trotline or making the no-return trip down the funnel of a crab pot. That crab would end up with its shell pried open by the deft actions of a picker, its meat plucked by swift fingers. Mike Ellison might be there observing as the crab met its fate. And Ammon Swartzentruber might select a bit of the victim's meat for a plane ride to Minneapolis. There, a microbiologist would examine it under a microscope, looking for taint.

But most likely the treasured meat from the blue fellow would go the commercial route, selling for as much as twice the value of porterhouse steak. And well worth it, the ultimate consumer might tell you.

Such is the stuff that resident inspection posts deal with.

*Roger W. Miller is editor of* FDA Consumer.

---

**A Consumer's Guide to Local FDA Offices**

*The following is a list of local Food and Drug Administration offices with full addresses and telephone numbers. The list is alphabetical by State with the two-letter State postal abbreviations in parentheses. There are 151 addresses: 21 district offices, 128 resident inspection posts, and 2 stations—Houston and San Antonio. Correspondence to the 128 should be addressed to "FDA Resident Inspection Post" and the appropriate address.*

**Alabama (AL)**

Suite 227
2112-11th Ave. S.
Birmingham 35205
(205) 254-1555

Ashbee Bldg.
2819 Springhill Ave.
PO Box 7354 Crichton Station
Mobile 36607
(205) 690-2161

Rm. 324, Aranow Bldg.
474 S. Court St.
PO Box 4974
Montgomery 36101
(205) 832-7116

**Alaska (AK)**

M/S 25
701 "C" St.
Anchorage 99513
(907) 271-5018

**Arizona (AZ)**

1314 N. Central Ave.
Phoenix 85004
(602) 261-3275/3280

301 W. Congress St.
Rm. 4P
Tucson 85701
(602) 792-6500

**Arkansas (AR)**

606 Garrison St.
Rm. 606
PO Box 1227
Fort Smith 72901
(501) 782-0911

Rm. 222, Tanglewood Office Bldg.
7509 Cantrell Rd.
Little Rock 72207
(501) 378-5257

**California (CA)**

2020 Milvia St.
Rm. 460
Berkeley 94704
(415) 486-3332

2202 Monterey St.
Suite 104E
PO Box 169
Fresno 93707
(209) 487-5321

Los Angeles district office
1521 W. Pico Blvd.
Los Angeles 90015
(213) 688-3776

Rm. 388, Fed. Office Bldg.
8th & "I" Sts.
PO Box 1115
Sacramento 95814
(916) 449-3274

880 Front St.
Rm. 4N34
San Diego 92188
(714) 293-5168

San Francisco district office
Rm. 526, Fed. Office Bldg.
50 U.N. Plaza
San Francisco 94102
(415) 556-0318

Rm. 108
586 N. 1st St.
San Jose 95112
(408) 275-7548

1600 N. Broadway
Rm. 220
Santa Ana 92706
(714) 836-2377

Rm. 106, Stockton Fed. Bldg.
401 N. San Joaquin St.
PO Box 1179
Stockton 95201
(209) 946-6306

300 S. Ferry St.
Terminal Island 90731
(213) 548-2332

### Colorado (CO)

Denver district office
500 Customhouse
Denver 80202
(303) 837-4915

United Bank Bldg.
Suite 470
201 W. 8th St.
Pueblo 81003
(303) 544-5277

### Connecticut (CT)

Rm. 339
US Courthouse & Fed. Bldg.
915 Lafayette Blvd.
Bridgeport 06603
(203) 579-5822

50 Founders Plaza
Rm. 308
East Hartford 06108
(203) 244-2529

### Delaware (DE)

Rm. 1202, Fed. Bldg.
300 South New St.
Dover 19901
(302) 674-0893

### Florida (FL)

Rm. G20, 400 W. Bay St.
PO Box 35069
Jacksonville 32202
(904) 791-3596

6501 NW 36th St.
Suite 200
Miami 33166
(305) 526-2920

Orlando district office
7200 Lake Ellenor Dr.
PO Box 118
Orlando 32802
(305) 855-0900

Suite 121, Taylor Bldg.
2574 Seagate Dr.
Tallahassee 32301
(904) 878-1735

Rm. 733
700 Twiggs Street Bldg.
Tampa 33602
(813) 228-2671

### Georgia (GA)

Atlanta district office
1182 W. Peachtree St.
Atlanta 30309
(404) 881-3162

OSHA Area Office, Suite "J"
400 Mall Blvd.
Savannah 31406
(919) 944-4106

Rm. 105, Tiff County Adm. Bldg.
PO Box 1709
Tifton 31794
(912) 382-5963

### Hawaii (HI)

Rm. 6320, P.J.K.K. Bldg.
300 Ala Moana Blvd.
PO Box 50061
Honolulu 96850
(808) 546-8387

### Idaho (ID)

1317 W. Idaho St.
PO Box 9246
Boise 83702
(208) 334-1319

PO Box 1672
Pocatello 83204

### Illinois (IL)

PO Box 418
301 W. Main St.
Benton 62812
(618) 439-9586

Rm. B-10
Fed. Bldg.
202 W. Church St.
Champaign 61820
(217) 398-5403

Chicago district office
1222 Post Office Bldg.
433 W. Van Buren St.
Chicago 60607
(312) 353-7379

Suite 105
2510 E. Dempster St.
Des Plaines 60016
(312) 297-1027

Rm. 204, US Post Office Bldg.
PO Box 536
Edwardsville 62025
(618) 656-3304

Suites A & B
297 E. Glenwood Lansing Rd.
Glenwood 60425
(312) 757-4828

Post Office Bldg., Rm. 320
100 NE Monroe St.
Peoria 61602
(309) 671-7293

Rockford Fed. Bldg., Rm. 120
211 S. Court St.
Rockford 61101
(815) 987-4324

Rm. 15, 600 E. Monroe St.
PO Box 1120
Springfield 62705
(217) 492-4095

2835 Belvidere Rd.
Rm. 115
Waukegan 60085
(312) 688-5462

### Indiana (IN)

Rm. 233, Fed. Office Bldg.
101 NW 7th St.
Evansville 47708
(812) 423-6871 (Ext. 286)

Suite 1111,
Anthony Wayne Bldg.
203 E. Berry St.
Fort Wayne 46802
(219) 422-6131

575 N. Penn St.
Indianapolis 46204
(317) 269-6500

Suite 418, Sherland Bldg.
105 E. Jefferson
South Bend 46601
(219) 232-4040

### Iowa (IA)

136 Fed. Bldg., B-09
1st St. at 1st Ave.
Cedar Rapids 52401
(319) 366-2411 (Ext. 507)

Rm. 302, US Post Office & Courthouse
4th & Perry Sts.
Davenport 52801
(319) 326-3323

631 Fed. Office Bldg.
210 Walnut St.
Des Moines 50309
(515) 284-4339

Rm. 219, US Post Office & Courthouse
6th & Douglas
PO Box 184
Sioux City 51102
(712) 233-3269

### Kansas (KS)

Rm. 112, Graham Bldg.
211 N. Broadway
Wichita 67202
(316) 267-6311 (Ext. 565)

### Kentucky (KY)

400 E. Vine St.
PO Box 1969
Lexington 40202
(606) 262-2312

Rm. 372B Fed. Office Bldg.
600 Federal Pl.
Louisville 40202
(502) 582-5237

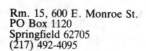

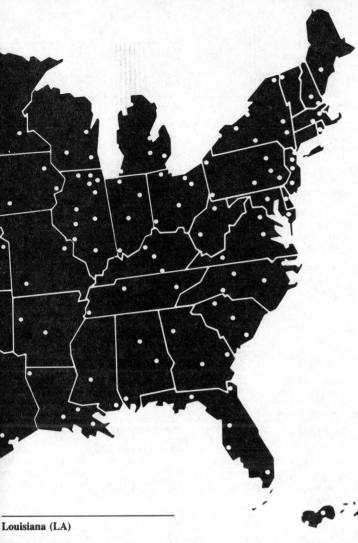

**Louisiana (LA)**

6130 Renoir Ave.
Baton Rouge 70806
(504) 389-0333

208 Eraste Landry Rd.
Lafayette 70506
(318) 235-2848

New Orleans district office
4298 Elysian Fields Ave.
New Orleans 70122
(504) 589-2401

Rm. 4A14
Fed. Bldg.
Shreveport 71101
(318) 226-5343

**Maine (ME)**

Rm. 401, Post Office & Fed. Bldg.
40 Western Ave.
PO Box 789
Augusta 04330
(207) 622-6171 (Ext. 268)

**Maryland (MD)**

Baltimore district office
900 Madison Ave.
Baltimore 21201
(301) 962-4012

1st Floor, Old Post Office Bldg.
PO Box 587
129 E. Main St.
Salisbury 21801
(301) 749-0540

**Massachusetts (MA)**

Boston district office
585 Commercial St.
Boston 02109
(617) 223-5066

Rm. 142, Post Office & Courthouse
436 Dwight St.
Springfield 01101
(413) 781-2420 (Ext. 397)

Waltham Fed. Center
Bldg. 104
424 Trapelo Rd.
Waltham 02154
(617) 894-2400 (Ext. 771)

WEAC
109 Holton St.
Winchester 01890
(617) 729-5700 (Ext. 32)

**Michigan (MI)**

Detroit district office
1560 E. Jefferson
Detroit 48207
(313) 226-6260

1007 N. Division
Grand Rapids 49503
(616) 456-2340

410 W. Michigan
Kalamazoo 49006
(616) 345-3203

Corr Bldg., Rm. 206
300 E. Michigan
Lansing 48933
(517) 377-1756

Rm. 73, Fed. Bldg.
100 S. Warren
PO Box 1228
Saginaw 48606
(517) 753-2061

**Minnesota (MN)**

Minneapolis district office
240 Hennepin Ave.
Minneapolis 55401
(612) 725-2121

**Mississippi (MS)**

Suite 302, Jefferson Bldg.
PO Box 1290
Biloxi 39533
(601) 435-4151

Suite 340
100 W. Capitol St.
Jackson 39201
(601) 960-4581

**Missouri (MO)**

Kansas City district office
1009 Cherry St.
Kansas City 64106
(816) 374-5646

St. Louis station
808 North Collins St.
St. Louis 63102
(314) 425-4189

Suite 312
300 S. Jefferson
Springfield 65806
(417) 865-3944 (Ext. 721)

**Montana (MT)**

Fed. Bldg. & Courthouse
Drawer 10035, Rm. 468
301 South Park
Helena 59626
(406) 449-5259

**Nebraska (NB)**

1619 Howard St.
Omaha 68102
(402) 221-4676

**Nevada (NV)**

Rm. 110
Fed. Office Bldg.
300 S. Las Vegas Blvd.
Las Vegas 89101
(702) 385-6361

1030 Bible Way, Rm. 8
Reno 89502
(702) 784-5770

**New Hampshire (NH)**

Rm. 342, Post Office & Courthouse
55 Pleasant St.
PO Box 488
Concord 03301
(603) 224-5511

**New Jersey (NJ)**

Suite 429-432
1800 Pavillion West
2101 Ferry Ave.
Camden 08104
(609) 757-5389

Newark district office
20 Evergreen Pl.
East Orange 07018
(201) 645-3023

Raritan Depot, Bldg. #17
Woodbridge Ave.
Edison 08817
(201) 494-5169

**New Mexico (NM)**

Rm. 4413, Fed. Bldg.
517 Gold Ave. SW
PO Box 1427
Albuquerque 87103
(505) 766-2022

**New York (NY)**

810 Leo W. O'Brien Fed. Bldg.
Clinton & N. Pearl Sts.
Albany 12207
(518) 472-6043

314 Fed. Bldg.
Henry St., PO Box 1093
Binghamton 13902
(607) 773-2752

New York district office
850 Third Ave.
Brooklyn 11232
(212) 965-5301

Buffalo district office
599 Delaware Ave.
Buffalo 14202
(716) 846-4478

Old Customs & Immigration
US Border Station
Meridan Rd.
Champlain 12919
(518) 298-8240

375 N. Broadway, Rm. 202
Jericho 11753
Mail to: PO Box 217
Main Branch
Hicksville 11802
(516) 433-2664

PO Box 3198
47 Route 17K
Newburgh 12550
(914) 562-2620

100 State St.
Rm. 328
Rochester NY 14614
(716) 263-6286

Rm. 1217, US Courthouse & Fed. Bldg.
100 S. Clinton St.
Syracuse 13260
(315) 423-5468

300 Hamilton Ave.
Rm. 309
White Plains 10601
(914) 683-9682

43

## North Carolina (NC)

Rm. DD-501
2500 E. Independence Blvd.
Charlotte 28205
(704) 371-6116

1207 W. Bessemer Ave.
Rm. 101
Greensboro 27408
(919) 378-5419

Fed. Bldg., Rm. 102
215 S. Evans St.
Greenville 27834
(919) 758-0738

Rm. 620
310 New Bern Ave.
PO Box 25730
Raleigh 27611
(919) 755-4474

## North Dakota (ND)

Rm. 242, Fed. Bldg.
657-2nd Ave.
Fargo 58102
(701) 237-5771 (Ext. 109)

## Ohio (OH)

Suite D
2872 W. Market St.
Akron 44313
(216) 375-5737

558 E. Main St.
PO Box 111
Canfield 44406
(216) 553-3919 (Ext. 3202)

Cincinnati district office
1141 Central Parkway
Cincinnati 45202
(513) 684-3504

Rm. 463
601 Rockwell Ave.
Cleveland 44114
(216) 522-4802

Rm. 231
New Fed. Bldg.
85 Marconi Blvd.
Columbus 43215
(614) 469-7353

Rm. 127, Fed. Bldg.
234 Summit St.
Toledo 43604
(419) 259-6347

## Oklahoma (OK)

PO Box 1683
722 N. Broadway
Magnolia Petroleum Bldg.
Oklahoma City 73101
(405) 231-4544

Rm. 3077, Fed. Bldg.
33 W. 4th St.
PO Box 3061
Tulsa 74101
(918) 581-7605

## Oregon (OR)

Rm. 694, Fed. Bldg.
511 NW Broadway
Portland 97209
(503) 221-2031

## Pennsylvania (PA)

Rm. 22, Fed. Bldg.
Sixth & State Sts.
PO Box 1796
Perry Square Station
Erie 16507
(814) 452-4518

Rm. 332, Fed. Bldg.
228 Walnut St.
PO Box 928
Harrisburg 17108
(717) 782-3414

Rm. 206, LBM Bldg.
Rts. 463 & 309
PO Box 576
Montgomeryville 18936
(215) 362-0740/41

Philadelphia district office
900 US Customhouse
2nd & Chestnut Sts.
Philadelphia 19106
(215) 597-4390

Suite 645
7 Parkway Center
Pittsburgh 15220
(412) 644-3394

US Post Office
Washington Ave. & Linden St.
Rm. 307
Scranton 18503
(717) 342-5699

## Puerto Rico (PR)

Cascade Salud Rm. 222
Regional Medical Center
Puerto Rico Department of Health
PO Box 3149
Marina Station
Mayaguez 00708

San Juan district office
Rm. 107, Post Office & Courthouse Bldg.
PO Box S-4427
San Juan 00905

## Rhode Island (RI)

152 Taunton Ave.
East Providence 02914
(401) 528-4613

## South Carolina (SC)

Rm. 210, US Customhouse
Market & E. Bay Sts.
PO Box 1077
Charleston 29402
(803) 724-4534

J. Marion Sims Bldg.
2600 Bull St.
Columbia 29201
(803) 765-5845

200 University Ridge
PO Box 171
Greenville 29602
(803) 235-3325

## South Dakota (SD)

US Fed. Bldg. & Courthouse
Rm. 212
515 9th St.
Rapid City 57701
(605) 348-3212

Rm. 107 Fed. Bldg.
400 S. Phillips Ave.
Sioux Falls 57102
(605) 336-2980 (Ext. 383)

## Tennessee (TN)

Suite 7002, 6300 Bldg.
Chattanooga 37411
(615) 894-3616

Rm. 4, 6415 Dean Hill Dr.
Knoxville 37919
(615) 584-9951

Rm. 56, Fed. Office Bldg.
167 N. Main St.
Memphis 38103
(901) 534-3684

Nashville district office
297 Plus Park Blvd.
Nashville 37217
(615) 251-5851

## Texas (TX)

Rm. 106
10 E. Elizabeth St.
PO Box 3267
Brownsville 78521
(512) 546-1123

Dallas district office
3032 Bryan St.
Dallas 75204
(214) 767-0317

Rm. 113-A
Paso Del Norte Bridge
1000 S. El Paso
El Paso 79901
(915) 543-7690

Bldg. 23, Rm. M-1
4900 Hemphill
PO Box 6398
Fort Worth 76115
(817) 334-5218

Houston station
1440 N. Loop
Suite 250
Houston 77009
(713) 226-5591

US Customs House
International Bridge
Rm. 102
PO Box 2593
Laredo 78041
(512) 724-2301

Rm. 415, Fed. Bldg.
US Courthouse
1205 Texas Ave.
Lubbock 79401
(806) 762-7648

Rm. 301
419 S. Main St.
San Antonio 78204
(512) 229-6735

## Utah (UT)

Rm. 1033, Adm. Bldg.
1745 W. 1700 South
Salt Lake City 84104
(801) 524-5285/86

## Vermont (VT)

Winston Prouty Fed. Bldg.
Rm. 210
Essex Junction 05452
(802) 951-6306

## Virginia (VA)

Rm. 309, 701 W. Broad St.
Falls Church 22046
(covers Washington, D.C. area)
(703) 285-2578

Federal Bldg., Rm. 821
200 Granby Mall
Norfolk 23510
(804) 441-3326

Rm. 11004, New Fed. Office Bldg.
7th & Marshall Sts.
PO Box 10048
Richmond 23240
(804) 771-2564

Rm. 757, Richard Poff Fed. Bldg.
210 Franklin Rd. SW
Roanoke 24001
Mail to:
PO Box 1554
Roanoke 24007
(703) 982-6380/81

## Washington (WA)

Rm. 314, Fed. Bldg.
104 W. Magnolia
PO Box 2783
Bellingham 98227
(206) 671-7150

Seattle district office
909 1st Ave.
Seattle 98174
(206) 442-5304

Rm. 773, US Courthouse
920 W. Riverside
Spokane 99201
(509) 456-4698

3803 W. Nob Hill Blvd.
Yakima 98902
(509) 575-5925

## West Virginia (WV)

Rm. 3201, Fed. Bldg.
500 Quarrier St.
Charleston 25301
(304) 924-1511 ask for 342-5772 (recorder)

Rm. 220, Fed. Bldg.
75 High St.
PO Box 909
Morgantown 26505
(304) 599-7410

## Wisconsin (WI)

828 Cherry St.
Green Bay 54301
(414) 433-3924

Rm. 306
212 East Washington Ave.
Madison 53703
(608) 264-5332

Lewis Center Bldg.
615 E. Michigan Ave.
Milwaukee 53202
(414) 291-3094

# Seizing Food And What To Do With It

*What happens when FDA finds food unfit for human consumption? It may be destroyed, it may be diverted to nonhuman use, it may be returned to the country of origin in the case of imported food, or it may—under certain circumstances—be reconditioned for use. This article looks at what happens to such foods.*

by Chris Lecos

One day not long ago Gerald E. Scholze left his home in the village of Ashwaubenon, Wis., and drove the 8 miles it takes to reach the one-room office he shares with Douglas Nelson in the football-obsessed city of Green Bay. Together, the two started to map out their plans for enforcing the Federal Food, Drug, and Cosmetic Act in an 11-county area of northeast Wisconsin. Scholze and Nelson are consumer safety officers—investigators, to be more precise—for the U.S. Food and Drug Administration.

Their schedule had been worked out a month ahead by FDA's district office in Minneapolis. But in the development of events this day, and several that followed, their carefully planned schedule started to come apart. The plan had called for a number of inspections to be made over the entire month, but a manufacturer's emergency recall of canned mushrooms contaminated by botulinal toxin—a dangerous food poison—took over top priority. It remained at the head of their priority list for most of the month. The mushroom recall covered 19 States.

Scholze and Nelson ascertained that four distributors in their 11-county area had obtained mushrooms from the Pennsylvania manufacturer; from the files and computers of each distributor they developed a list of some 300 food establishments, mainly restaurants, that bought the mushrooms. They alerted each food operator about the recall by phone or by direct visit, personally going to one-fourth of the places to verify that the mushrooms had been removed. In one restaurant, says Scholze, the manager pulled the mushrooms from the storage area but had "missed one that an employee had under the shelf."

The two FDA men, like their colleagues in field offices and resident posts throughout the country, have performed a routine task to help protect consumers from a contaminated food product. It is not unusual for investigators to concentrate their time and effort on what Agency officials describe as "high-risk work"—food, drug, and other problems with a high degree of risk or hazard to the public. It often means other inspections must be delayed or deferred.

The Food, Drug, and Cosmetic Act is intended to assure that the public gets food that is wholesome and safe to eat and that is stored and produced under sanitary conditions. FDA uses its regulatory and investigative powers to move against foods that the law defines as adulterated or misbranded. This includes food produced in the United States or shipped from foreign countries. Periodic inspections, voluntary recalls, and corrective actions by industry, educational programs, and legal proceedings, when deemed necessary, are employed by FDA to obtain compliance. But, say FDA officials, it is a function not always understood by the public.

Many people, says Scholze, believe FDA guarantees that foods on the market are safe and good to eat. Thus, when something goes wrong, they think FDA did something wrong. "They fail to realize that we can't get into a [food] plant that often, that we are only a regulatory agency that acts when it finds a violation. They also don't understand the enormous responsibilities we have in so many different areas." He is alluding to the fact that there are 150,000 food, drug, cosmetic, medical device, radiological, and other firms whose operations are within FDA's jurisdiction. The size of the food industry alone illustrates the difficulty of FDA's inspection responsibilities.

FDA currently is concerned with the operations of an estimated 79,000 food establishments. Most of these manufacture, process, ship, pack, label, and store a substantial portion of the food Americans eat each day. When one considers the sheer volume of food involved, one begins to understand why many FDA officials feel they face an almost impossible task in trying to monitor the safety of all foods, and why, in the view of many investigators, they can only scratch the surface of the total amount of food involved.

Donald C. Healton, executive director of regional operations (EDRO), which oversees FDA's field offices and activities from FDA headquarters in Rockville, Md., said recently that inspections of food establishments are being conducted on the average of once every 6 years. In 1976, it was once every 4½ years. During the 12-month period ending September 30, 1979, FDA conducted 12,855 inspections of food operations.

When a food contamination situation is considered serious or widespread, FDA has the option of filing a complaint and obtaining the ap-

*October 1980 / FDA Consumer*

45

*Random samples of bananas from Panama are removed by Import Inspector Mascari from boxes being unloaded via conveyor belt from a freighter docked in a slip off the Mississippi River. The samples will go back to the laboratory to be tested for pesticide residues.*

proval of a U.S. district court for a mass seizure of food. In this fashion food, good and bad in a single location, can be brought under Federal control to prevent contaminated foods from being sent into the marketplace.

Because the food standards and laws of other countries are not always as strict as those here, imported foods are a major concern to FDA. Again, the Agency's problem is one of dealing with volume. Some products from certain foreign outlets and countries are detained automatically because of past violation histories and poor standards. Fresh fish and seafoods of various types are spot checked carefully. During the 1979 fiscal year, ending September 30, there were nearly 5,977 detentions of imported food, resulting in more than $205 million worth of food that was either reconditioned under FDA supervision or was refused entry into the country, according to Richard R. Klug, assistant for import operations. For the 8-month period October 1979 through May 1980, Klug said, FDA district officers had detained 4,306 shipments of food valued at $185 million.

Under the Food, Drug, and Cosmetic Act, foods shipped to the United States are subject to inspection at the point of entry through U.S. Customs and can be detained by FDA and denied entry if they fail to comply with requirements of the law. Products that fail FDA's inspections sometimes can be brought into compliance through reconditioning if the methods to be used are first approved by FDA. If the detained shipments are not reconditioned satisfactorily, they must be exported or destroyed. The approval of a Federal court is not needed in an import detention because the product is technically not yet in the United States.

In the case of foods produced in the United States, a court-approved seizure is an involved, cumbersome process that begins with a detailed inspection of a food establishment, the preparation of a large volume of reports that spell out the violations, the gathering of evidence to back up FDA findings of adulteration of food, a series of administrative approvals at district and headquarters levels, and finally, the preparation of legal documents by a U.S. attorney who will file FDA's complaint in a U.S. district court. If a food seizure is approved by a Federal judge, U.S. marshals carry out the court's edict. The owner of the goods can contest the charges and demand a trial or, if he does not want to fight it, can enter into a consent decree that would permit condemnation. The owner then has the option of agreeing to the destruction of the seized goods or requesting an opportunity to recondition at least some of the foods under FDA's supervision so they can be used for human and/or animal consumption.

Unlike drug firms, which are required to register with FDA and which are subject to inspection once every 2 years, there are no registration requirements for most food operations in the United States. FDA's surveillance of domestic food operators is concentrated on manufacturing and processing plants, packing and labeling plants, grain elevators, and warehouses. Warehouses that store and distribute foods alone comprise about 24,600 of the 79,000 food establishments FDA has on its inventory of places deemed within its jurisdiction. The only exceptions to the nonregistration requirement are commercial producers of low-acid foods that are heated and packed in hermetically sealed metal, glass, or plastic containers, and acidified foods—those to which acid or acid food is added (beans, cucumbers, cabbage, peppers, etc.) who must register with FDA. This applies to U.S.-based companies and those in other countries which export their goods to the United States. The registration requirements were adopted by FDA in 1973 after Agency investigations revealed lax practices in canned food processing and inadequate safeguards against botulism, an often fatal form of food poisoning.

A comprehensive June 1976 study provides an insight into the Agency's surveillance and compliance activities. The study was prepared at a time when FDA was asked to re-examine its policy on "blending"—the mixing of adulterated with unadulterated foods—a practice which FDA currently prohibits. The 1976 study was based on a review of more than 1,000 court-approved food seizures, 195 industry recalls of foods, 1,985 "corrective action" reports by industry that resulted in the destruction and/or diversion of foods to nonhuman food channels, and 604 import detentions, selected to represent a statistical sample of 12,000 detentions. The data covered the fiscal years 1973 through 1975.

During the 3-year study period the report revealed that:

• Nearly 47 million pounds of food produced in the United States was either destroyed or diverted to nonhuman food uses as a result of FDA enforcement activities—seizures, recalls, and voluntary destruction—and another 115 million pounds of food offered for import was refused entry and exported or destroyed.

• Sanitation violations—those caused by rats, mice, insects, and birds along with nontoxic molds, and other filth and foreign materials—were responsible for the largest quantities of destructions and diversions from human food channels. More than 32.3 million pounds (69 percent of the total) of the domestically produced foods and more than 50.3 million pounds (nearly 44 percent) of detained imports resulted from such violations.

• Botulism, salmonellosis, and other microbiological contaminants were responsible for the destruction or diversion of 8.6 million pounds of U.S.-produced foods and 36.2 million pounds of imported foods.

• The rest, as illustrated by the table, were foods destroyed or diverted because of contamination by chemicals, metals, natural poisons, and such miscellaneous adulterants as glass particles in food.

The report noted that court-approved seizure actions had a big impact on domestically produced foods. During the 3-year period, the report said, 20.1 million pounds of seized foods had to be destroyed or diverted to nonhuman use. Other seized foods were reconditioned and some foods may have been distributed before the necessary legal action could be taken. FDA investigators point out that some food gets past them into normal food channels between the times a plant or warehouse is inspected and a court order is obtained to seize the food. The report also indicated that there is a direct relationship between the amount of food destroyed or diverted each year and the frequency with which the Agency is able to inspect food operations.

In an evaluation of 50 general categories of food commodities, the impact of enforcement actions was greatest on cereal and grain products produced in this country. Of the 47 million pounds destroyed or diverted, 18.6 million pounds—or more than 40 percent—were in this cereal/grain category.

Out of some 40 food categories of imported products, fresh fish and shellfish accounted for 26 percent of the foods refused entry. Most operators whose foods are denied entry generally choose to re-export to other countries. The report estimated that "probably less than five percent" of the detained imports actually were destroyed.

The 1976 report was prepared at a time when there was widespread concern over a projected food shortage in the world and FDA officials were asked to determine whether more food could be made available if FDA eased its prohibitions against blending. At that time, most FDA officials were generally opposed to any lowering of the standards, a policy that is unchanged today.

Said Healton recently: "We don't think people should take garbage and mix it with good food and make lots more garbage—which is really what you are talking about, because that would be deliberately hiding a contaminated product through the blending process." When food is seized, FDA requires that the contaminated products be segregated and either destroyed or reconditioned under FDA supervision, Healton explained.

The Federal Food, Drug, and Cosmetic Act contains no explicit language dealing with the issue of blending. In general, the law is a prohibition against the interstate distribution to humans and animals of foods that are adulterated or misbranded. Adulterated foods are defined as those that are defective, unsafe, unfit, decomposed, filthy, or produced under insanitary conditions. A food is misbranded if the label contains false, misleading, or incorrect statements. The law also concerns itself with prohibitions against the addition of "poisonous or deleterious" substances; it requires action if pesticide residues in foods exceed tolerances set by the Federal Government.

Court decisions back up the basic position that Government is not required to establish that an adulterated food is unfit and dangerous to health before it can take action. As one court put it: "There is no room for controversy over percentages of filth under the statute itself, for it excludes all." In another case, the court stated: "There can be no doubt that this section of the act was designed to protect the esthetic tastes and sensibilities of the consuming public and that the visible presence of such material in food would offend both." In this particular case, the defendant had argued that the statute was directed only to filth that was perceptible to the consumer, but the court rejected that argument, adding:

"To so interpret this section of the statute would largely deprive the people of the protection it seeks to give. The consumer ordinarily requires no governmental aid to protect him from the use of food products, the filthy adulteration of which he can see, taste or smell. What he really needs is governmental protection from food products, the filthy contamination of which is concealed within the product."

*Chris Lecos is a member of FDA's public affairs staff.*

An FDA
# CONSUMER MEMO

## Food Poisoning: The 'Infamous Four'

You don't feel good. Your stomach is upset. You have abdominal pains, diarrhea, perhaps a headache.

You may think you are coming down with the flu. That's possible. But, it's also possible you have fallen victim of one of the "infamous four."

These are four common sources of food poisoning—*Salmonella, Clostridium perfringens, Staphylococcus,* and *Clostridium botulinum.* Don't be floored by the big names. They are simple, little bacterial organisms—one-celled organisms that multiply by dividing. To divide they need food, warmth, and moisture. These bacteria may be found in all kinds of foods because they are everywhere in the environment.

The "infamous four" can cause illness or, in extreme cases, death.

Many people—from the farmer who produces food through the retailer who sells it—work to minimize the danger from these bacteria. The U.S. Department of Agriculture, or a State agency, inspects meat and poultry for wholesomeness. The Food and Drug Administration inspects plants where all other foods are produced.

But from the supermarket shelf to the dinner table you are the one responsible for protecting your food. Your responsibility begins in the supermarket. Don't buy any food in containers—cans or packages—that are outdated, broken, bent, or leaky. Especially avoid bulging cans—the food could contain lethal botulinal toxin. Let the store manager know there is something wrong so that he can take such packages or cans off the shelf.

Another thing you can do is make sure meat, poultry, and frozen foods are kept cold. Buy them last, if possible, so they don't warm up or defrost in the cart while shopping.

Make the grocery store your last shopping stop so you can go right home. At home, immediately put your meat and poultry in the refrigerator, which should be at 40 degrees F., or in the freezer at 0 degrees F. Put frozen foods in the freezer right away. This prevents or inhibits the growth of any bacteria in the food.

When you take fresh meat, poultry, or frozen foods out to cook, don't do any favors for the "infamous four." Thaw foods in the refrigerator, or cook them frozen. If you must thaw in a hurry, do so in a watertight plastic bag submerged in cold water.

Never thaw frozen food uncovered on a kitchen counter at room temperature. Remember, there are bacteria everywhere, and room temperature lets them grow.

The way the "infamous four" usually strike in the home is through cross-contamination. Since most bacteria are killed under proper cooking temperatures, an illness from cross-contamination can occur only if you've failed to take the right precautions.

Here's how that can happen:

Your chicken has been thawed and you have cut it up on your cutting board. You put the chicken into the frying pan and begin preparing other foods. If you use the same knife and cutting board without first washing the utensils and your hands in soap and hot water, you could contaminate whatever else you prepare.

For example, say you have cooked potatoes for potato salad. You slice them with the same knife on the same cutting board you used for the raw chicken without washing either the board, the knife, or your hands. You could easily spread *Salmonella* bacteria to the potatoes. If you take the salad on a picnic where temperatures are warm enough for bacterial growth, they will multiply rapidly to the levels where infection can occur.

Good habits in the kitchen can help prevent food contamination. For example, don't keep main dishes at room temperature for longer than 2 hours before serving. Bacteria grow rapidly at room temperature.

Follow simple hygienic practices. Use rubber gloves if you have cuts or sores on your hands. The gloves not only keep bacteria out of the sores, but also keep bacteria from the sores from getting on the food.

Put leftovers into the refrigerator as soon as the meal is over. Refrigerate them in small or shallow containers so they will cool quickly.

Pets also are carriers of bacteria. After handling pets, wash your hands before working with food. Don't allow pet feeding dishes, toys, or bedding in the kitchen. Don't let pets touch food, utensils, or food-working surfaces.

Here are some details about the "infamous four":

*Salmonella* (sal' mo-nel' a) is one of the most common causes of food poisoning. While it is not often fatal, is is widespread. More than two million cases of illness from *Salmonella* poisoning are believed to occur in the United States each year.

*Salmonella* is most commonly found in raw meats, poultry, eggs, milk, fish, and products made from them. Processed foods such as chocolate, yeast, casein, and spices have also been known to harbor this pathogen. Other sources can be pets, such as dogs, cats, turtles, birds, and fish.

There is no way to tell by looking at, tasting, or smelling the food whether *Salmonella* germs are present. To avoid them, however, do not handle food excessively and keep it below 40 degrees F. or above 140 degrees F.

*Salmonella* germs in food are destroyed by heat. So, always use a meat thermometer and cook foods thoroughly.

Also, heat leftovers thoroughly. Broth and gravies should be brought to a rolling boil for several minutes when reheating.

Symptoms of *Salmonella* food poisoning are fever, headache, diarrhea, abdominal discomfort, and occasionally vomiting. These appear in 24 hours after eating contaminated food. Most people recover in 2 to 4 days. Children under 4, elderly people, and people already weakened by disease could become seriously ill.

*Clostridium perfringens* (klos-trid' i-um per-fringe'-ins) is widely distributed in nature—in the soil, dust, on food, and in the intestinal tracts of man and other warm-blooded animals. It is more widespread over the earth than any other disease-causing microorganism.

Disease outbreaks frequently occur when foods are held in large quantities at improper temperatures for several hours or overnight. Perfringens outbreaks are closely associated with restaurants or other large feeding establishments where foods are held for long periods of time on steam tables or other warming devices.

To avoid *Clostridium perfringens* food poisoning, meats should be properly cooked, held hot (above 140 degrees F.) and served hot. If you cook meat for later use, cool the meat rapidly in small containers in a refrigerator to 40 degrees F. or below. Thoroughly reheat leftover meats or meat dishes before serving—and bring leftover gravy to a rolling boil before serving. Maintain cold sliced meats and other cold cuts below 40 degrees F. and serve them cold.

Large numbers of perfringens bacteria can cause diarrhea and abdominal pain in 4 to 22 hours—usually in about 12 hours.

*Staphylococcus* (staf' y-lo-kok' us)—known as staph—is quite common. Staph organisms are in your respiratory passages and on your skin. They usually enter food from a human or an animal.

Staph germs grow in a wide variety of foods—all meats, poultry and egg products, egg, tuna, chicken, potato or macaroni salads, cream filled pastries, and sandwich fillings. If staph germs are allowed to multiply to high levels, they form a toxin that cannot be boiled or baked away.

Symptoms of staph food poisoning are diarrhea, vomiting, and abdominal cramps. They occur 1 to 7 hours after eating and may persist for 24 to 48 hours. Staph food poisoning is rarely fatal.

*Clostridium botulinum* (klos-trid' i-um, bot'u-li'num), while rare, is often fatal. A tiny amount of the toxin from botulinal microorganisms can kill you. Botulinal spores are found throughout the environment and are harmless. However, in the proper environment, and when not destroyed by heat, the spores divide and produce a poisonous toxin. High heat makes the toxin harmless.

Only seven deaths from botulism in U.S. commercially canned foods have been reported since 1940. But poisoning from home-canned foods happens more often. About 700 people have died of botulism since 1925 from eating contaminated home-canned products.

Undercooking is the real culprit in home canning. In general, high-acid foods may be canned by boiling; but all others, including meats and poultry, should be canned in pressure cookers at the appropriate heat level for the required length of time.

Signs of botulism begin 12 to 36 hours after eating the contaminated food. The signs include double vision, inability to swallow, speech difficulty, and progressive paralysis of the respiratory system.

Medical help must be obtained fast. If botulism is suspected, a doctor should be called immediately.

To guard against botulism, inspect all food containers before opening and serving—whether commercially canned or home canned. Avoid food in containers with leaky seals, and bent, broken, or bulging cans. DO NOT taste the contents! Do not even open a bulging can. It takes only a small amount of botulinal toxin to be fatal.

*Salmonella, Clostridium perfringens,* and *Staphylococcus* are invisible and can be detected only by a bacteriological test. Botulism may provide its own clues in canned foods. Bulging cans, an off color, spoilage or other unusual odor, or suspect appearance are the tip-off. If in doubt: DON'T EAT IT! DON'T EVEN TASTE IT! If it's home canned, dispose of it where no person or animal can get to it. Save any commercially canned product you suspect and report it to the Food and Drug Administration, or your local public health officials.

The "infamous four" are everywhere, but with proper food handling practices, you can prevent them from causing illness in your home.

## FOR MORE INFORMATION

The Food and Drug Administration publishes considerable information about food safety. For more information about this subject, you can write your nearest FDA office, or Food and Drug Administration, HFJ-10, 5600 Fishers Lane, Rockville, Maryland 20857.

U.S. DEPARTMENT OF HEALTH,
EDUCATION, AND WELFARE
Public Health Service
Food and Drug Administration
5600 Fishers Lane
Rockville, Maryland 20857
HEW Publication No. (FDA) 80-2046  Revised October 1979

# ADDITIVE AVALANCHE

# Section Two

## ADDITIVE AVALANCHE: Turning the Tongue Against the Brain

The food processing giants generate more sales and profits whenever they can induce you to buy food because it is soft, tasty, colorful and convenient to prepare. But buying food for those characteristics in no way assures that such foods are also nutritious, sanitary, free from harmful additives and fairly priced. Often, just the opposite is assured.

The modern chemical technology of food too frequently enriches the food sellers at the expense of consumers, especially vulnerable children. The article "Let Them Eat Junk," by New York Times reporter William Serrin, in *The Saturday Review,* provides a good overview of what the food conglomerates are doing for disease and appetite.

What purposes do these food additives serve? This question is addressed by Phyllis Lehmann's graphic article "More Than You Ever Thought You Would Know About Food Additives. . . Part III," which appeared in *FDA Consumer*. To color, to flavor, to enhance flavor, to sweeten, to stabilize and thicken are a few of the purposes pursued by additives. Salt or sodium intake is analyzed for its adverse health effects in an interview with Dr. Alan L. Forbes of the FDA's Bureau of Foods.

The debate over which additives cause cancer rages year in and year out and can be confusing. The National Cancer Institute has produced a question and answer booklet titled "Everything Doesn't Cause Cancer," which discusses the oft-asked inquiry: "How do we know what substances cause cancer and are animal tests reliable?"

We thought you would enjoy a debate on the subject of animal tests, human cancer and the Delaney amendment of 1958 to the Food and Drug laws prohibiting the use of any food additive that induces cancer in man or animals. This debate appeared in *Chemical and Engineering News* and draws out many of the "pros and cons" that will be made in the coming months during the food industry's drive in Congress to cripple the Delaney safeguard.

An explanation of the Delaney amendment and its importance as a bastion against cancerous food additives is the contribution of Public Citizen's Health Research Group report, "Cancer Prevention and the Delaney Clause." This report is a standard reference for consumer groups and individuals engaged in defending the food supply. More facts and recommendations come from the lively Center for Science in the Public Interest (CSPI) in its articles "The Delaney Clause Draws Fire" and "The Invisible Additives." A page on CSPI's experience with the notorious Hot Dog may stay with you the next time you're tempted to ingest one of those pink missiles.

Reprinted from Saturday Review, February 2, 1980.

# LET THEM EAT JUNK

by William Serrin

## The Triumph of Food Processing

HUNDREDS OF MEN AND WOMEN laden with plastic shopping bags wander a vast midway of booths. Some wear straw hats, party favors from a turn-of-the-century gala the night before. All around them, on placards hanging from the roof of the Cervantes Convention Exhibition Center in St. Louis, are signs hawking their companies' products.
- Prefabricated pork chops triumph with H & R natural and artificial flavors.
- VD has the scoop on True-to-Life flavors.
- Nutrifox VDD sodium tripolyphosphate is the one for meat-curing applications.
- Now Amazio adds more eye and mouth appeal to processed foods faster than you can say "pregelatinized starches."

These men and women are food scientists. More than farmers, ranchers, or nutritionists, they determine what Americans eat. At a recent convention of the Institute of Food Technologists (IFT), they bustled up and down the corridors between display booths, nibbling on the products that the large food companies hope to place on the nation's dining tables: low-calorie watermelon punch, imitation vanilla cookies, fudge made from artificial chocolate, imitation cream cheese, imitation mozzarella cheese, imitation provolone cheese, sausages made with artificial meat flavor, popcorn flavored with imitation butter, freeze-dried raspberry yogurt chunks, peppermint-flavored mints.

Until the past generation, food scientists concerned themselves largely with techniques of preservation—canning, refrigeration, cellophane wrapping. The technology explosion that occurred during World War II and the decades afterward helped to create an industry more concerned with producing new foods in new forms than with preserving and transporting the existing ones economically and conveniently. Two new practices that had enjoyed immense growth in the 20 years before the war, market research and national advertising, were of great assistance in the birth of what amounted to a whole new enterprise—the modern food industry.

Today Americans spend $260 billion annually on food, almost half of which goes toward the purchase of highly processed items, including convenience and snack foods. We also eat about 40 percent of our meals away from home, spending an additional $105 billion on what is known as the food-service business. Food science is the backbone of this industry as well.

Not long ago the food business, as well as the food-service business, consisted of local entrepreneurs with their own factories, or warehouses, or shops. Today not only the frozen, wrapped, uniform food at the grocery store but also the fast-food restaurants, many of the farms, the food research groups, and even the seeds are controlled by a handful of huge conglomerates with names like General Foods, General Mills, Procter and Gamble, Kelloggs, along with such non-food companies as ITT and General Electric. Many of these firms sell literally hundreds of products and spend vast

**The cereal-makers, with no new mouths to conquer at home, have turned to the Third World.**

amounts of money on marketing and advertising to convince the public that the new items are somehow different from their predecessors. The lack of competition among these firms has led to charges that food prices are artificially high, as well as, in one case, an FTC antitrust suit. Companies have also been charged with manipulative and deceptive advertising. Critics of the food business are also distressed that the industry has managed almost to saturate the American market and is now looking abroad, especially to the developing nations of the Third World, for new consumers. This internationalization has raised the fear that America will soon be blanketing the world with frozen french fries. But all of these criticisms pale before the essential one: that food-processing companies have succeeded in debasing the American diet.

JUST WHAT HAVE technologists done to our food? The answers can be found not only in the foods they put on American plates but in the ideas they put into the minds of American consumers by means of marketing and sales promotion.

"Today's consumers really don't need anything in the way of new products," the research director for Libby, McNeill & Libby, the large Chicago-based food company, told the IFT a few years ago. Nonetheless, he contended, "they are constantly searching for something just a little better or different...." An aggressive company, he pointed out, must capitalize on this. Though one firm might look at a proposed orange-juice product and ask, "Who needs it?" a go-getter, the marketing man suggested, would know better. "A company could come along with an orange-juice product to which an additional color or sweetness ... or ... an important nutritional component has been added, and do well in the marketplace."

Enter the food technologist. He can add ingredients that extend the shelf life or keep processed foods stable so that, for example, chocolate pudding doesn't turn into a mess of separated layers of goo, all different colors. He can add substances that produce what the industry calls "fine surface gloss." He can simplify production, saving labor and its costs. Raw foods—milk, meat, eggs, grains, fruits, vegetables—are not important. They exist only to be simulated in the laboratory, with the manufactured copies transformed into new foods, those known in the trade as "fun foods" or "consumer hot buttons." Companies can capitalize on almost any characteristic of these products—the novelty, the taste, the cooking time, a new or unusual container—whatever might pique the fancy of the consumer thumbing through the newspaper food section or strolling the aisles of the supermarket.

"The technology is skewed toward anything that can make a buck and away from anything that improves quality," says James S. Turner, a Washington, D.C. attorney and author of *The Chemical Feast*, a study of the regulatory policies of the Food and Drug Administration. "Nothing is heard from the scientific community about quality....The scientists say to the companies, 'We can improve your sales,' and then they come up with the flavors, colors, and extenders that are added to food substances to make them appear to be food. But it's not food. We don't even know what food is in our society."

The overwhelming dominance of manufactured foods in the American diet belies the idea that we are experiencing a revolution in taste and cooking, a notion largely advanced by food editors and writers. A casual inspection of big-city newspapers and magazines suggests that everyone is puttering around a brick-lined kitchen, a gourmet cookbook in hand, the food processor purring, turning out some delicate, exotic dish. But the reality is that consumers have accepted food technology, and our ever-more homogenized diet is destroying the nation's rich culinary traditions.

Regional and ethnic distinctions are disappearing from American cooking. Food in one neighborhood, city, or state looks and tastes pretty much like food anywhere else. Americans are sitting down to meals largely composed of such items as instant macaroni and cheese, soft white bread, oleomargarine, frozen doughnuts, and Jell-O. "Today it is possible to travel from coast to coast, at any time of year, without feeling any need to change your eating habits," according to a brochure printed by the IFT. "Sophisticated processing and storage techniques, fast transport, and a creative variety of formulated convenience-food products have made it possible to ignore regional and seasonal differences in food production...."

The success of food scientists can be charted by the radical transformation of the American diet in recent years. Consumption of fluid milk, for instance, has slipped more than 30 percent since 1960; soft drinks are now the nation's number-one beverage. Potato products, largely frozen french fries, became 464 percent more popular between 1910 and 1976, a year in which more fries, in tons, were sold than any other frozen vegetable. In 1940, homemakers bought 50 percent of the flour sold in the United States; today they buy 11 percent. Most flour now goes into processed breads, cakes, pies, and buns sold at grocery stores and fast-food restaurants (40 percent of the beef consumed here is hamburger).

One of the great recent triumphs of food science is Procter and Gamble's potato-chip-like product, Pringles. The baby of several years of research, Pringles are made not from potato slices, like traditional chips, but from a potato paste. They have a salt taste on only one side: Why put taste on both sides, the manufacturer reasoned, when only one will touch the tongue? Without food science, all milk shakes would still contain milk. Many are now made instead from vegetable oils such as coconut or palm, milk-food solids, emulsifiers, flavors, and sugars.

Fast-food restaurants, a booming sector of the economy, depend upon the miracles of modern food science no less than do the drinks and snacks you buy in the grocery store. A firm like Denny's, for example, has some 700 outlets in 41 states, and others scattered across Great Britain and Japan. With labor turnover in the industry running at about 350 percent a year, Denny's must serve foods that can be prepared by almost anyone. Their cooks are not likely to have high culinary skills. Even though Denny's puts pictures on its food containers to make sure employees with limited reading ability can manage the cooking, the chain still relies on food science to make preparation even easier. The foods Denny's chooses are ready for eating when heated to 90 degrees, and are palatable at temperatures between 75 and 105. Though the chain used frozen foods heavily at first, the ballooning costs of refrigeration are dictating a switch to dehydrated foods. In the future, Denny's plans to use foods packaged in

**A crucial tool in the growth of the great food conglomerates has been television advertising.**

retort pouches—flexible aluminum cans whose contents can be quickly warmed under the kitchen faucet. The firm is looking for preportioned servings that will keep for the longest possible time.

Among the long list of accomplishments of modern food science, advances in nutrition cannot be said to rank very high. A change in public taste has led to a rash of "natural" items, products with some crucial vitamins restored. But food scientists and the companies that they represent do not feel that the diet they have created is deficient in any fundamental way. They do not take well to criticism of it—as one scene at the IFT convention demonstrated. In one part of the program, 400 scientists gathered to watch movies about the health aspects of their business.

Murmurs of approval ran through the audience during the screening of *Chemicals: A Fact of Life,* filmed through the courtesy of the Monsanto Company. *Adventures in Packaging,* by the Package Research Laboratory, enjoyed a similar reception. But the background commentary changed tone with the appearance of the "controversial" films. Food scientists snickered as they watched *The Junk Food Man* and *It's Easy as Selling Candy to a Baby*, made by groups critical of the food-processing industry. In one film, featuring Fat Albert and the Bill Cosby Kids, while Fat Albert wolfs down sweets the viewers are instructed in the dangers of eating sugary foods as well as of snacking. No discussion followed the films. The movies were uniformly ridiculed from the beginning.

Why show them, then? I. D. Wolf, the program's moderator, and a professor of food science from the University of Minnesota, said that the controversial movies were full of "inaccuracies and exaggerations," but that they were representative of the sort of information about processed food now reaching the public. Since laymen might take such charges seriously, she continued, it was up to food scientists to familiarize themselves with them and rebut them.

They have a great deal of rebutting to do. As Americans have become more concerned about their health, and somewhat more circumspect about what they put into their mouths, allegations like those in the "controversial" films have become common coin. Dr. Michael Jacobson, a well-known food activist and author, claims that the American diet "promotes high blood pressure, strokes, heart disease, obesity, tooth decay, diabetes, and probably certain forms of cancer—surely bowel cancer and breast cancer. Diet is not the only cause of these afflictions, but it is significant.... And if you add them all up, they cause half the deaths Americans succumb to annually."

More bad news comes from the Senate Select Committee on Nutrition and Human Needs, which points out that processed food is exceptionally high in sugar and salts. The committee reported that six out of the 10 leading causes of death are linked to over-consumption of fats, cholesterol, sugar, salts, and alcohol.

Food executives often don't take the comments about overweight Americans too seriously. They like to joke that their industry has been so successful at providing wholesome, good-tasting food that the nation's major nutrition problem is obesity. And they are attacking even that condition, the executives say, through edibles that are low in calories.

But food executives do not usually joke when confronted with the allegations about the hazards of their products. Dr. Jack Francis, president-elect of the IFT, concedes that sugar contributes to tooth decay and that high consumption of fats "probably" leads to obesity; but he testily labels other charges "a lot of speculation." Francis calls for more research. And what if further investigation confirms the dangerous consequences of eating processed food? "What is the alternative?" he asks. "Go back to the diet we had 100 years ago? ... People aren't going to go back to the salt pork, cabbage, carrots, and potatoes that they ate in New England a century back. We have to feed people, and we have to transport food, and we have to make food delivery as efficient as we possibly can.... That means more processing, not less."

**T**HE BEWILDERING diversity of canned, frozen, and dehydrated foods on the grocer's shelves gives the mistaken impression that the food business is teeming with small, competitive firms as it was in Francis's "old days," when food meant fresh produce. But as the focus of the food industry has shifted from farm to factory, conglomerates have grown bigger and bigger at the expense of small, independent operators. The number of food companies, according to Carol Tucker Foreman, Assistant Secretary for Food and Consumer Services for the Department of Agriculture, has declined from 44,000 in 1947 to 22,000 in 1972 and is still dropping. Researchers J.M. Connor and Russell C. Parker say that 200 corporations control 63.5 percent of food and tobacco processing sales. They have estimated that monopoly overcharges in the food manufacturing industries totaled $10 billion, or about 6 percent of food sales, in 1975.

Indeed, concentration has become the rule throughout the food business. Farms, the equipment that they use, as well as the crops that they produce, have come more and more under the domination of a shrinking number of conglomerates. Regional supermarket chains have all but eliminated substantial local competition in many cities. The processed-food market is no less concentrated. Campbell's makes 90 percent of our soup. And only four other firms sell us a large majority of many canned fruit and vegetable products.

The billion-dollar food-processing companies not only manufacture hundreds of consumer goods but they have extended their grasp throughout the food business. For example, General Mills sells a long list of cereals, including Wheaties and Cheerios, Betty Crocker cake mixes, Saluto pizza—and owns Red Lobster inns. General Foods has Burger Chef restaurants and markets Maxwell House coffee, Post cereals, Kool-Aid, Jell-O, and Tang. Ralston Purina sells Chicken of the Sea tuna, Chex cereals, and a long list of pet food, and owns the Jack-in-the-Box restaurant chain. Del Monte, which is itself part of the R.J. Reynolds tobacco company, has seed farms and engages in food and agricultural research, agricultural production, transportation, storage and distribution, as well as food processing and service.

One of the crucial tools in the growth of the great food conglomerates has been television advertising. Since many of their products are essentially alike, an advertised image may provide the critical difference in the mind of the consumer. Lay's potato chips, for example, may not differ intrinsically from Wise potato chips; but who could fail to be moved by the sight of Bert Lahr gazing rapturously at Lay's finest? With this kind of clout in mind, food companies have turned heavily to television advertising.

Here, too, government regulatory agencies have stepped in to halt what they consider systematic abuses. Spurred by a series of petitions from public-interest groups, the Federal Communications Commission tried to persuade manufacturers of presweetened cereal, candy, and toys to reduce the numbers of their television commercials to children and to include messages that might mitigate some of the ill effects of their products. Broadcasters and advertisers agreed to decrease slightly the hourly time allotted for advertising on children's TV but have otherwise proved intransigent. The cereal industry has lobbied intensely on the issue, reportedly spending $30 million (peanuts compared to the $600 million spent on children's TV advertising), and has so far staved off further reform.

Action for Children's Television (ACT), a Boston-based group, continues to pressure the FCC. Peggy Charren, the president, argues that children should "be protected from deception in the marketplace the same way adults are....Nobody's ever told a child that Milky Way causes cavities; the message is 'Milky Way at work, rest, or play.' After 10 years of this argument they could at least say, 'at work, rest, or play, and if you don't brush your teeth you're going to get a cavity.'" ACT would prefer, however, that commercials for sweets be removed from children's TV entirely.

WHATEVER THE MARKETING strategies and advertising finesse, the food industry faces one apparently intractable problem: America can eat just so many Pringles. The food-processing business has already responded to the problem of satiation by inventing new packages, new slogans, and yet more exotic products. But the national population is stabilizing, and the food companies must look for growth opportunities elsewhere—to pets, for example: Americans now feed 40 million dogs and 30 million cats, whose taste buds remain more or less unexplored territory. Grocery shelf space devoted to pet food has expanded dramatically in the last few years, and dogs now enjoy a variety of comestibles not much less impressive than their masters'.

The biggest growth area of the future is neither new products nor new species, but new countries. General Foods now sells coffee and powdered drinks in Europe, ice cream in Brazil, candy and gum in Mexico, and Tang, that venerable pick-me-up of astronauts, to Japan. Borden's foreign sales account for 20 percent of its total volume. The cereal-makers, with no new mouths to conquer at home, have turned increasingly to the Third World nations of South America and the Middle East. Indeed, the upper crust of the Third World, a vast market full of people who have never so much as seen a frozen apple pie, represents the next great market for many food-processing companies. "The most compelling job," said Kellogg International's Vice-President Charles Tornabene in a *Business Week* interview, "is to change people's food habits." Converting the entire world to the processed-food gospel may prove a daunting task.

In his closing speech to the IFT delegates, William Beers, retired chairman of Kraft, Inc., proposed that underdeveloped nations be encouraged to revamp their food systems into miniature versions of our own. Third World nations, said Beers, "must acquire modern infrastructures—industrial, technological, and marketing." Farmers, said Beers to the cheering scientists, "must be motivated to adopt and use modern agricultural technology." And finally, developing nations must pattern their food marketing systems after our own. Then, he concluded, "there will be demands for new types of processed foods, packaging, transportation, and distribution."

The consequences of selling and aggressively advertising sophisticated food products to Third World consumers have been painfully illustrated in the case of infant formula milk. Most manufacturers of infant formula—Nestlé, Borden, Carnation, and Bristol-Myers, among others—market their product to mothers in developing nations. The companies sent "milk nurses"—employees dressed up as nurses—into maternity wards, where they handed out free samples of the formula. Then the companies launched highly effective advertising campaigns. Mothers in the Ivory Coast were found feeding their children Nescafé after a radio message proclaimed that "Nescafé makes men stronger, women more joyful, and children more intelligent."

Besides taking the message too literally, mothers often do not understand how or simply are not able to use the formula: Once back in the village, clean water may not be available to mix with the formula powder. The instructions may be written in the wrong language, or the mother may be illiterate. Money runs out, they can no longer afford the formula, and they have lost their own ability to nurse. For these reasons, according to INFACT, an anti-formula group, as many as 10 million infants die annually from diseases related to bottle-feeding. But the milk companies have done little to counteract the damage they may have caused.

If the past is any indication, however, the companies will survive these complaints and controversies. The food-processing industry has shown itself to be extraordinarily adaptive. When the American consumer began clamoring for "natural foods," the industry stuck "natural" on its labels. While its very success depends on its ability to abolish the past, the industry suggests that its goods have the taste and quality of old-fashioned foods. Running out of Americans to feed, the food business has moved into the Third World, proclaiming itself a force of economic liberation rather than a purveyor of dubious goods.

Of course, the industry is not wholly responsible for the decay of the American diet. After all, no one is forced to choose the laboratory's products over the farm's; no one has to buy orange drink instead of oranges. The industry has found willing customers in the American public.

Yet the questions remain: How do we deal with the technology we have spawned? What kind of food system should we move toward in the future, not only for ourselves but for the rest of the world? Are world hunger and malnutrition to be combated with massive doses of instant junk foods? Unless we choose to confront these questions, we can be assured a steady diet of frozen pizza and cupcakes.

*William Serrin is a national correspondent for the* New York Times.

# Bread and Factories
*Saturday Review, 2/2/80*

*Enriched flour (barley malt, ferrous sulfate, niacin, thiamine mononitrate, riboflavin), water, corn syrup, partially hydrogenated vegetable shortening (soybean and/or cottonseed and/or palm oil), yeast, salt, soy flour, calcium sulfate, sodium stearoyl-2-lactylate, mono- and diglycerides, whey dicalcium phosphate, potassium bromate, calcium propionate, potassium bromate.*

—Ingredients listed on a loaf of bread

Oxides of nitrogen, chlorine, nitrosylchloride, chlorine dioxide, benzoyl peroxide, acetone peroxide, azodicarbonamide, plaster of paris.

—Some of the flour additives and processing chemicals that need not, according to the Code of Federal Regulations, be listed on the package.

Civilization was built on grain. Western civilization used it mainly in the form of bread made of flour, water, a little leaven (yeast), and salt. For the basic loaf, anything else was an adulteration, and the perpetrator was subject to being pilloried in the marketplace.

When bread was well made, the staff of life asked little of the earth. All agriculture was of course organic, and manures and rotation restored the soil. The millstones were turned by people, then by beasts, then by water or wind. The ovens were heated by such fuel as came to hand.

Until the late industrial age, it was common knowledge that bread, like wine, varied in taste and quality with the soil and conditions in which the wheat was grown. Platina, a 15-century Vatican epicure, wrote that the best bread was made from wheat that grew on hillsides (and indeed the most delicious bread that I ever ate was in a Greek mountain village where the wheat was still sown by hand, nourished with donkey manure, cut by scythes, flailed, and stone-ground. In the mid-19th century, Frederick Law Olmsted observed that the closer he got to the new lands of the frontier, the better the bread tasted. Sylvester Graham, the prophet of whole grain, wrote:

*They who have never eaten bread made of wheat recently produced by a pure virgin soil, have but a very imperfect notion of the deliciousness of good bread, such as is often to be met with in the comfortable log houses in our western country.*

By 1837, the large millers of the Atlantic Coast were speeding up their mills and shipping tight-packed flour in barrels to all ports. Mary Hooker Cornelius in *The Young Housekeeper's Friend*, published in 1846, wrote:

*Newly ground flour which has never been packed is very superior to barrel flour, so that the people in Western New York, that land of finest wheat, say that the New England people do not know what good flour is.*

Mrs. Cornelius was already observing the gradual abandonment of wheat-growing in the older regions along the coast. The industry settled in the Great Plains and gradually expanded into vast, single-crop, mechanized spreads. As the rich soil thinned, increasing amounts of fertilizer and herbicide and pesticide, all synthesized from natural gas and petroleum, had to be applied each year. The result is an enormous crop, a source of national wealth and power but also a poor-tasting flour.

Already in Graham's time, the mills were bolting out of the wheat all of the bran and much of the germ, which would gum up and scorch if the millstones turned fast. In the next 40 years, steel shears and rollers completed the transformation of flour from a golden, fat, and nourishing food to a lifeless chalk dust. Yeast could barely live on it, so bakers added sugar. From decade to decade, cookbooks added more and more sugar to replace the rich flavor of true flour, creating the addiction to sweetness that now afflicts the American palate. Yeast itself was often replaced by faster chemical leavens, which brought about a national scandal at the turn of the century, involving charges that the public was being poisoned and that officials were being corrupted to let it happen.

Great new strides were achieved in this century. Flour mills were concentrated in a few major grain centers. There they ground the wheat, employing a score of chemicals to keep it from spoiling, packaged it, and shipped it long distances to market.

The neighborhood baker went the way of the town mill. A dwindling number of bread factories dominated the market with heavy advertising, delivering the product in huge vans. Fuel was cheap. The new factory loaf was "improved" to its present condition of wrapped, sliced Styrofoam, dosed with fungicide to prevent mold and with polysorbates to keep it from drying (the TV commercial for one packaged mix cries "Super-moist!" as if that were a virtue.). Permanent shelf life was achieved.

The most advanced bakeries now resemble oil refineries. Flour, water, a score of additives, and huge amounts of yeast, sugar, and water are mixed into a broth that ferments for an hour. More flour is then added, and the dough is extruded into pans, allowed to rise for an hour, then moved through a tunnel oven. The loaves emerge after 18 minutes, to be cooled, sliced, and wrapped.

They call this bread.

A century of complaint about the impoverishment of the staff of life has led the industry to "enrich" it by adding a few of the nutrients it has removed — only a few — and none of the rich array of earthy flavor and body that our forebears loved.

Clinicians recently discovered what the ancients well knew, that roughage was an important element of diet. ITT Continental Baking Company met this need with a loaf that promised added fiber. The government has insisted that the company identify the ingredient more plainly. It is sawdust. We have come to that.

*John Hess is a syndicated columnist, and the co-author of* The Taste of America.

# More Than You Ever Thought You Would Know About Food Additives... Part III

by Phyllis Lehmann

*"Not the same thing a bit!" said the Hatter. "Why, you might as well say that 'I see what I eat' is the same as 'I eat what I see'!"*
—From ALICE IN WONDERLAND

*"In sight, it must be right."*
—Advertising slogan of a Midwest-based fast food chain

The Mad Hatter lived more than a hundred years ago in the imagination of Lewis Carroll. Yet his seemingly nonsensical remark about eating what one sees and seeing what one eats makes a lot of sense today. That's because today food additives are used to make foods look, taste, and even feel the way we want them to look, taste, and feel.

Some of the additives are put into foods simply to make them more appealing. Others are there to aid in the processing and preparation of the foods. These additions rouse more than a little suspicion among eaters. That's why the fast food chain uses the slogan about preparing its food out in the open so everyone can see it.

This article, the third in a series on food additives, will take a look at those two groups of additives: substances used to spark the color or taste of foods, and those that make foods behave the way we expect them to even after they've left the manufacturing plant.

**Making Food More Appealing**

Four classes of additives are used to heighten the appeal of foods: colors, flavors, flavor enhancers, and sweeteners.

**Colors**

Coloring agents add controversy along with color. They add controversy because they are used solely to improve appearance. They contribute nothing to nutrition, taste, safety, or ease of processing. And some consumer advocates argue that food is often made to look more appetizing at the risk of increasing health hazards.

Today food colors are used in virtually all processed foods. While their use is not restricted, per se, they cannot be used in unnecessary amounts or to cover up unwholesome products. Artificial colors must be listed as ingredients in all foods except butter, ice cream, and cheese.

There are 35 colors currently permitted for use in food. Nearly half of them are synthetic colors, which are created in laboratories. The manmade colors find the widest use because they are stronger than natural colors and thus can be used by manufacturers in smaller quantities and at less cost.

The first food colors were generally harmless vegetable dyes. In the 19th century days of the Mad Hatter, toxic mineral pigments, such as lead and copper, were used to change the color of foods.

Testing and certification of food colors were first required in the original Food and Drugs Act of 1906. Today food colors are regulated under 1960 legislation that makes the food industry responsible for proving the safety of the additives.

Synthetic food colorings usually refer to the coal-tar dyes, which are laboratory creations that are chemically different from anything found in nature. (They are now derived from petroleum rather than from coal tar.) For identification, the colors are assigned initials, the shade, and a number. FD&C Red No. 40, for example, indicates that it is a red coloring used in foods, drugs, and cosmetics.

In the past half dozen years, FDA has prohibited four colors from use in foods. A violet used to stamp meats; Red No. 2, which was suspected as a carcinogen; Red No. 4, used in maraschino cherries and shown to cause bladder lesions and damage to adrenal glands in animals; and Carbon Black, used in candies such as licorice and jelly beans.

FDA also proposed to prohibit Orange B, used in sausage and hotdog casings, because of possible contamination with a carcinogen. The manufacturer voluntarily stopped producing it in 1978.

The two most widely used food colors now are Red No. 40 and Yellow No. 5, and both are under fire because of reports of possible health risks. Red No. 40 is suspected of causing premature malignant lymph tumors when fed in large amounts to mice. In 1977, the Health Research Group petitioned FDA to prohibit its use along with several other colors. FDA denied the petition, but tabled a final decision on Red No. 40 pending a study review.

The problem with Yellow No. 5 is that it causes allergic reactions—mainly rashes and sniffles—in an estimated 50,000 to 90,000 Americans. The reactions are usually minor but in some instances can be life threatening. Because of its relatively narrow effect, FDA has proposed a regulation to require manufacturers to list Yellow No. 5 on the labels of any food products containing it.

### Flavors

Some 1,700 natural and synthetic substances are used to flavor foods, making flavors the largest single category of food additives. Most of the flavors are synthetic because they are cheaper than the real McCoy and because there probably would not be enough of the real McCoys—strawberries, for example—to produce all the strawberry flavoring desired by diners. Artificial flavorings are usually not derived from a single chemical but are the result of a complex process that involves analyzing the individual chemicals present in a flavor, reproducing those in a laboratory, and synthesizing them to create a taste approximating the real thing.

Flavors are listed on food labels in general terms, such as "artificial flavor" or "spices." If a product contains any added flavoring, either natural or synthetic, that fact must be noted on the label. For example, a label that says "strawberry yogurt" means that the product contains all natural strawberry flavor. "Strawberry-flavored yogurt" indicates that it contains natural strawberry flavor plus other natural flavorings. "Artificially flavored strawberry yogurt" means that it contains only artificial flavorings or a combination of artificial and natural flavors.

Flavors have come under less criticism than colors, perhaps because they serve a more direct purpose in foods. Still, some consumer groups question the necessity of using artificial flavors. FDA scientists maintain that anyone sensitive to artificial flavors would be likely to react to natural ones as well because of the chemical similarities.

A few flavorings have been prohibited in the past 10 to 15 years because of health hazards. Probably the best known outcast was safrole, the principal flavoring in sassafras root, once widely used in root beer. FDA banned safrole after tests showed it caused liver cancer in rats. Coumarin, used as an anticoagulant drug, was once present in imitation vanilla extract and other flavorings but was banned for food use because large amounts could cause hemorrhaging.

### Flavor Enhancers

These compounds magnify or modify the flavor of foods and yet do not contribute any flavor of their own. Some of them work by temporarily deadening certain nerves—those responsible for perception of bitterness, for example—thereby increasing the perception of other tastes.

The best known flavor enhancer is the amino acid, monosodium glutamate (MSG), widely used in restaurants and in prepared foods. Scientists are not sure exactly how it works, but suspect that it increases the nerve impulses responsible for perception of flavors. Several years ago, public pressure persuaded manufacturers to stop using MSG in baby foods after studies showed that large amounts had destroyed brain cells in young mice.

MSG also produces the so-called "Chinese restaurant syndrome," which causes some people to have a burning sensation in the neck and forearms, tightness in the chest, and headache after they consume the relatively large amounts of MSG often used in food served in Chinese-style restaurants.

### Sweeteners

Though technically flavors, sweeteners are generally considered a separate category. They are among the most commonly known food additives. Who has not heard of saccharin?

Sweeteners are classified as nutritive and non-nutritive. The nutritive ones, metabolized by the body to produce energy, include the natural sugars such as sucrose (common table sugar), glucose, and fructose, as well as sugar alcohols, such as sorbitol and mannitol. Non-nutritive sweeteners, which are not metabolized and therefore contribute no calories to the diet, include cyclamate, which is currently prohibited from use in food, and saccharin.

Natural sugars are widely used in foods, not just as sweeteners but also to create a heavier mouth feel in soft drinks and as browning agents in baked goods. Some consumer groups oppose adding sugar to food because they say it represents "empty" calories devoid of vitamins, minerals, or protein. They also argue that sugar contributes to tooth decay.

The sugar alcohols, chemical variants of natural sugars, have been around for decades but have been promoted in recent years as "low-cal" alternatives to sugar and as less likely to cause tooth decay. Foods containing these sweeteners, are not truly low-calorie. Nor can they be considered completely "free" foods for diabetics, as they can and do lead to the production of some blood sugar. An FDA regulation scheduled to become effective in 1980 will require manufacturers to state on labels that inclusion of these nonsugar sweeteners does not mean the product is "low-cal" or "reduced calorie."

The most widely used sugar alcohol is sorbitol, put in chewing gum, mints, candies, and dietetic ice cream. Though safe, it does have a laxative effect in large amounts. Mannitol, on the other hand, may cause diarrhea in relatively small amounts because it accumulates water during its extremely slow passage through the intestinal tract. This side effect has caused mannitol's use to be limited to the powdery coatings on some chewing gums.

Xylitol, another sugar alcohol promoted several years ago for sugarless gum and dietetic food, fell from favor following several negative health reports and studies. Some manufacturers have voluntarily stopped using it.

The non-nutritive sweeteners—cyclamate and saccharin—have proved to have a taste for controversy. FDA banned cyclamate in 1969 after studies indicated it caused cancer in animals. The manufacturer has sought reinstatement, citing new studies, but so far it remains off the store shelves.

Saccharin was originally on the GRAS (Generally Recognized As Safe) List but was removed in the early 1970's when evidence of health hazards began to mount. In April 1977, FDA proposed to ban saccharin as an additive in food (it's primarily used in diet sodas) following a Canadian government study that showed it caused bladder tumors in rats. The ban would have allowed the continued use of saccharin as a tabletop sweetener if the industry could prove it was beneficial to diabetics or to people on weight-reduction diets.

The public outcry was such you would have thought FDA was trying to take real candy from all the Nation's babies. Congress heard the outcry and passed an 18-month moratorium on the FDA ban proposal. The moratorium ended May 23, 1979, and FDA has announced its intention to repropose the ban—a regulation-making process that could take 15 to 18 more months.

## Preparing and Processing Foods

The final category of food additives consists of those used in the preparation and processing of foods. These additives are used by manufacturers to get desired effects during processing and beyond. To the consumer, the additives give the food some of the characteristics that are associated with the products.

The functions of these additives are many. Some cause baked goods to rise. Others prevent ice crystals from forming in ice cream and keep peanut butter from separating into oily and dry layers. Because of such additives, shredded coconut stays fresh and moist in the can.

Of the four major categories of food additives, these are the least clouded by controversy. Consumer groups caution about using some specific compounds, but there is nothing like the furor generated by other additives such as saccharin and nitrites.

There are seven major groups of additives that are considered aids in processing or preparation of foods:

### Emulsifiers (Mixers)

Some liquids don't mix unless there is an emulsifier around. In salad dressing, for example, oil and vinegar normally separate as soon as mixing stops. When an emulsifier is added, the ingredients stay mixed longer. In pickles, beverages, and candies, emulsifiers help disperse flavors and oils that otherwise would not be soluble in water. Without these compounds, ice cream and other frozen desserts would separate and lose their creamy texture. In baking, emulsifiers improve the volume and uniformity of breads and rolls as well as make batter and dough easier to handle.

Many emulsifiers come from natural sources. Lecithin, naturally present in milk, keeps fat and water together. Egg yolks, which also contain lecithin, improve the texture of ice cream and mayonnaise. The mono- and diglycerides come from vegetables or animal tallow and make bread soft, improve the stability of margarine, and prevent the oil and peanuts in peanut butter from separating.

Several emulsifiers have elicited concern about health risks. The Center for Science in the Public Interest, in its poster "Chemical Cuisine," advises consumers to avoid brominated vegetable oil (BVO), used to maintain the characteristic cloudy appearance of citrus-flavored drinks by keeping oils in suspension. The Center warns that residues from the additive accumulate in body fat. Because of this problem, FDA has removed BVO from the GRAS List and set specific levels at which it may be used.

### Stabilizers and Thickeners

These compounds "improve" the appearance of food and the way it feels in the mouth by producing a uniform texture. They work by absorbing water. Without stabilizers and thickeners, ice crystals would form in ice cream and other frozen desserts and particles of chocolate would settle out of chocolate milk.

Stabilizers also are used to prevent evaporation and deterioration of the volatile flavor oils used in cakes, puddings, and gelatin mixes.

Most stabilizers and thickeners are natural carbohydrates. Gelatin—made from animal bones, hooves, and other parts—and pectin—from citrus rind—are used in home and commercial food processing. Extra pectin, for example, is added to thicken jams and jellies.

In the past 30 years, vegetable gums—from trees, seaweed, and other plants—have become widely used as thickeners. These are so effective that 0.1 percent can produce the same degree of thickness in water as a high concentration of starch. One problem is that some, such as tragacanth gum and gum arabic, cause allergic reactions in a few susceptible people. The Center for Science in the Public Interest warns consumers especially about tragacanth gum, which it says has caused some severe allergies. FDA does not believe this problem is any more prevalent than allergies from eggs, chocolate, milk, or other natural foods.

### pH Control Agents

These affect the texture, taste, and safety of foods by controlling acidity or alkalinity. Acids, for example, give a tart taste to such foods as soft drinks, sherbets, and cheese spreads. A more important use is to insure the safety of low-acid canned foods, such as beets. Normally, these low-acid foods have to be cooked longer at higher heat than acidic foods to render them sterile because they are more receptive to the bacteria that cause botulism. By adding acids, manufacturers can eliminate the need for extra heat that might detract from the marketable quality of the food. Natural organic acids, such as citric, fumaric, tartaric, and malic acids, are used in canned foods, although mineral acids, such as hydrochloric, are preferred in some cases.

Alkalizers alter the texture and flavor of many foods, including chocolate. After cocoa beans are picked, they are allowed to dry and ferment before they are made into chocolate. During processing, alkalizers are added to neutralize the acids produced during fermentation and to provide a darker, richer color and milder flavor in the finished product.

### Leavening Agents

Although air and steam help create a light texture in bread and cake, carbon dioxide is the key to making baked goods rise properly. Without leavening agents that produce or stimulate production of carbon dioxide, we would not have light, soft baked goods.

The earliest leavening agent was yeast, which produces carbon dioxide through fermentation. Today two other leavening agents are found in home kitchens and commercial bakeries. One, baking soda (sodium bicarbonate), releases carbon dioxide when heated. Usually an acid ingredient, such as sour milk, is used along with baking soda to eliminate the soapy-tasting byproduct of this chemical reaction. The second is baking powder, a combination of sodium bicarbonate and acid salts that react in the presence of water to produce carbon dioxide.

### Maturing and Bleaching Agents

Maturing and bleaching agents are used primarily to get flour ready for baking because natural pigments give freshly milled flour a yellowish color.

Flour also lacks the qualities necessary to make a stable, elastic dough. When aged for several months, it gradually whitens and matures to become useful for baking.

In the early 1900's scientists discovered they could hasten bleaching and maturing—and eliminate costly storage—by adding certain chemicals. These agents do not remove anything from the flour and leave little residue. They simply change the yellow pigments to white and develop the gluten characteristics necessary for baking.

Bleaching agents, such as benzoyl peroxide, also are used to whiten milk used for certain cheeses known for their whitish curd, such as blue cheese and gorgonzola. Bleaching is considered necessary because the grass that cows eat causes them to yield buff-colored milk.

### Anti-caking Agents

Compounds such as calcium silicate, iron ammonium citrate, and silicon dioxide are used to keep table salt, baking powder, confectioner's sugar, and other powdered food ingredients free flowing. By absorbing moisture, these chemicals prevent caking, lumping, and clustering that would make powdered or crystalline products inconvenient to use.

### Humectants

Humectants are substances that retain moisture in shredded coconut, marshmallows, soft candies, and other confections. One of the most common is glycerine. The sweetener sorbitol also is used for this purpose.

Although these are the major additives used in processing and preparation, there are other additives with specialized uses, such as clarifying agents, which remove small mineral particles that cloud such liquids as vinegar; firming agents that help coagulate certain cheeses and improve the texture of pickles, maraschino cherries, canned peas, tomatoes, potatoes, and apples; foam inhibitors that prevent foam formation on pineapple juice or on other foods during washing, cooking, or processing; and sequestrants that chemically "hold" minerals in soft drinks that might otherwise settle out and cloud the beverage.

*Phyllis Lehmann is a freelance writer. Charts for these articles were developed by Sandy Barwick, CAO at Grand Rapids.*

# ADDITIVES: What, Where, Why They Are ...

## PURPOSE: To Aid in Processing or Preparation
## CLASS: Emulsifiers

| Some Additives | Where You Might Find Them | Their Functions |
|---|---|---|
| Carrageenan | Chocolate milk, canned milk drinks, whipped toppings | Help to evenly distribute tiny particles of one liquid into another, e.g., oil and water; modify surface tension of liquid to establish a uniform dispersion or emulsion; improve homogeneity, consistency, stability, texture. |
| Lecithin | Margarine, dressings, chocolate, frozen desserts, baked goods | |
| Mono/diglycerides | Baked goods, peanut butter, cereals | |
| Polysorbate 60, 65, 80 | Gelatin/pudding desserts, dressings, baked goods, nondairy creams, ice cream | |
| Sorbitan monostearate | Cakes, toppings, chocolate | |
| Dioctyl sodium sulfosuccinate | Cocoa | |

## PURPOSE: To Aid in Processing or Preparation
## CLASS: Stabilizers, Thickeners, Texturizers

| Some Additives | Where You Might Find Them | Their Functions |
|---|---|---|
| Ammonium alginate<br>Calcium alginate<br>Potassium alginate<br>Sodium alginate | Dessert-type dairy products, confections | Impart body, improve consistency, texture; stabilize emulsions; affect appearance/mouth feel of the food; many are natural carbohydrates which absorb water in the food. |
| Carrageenan | Frozen desserts, puddings, syrups, jellies | |
| Cellulose derivatives | Breads, ice cream, confections, diet foods | |
| Flour | Sauces, gravies, canned foods | |
| Furcelleran | Frozen desserts, puddings, syrups | |
| Modified food starch | Sauces, soups, pie fillings, canned meals, snack foods | |
| Pectin | Jams/jellies, fruit products, frozen desserts | |
| Propylene glycol | Baked goods, frozen desserts, dairy spreads | |
| Vegetable gums: guar gum, gum arabic, gum ghatti, karaya gum, locust (carob) bean gum, tragacanth gum, larch gum (arabinogalactan) | Chewing gum, sauces, desserts, dressings, syrups, beverages, fabricated foods, cheeses, baked goods | |

## PURPOSE: To Aid in Processing or Preparation
## CLASS: Leavening Agents

| Some Additives | Where You Might Find Them | Their Functions |
|---|---|---|
| Yeast | Breads, baked goods | Affect cooking results; texture and increased volume; also some flavor effects. |
| Baking powder, double-acting (sodium bicarbonate, sodium aluminum sulfate, calcium phosphate) | Quick breads, cake-type baked goods | |
| Baking soda (sodium bicarbonate) | Quick breads, cake-type baked goods | |

## PURPOSE: To Aid in Processing or Preparation
## CLASS: pH Control Agents

| Some Additives | Where You Might Find Them | Their Functions |
|---|---|---|
| Acetic acid/sodium acetate | Candies, sauces, dressings, relishes | Control (change/maintain) acidity or alkalinity; can affect texture, taste, wholesomeness. |
| Adipic acid | Beverage/gelatin bases, bottled drinks | |
| Citric acid/sodium citrate | Fruit products, candies, beverages, frozen desserts | |
| Fumaric acid | Dry dessert bases, confections, powdered soft drinks | |
| Lactic acid | Cheeses, beverages, frozen desserts | |
| Calcium lactate | Fruits/vegetables, dry/condensed milk | |
| Phosphoric acid/phosphates | Fruit products, beverages, ices/sherbets, soft drinks, oils, baked goods | |
| Tartaric acid/tartrates | Confections, some dairy desserts, baked goods, beverages | |

## PURPOSE: To Aid in Processing or Preparation
## CLASS: Humectants

| Some Additives | Where You Might Find Them | Their Functions |
|---|---|---|
| Glycerine | Flaked coconut | Retain moisture. |
| Glycerol monostearate | Marshmallow | |
| Propylene glycol | Confections, pet foods | |
| Sorbitol | Soft candies, gum | |

## PURPOSE: To Aid in Processing or Preparation
## CLASS: Maturing and Bleaching Agents, Dough Conditioners

| Some Additives | Where You Might Find Them | Their Functions |
|---|---|---|
| Azodicarbonamide | Cereal flour, breads | Accelerate the aging process (oxidation) to develop the gluten characteristics of flour; improve baking qualities. |
| Acetone peroxide<br>Benzoyl peroxide<br>Hydrogen peroxide | Flour, breads & rolls | |
| Calcium/potassium bromate | Breads | |
| Sodium stearyl fumarate | Yeast-leavened breads, instant potatoes, processed cereals | |

## PURPOSE: To Aid in Processing or Preparation
## CLASS: Anti-caking Agents

| Some Additives | Where You Might Find Them | Their Functions |
|---|---|---|
| Calcium silicate | Table salt, baking powder, other powdered foods | Help keep salts and powders free-flowing; prevent caking, lumping, or clustering of a finely powdered or crystalline substance. |
| Iron ammonium citrate | Salt | |
| Silicon dioxide | Table salt, baking powder, other powdered foods | |
| Yellow prussiate of soda | Salt | |

## PURPOSE: To Affect Appeal Characteristics
## CLASS: Flavor Enhancers

| Some Additives | Where You Might Find Them | Their Functions |
|---|---|---|
| Disodium guanylate | Canned vegetables | Substances which supplement, magnify, or modify the original taste and/or aroma of a food—*without* imparting a characteristic taste or aroma of its own. |
| Disodium inosinate | Canned vegetables | |
| Hydrolyzed vegetable protein | Processed meats, gravy/sauce mixes, fabricated foods | |
| MSG (monosodium glutamate) | Oriental foods, soups, foods with animal protein | |
| Yeast-malt sprout extract | Gravies, sauces | |

## PURPOSE: To Affect Appeal Characteristics
## CLASS: Flavors

| Some Additives | Where You Might Find Them | Their Functions |
|---|---|---|
| Vanilla (natural) | Baked goods | Make foods taste better; improve natural flavor; restore flavors lost in processing. |
| Vanillin (synthetic) | Baked goods | |
| Spices and other natural seasonings and flavorings, e.g., clove, cinnamon, ginger, paprika, turmeric, anise, sage, thyme, basil | No restrictions on usage in foods—found in many products | |

## PURPOSE: To Affect Appeal Characteristics
## CLASS: Natural/Synthetic (N/S) Colors

| Some Additives | Where You Might Find Them | Their Functions |
|---|---|---|
| N Annatto extract (yellow-red) | No restrictions | Increase consumer appeal and product acceptance by giving a desired, appetizing, or characteristic color. Any material which imparts color when added to a food. Generally *not* restricted to certain foods or food classes. May *not* be used to cover up an unwholesome food, *or* used in excessive amounts.<br>*Must* be used in accordance with FDA Good Manufacturing Practice Regulations. |
| N Dehydrated beets/beet powder | No restrictions | |
| S Ultramarine Blue | Animal feed only .5% by wt. | |
| N/S Canthaxanthin (orange-red) | Limit = 30 mg/lb of food | |
| N Caramel (brown) | No restrictions | |
| N/S Beta-apo-8' carotenal (yellow-red) | Limit = 15 mg/lb of food | |
| N/S Beta carotene (yellow) | No restrictions | |

| | | |
|---|---|---|
| N Cochineal extract/carmine (red) | No restrictions | |
| N Toasted partially defatted cooked cottonseed flour (brown shades) | No restrictions | |
| S Ferrous gluconate (turns black) | Ripe olives | |
| N Grape skin extract (purple-red) | Beverages only | |
| S Iron oxide (red-brown) | Pet foods only .25% or less by wt. | |
| N Fruit juice/vegetable juice | No restrictions | |
| N Dried algae meal (yellow) | Chicken feed only | |
| N Tagetes (Aztec Marigold) | Chicken feed only | |
| N Carrot oil (orange) | No restrictions | |
| N Corn endosperm (red-brown) | Chicken feed only | |
| N Paprika/paprika oleoresin (red-orange) | No restrictions | |
| N/S Riboflavin (yellow) | No restrictions | |
| N Saffron (orange) | No restrictions | |
| S Titanium dioxide (white) | Limit = 1% by wt. | |
| N Turmeric/Turmeric oleoresins (yellow) | No restrictions | |
| S FD&C Blue No. 1 | No restrictions | Synthetic color additives subject to certification: inspected and tested for impurities. |
| S Citrus Red No. 2 | Orange skins of mature, green, eating-oranges. Limit = 2 ppm. | |
| S FD&C Red No. 3 | No restrictions | |
| S FD&C Red No. 40 | No restrictions | |
| S FD&C Yellow No. 5 | No restrictions | |

## PURPOSE: To Affect Appeal Characteristics
## CLASS: Sweeteners

| Some Additives | Where You Might Find Them | Their Functions |
|---|---|---|
| Nutritive Sweeteners:<br>Mannitol—sugar alcohol<br>Sorbitol—sugar alcohol | Candies, gum, confections, baked goods | Make the aroma or taste of a food more agreeable or pleasurable. |
| Dextrose<br>Fructose<br>Glucose<br>Sucrose (table sugar) | Cereals, baked goods, candies, processed foods, processed meats | |
| Corn syrup/corn syrup solids<br>Invert sugar | Cereals, baked goods, candies, processed foods, processed meats. | |
| Non-nutritive sweeteners:<br>Saccharin | Special dietary foods, beverages | |

# The Case For Moderating Sodium ~~Salt~~ Consumption

Americans consume sodium equivalent to some 2 to 2 1/2 teaspoons of salt a day, and around 60 million Americans suffer from high blood pressure to one degree or another. For many people, there may be a connection. Allan L. Forbes, M.D., associate director for nutrition and food sciences in FDA's Bureau of Foods, discusses the link and efforts to help consumers moderate their salt or sodium intake. Dr. Forbes was interviewed by Roger W. Miller, editor of FDA CONSUMER.

**Q.** *Dr. Forbes, why is too much sodium a health problem?*

**A.** The basic reason is that sodium intake is interrelated with hypertensive diseases or high blood pressure. This has been studied for many years and, when you really come down to it, what we know is that population groups around the world who have very low sodium intakes have virtually no hypertension, and population groups that have very high sodium intakes have more hypertension. The thing we don't know with the precision that we wish we did know is if there is an "ideal level" of sodium intake for a population. We certainly know—and we emphasize the point—that very low intakes likely protect against hypertension, and very high intakes tend to predispose some individuals to the development of hypertensive disease. And certainly there is no doubt that, in many individuals with high blood pressure, sodium restriction, best coupled with weight reduction, lowers blood pressure.

**Q.** *What are some examples of the population groups that you refer to?*

**A.** Well, for example, in parts of the Orient you find that the biggest killer is hypertension, particularly in northern Japan but also in Thailand and China. There are some population groups scattered around the world—Polynesia, some of the islands off Southeast Asia, northern Brazil and in Africa—where the prevalence of hypertensive disease is very, very low and the salt/sodium intakes are also very low.

**Q.** *How does hypertension kill?*

**A.** Hypertension is the most common cause of serious stroke—that is, a hemorrhage into the brain from an artery.

**Q.** *And pressure on the artery causes it?*

**A.** We don't fully understand even today what occurs in the wall of the artery. But we know that what happens is that a leak occurs in the artery and a hemorrhage develops into the brain. That is a stroke and a particularly bad type of stroke. Hypertension also is associated with heart attacks. The basic cause of most heart attacks is another disease called atherosclerosis in which there is a blockage inside the artery from cholesterol deposits on the wall of the artery. When an individual has atherosclerosis, with the resulting narrowing of the coronary artery, high blood pressure makes the possibility of having a heart attack even greater. So the two diseases are synergistic in terms of producing adverse effects, particularly relative to the heart. The same is true of kidney disease. When hypertensive disease gets very bad, a lot of troubles arise in the kidneys. Over time, the untreated patient will suffer a decline in kidney function—the ability of the kidneys to clean blood and to excrete products into urine that the body wants to get rid of. So when we're talking about hypertension, we're talking about the possibility of serious brain disease, serious heart disease and serious kidney disease. Hypertension is directly or indirectly contributory to the top four or five causes of death in the United States.

**Q.** *What does sodium do in the body to cause high blood pressure?*

**A.** First, I want to stress that sodium is absolutely essential for life itself. In other words, you would not survive on a sodium-free diet. Sodium helps maintain what is called the osmotic pressure of the blood, and therefore is absolutely essential for maintaining blood pressure. Sodium attracts water into the blood vessels, thereby maintaining

"... most of the leading health and medical experts ... have concluded that it is reasonable to moderate the amount of sodium intake for everybody."

proper blood volume and keeping the pressure within the blood vessels more or less constant. Sodium also is intimately involved in the production of hormones which regulate blood pressure. Another thing that sodium does—along with other ions like bicarbonate and potassium and chloride—is that it maintains what we call acid-base balance. The body must be adjusted with tremendous precision as to how acid it is and how alkaline it is. This is done, in large measure, by the kidney.

**Q.** *Do non-hypertensives have to watch their sodium intake?*

**A.** That is a very good question, and a rather difficult one to answer. We know that there are over 30 million people in the United States who have hypertension, and we know that there are a lot more—perhaps another 30 million—who have what is called borderline hypertension or are in the high normal range. A large proportion of these borderline people are highly likely to develop hypertension later in life, if it goes undiagnosed and untreated. Part of the problem is that there are some people who are particularly sodium sensitive and some people who are not. It is very difficult, given the current state of the medical art, to determine in advance who is and who isn't. So there is a lot of research right now to try to develop methods that will allow the medical profession to determine in advance whether an individual is sodium sensitive or whether he is a part of the population that isn't. The result is that most of the leading health and medical experts in the United States and in a lot of other parts of the world have concluded that it is reasonable to moderate the amount of sodium intake for everybody.

**Q.** *Aren't some people more likely to be hypertension candidates...*

**A.** Indeed...

**Q.** *...blacks, for example?*

**A.** Indeed, that's absolutely right. It is clearly known that blacks have a higher prevalence of hypertension than whites do.

**Q.** *Is it known why they do?*

**A.** This point is not clear. There definitely appears to be a strong genetic component. But that applies to white people as well, and applies to people in general. Exactly why the blacks are somewhat more likely to develop hypertension is still not clear. There have been studies that have demonstrated that it isn't simply because blacks may consume a bit more sodium than whites generally do. So with hypertension, I think we have to emphasize—in communicating to the public—that like so many other diseases, its cause is multi-factorial. Hypertension is due to a number of factors interacting together: sodium intake, genetic susceptibility to the disease, the presence of obesity which aggravates hypertension, smoking, stress and probably other factors that are still unknown.

**Q.** *When we're talking about sodium aren't we talking mainly about salt, which is 40 percent sodium?*

**A.** Well, yes, in a way, but I think it is important to point out that sodium comes into our diet in a lot of ways. We don't know for sure how much of our total sodium comes from salt itself, and it varies a lot from person to person. Salt is by far the major source, but I think it is very important to bring to the public's attention that sodium comes from many other sources, including other ingredients added to foods. Baking soda is sodium bicarbonate. Oftentimes vitamin C, which we call ascorbic acid, is added to a food as sodium ascorbate. Lots of people are aware of monosodium glutamate because it has been very much publicized relative to the Chinese restaurant syndrome.

**Q.** *Known popularly as MSG?*

**A.** MSG, exactly. About 10 percent of MSG is sodium, so sodium is throughout our food supply. However, much of the sodium in foods is put there by nature. It is present in all parts of plants and ani-

"... there has been a gradual development of a national medical and biological consensus about the association between sodium and blood pressure."

mals. So the sodium in foods comes from many sources, and they are all important because they all contribute to the total amount in the diet. And, once it is in the body and circulating in the blood, the body cannot distinguish where the sodium came from.

**Q.** *And where it comes from is not just the salt shaker or Mother Nature but it is put into food. It is processed into food. How much is added in that way?*

**A.** You know, in this country we have the most extraordinarily diversified diet. There are some 18,000 products in the supermarkets, and practically all of them contain sodium to one degree or another. And dietary patterns are so different between individuals that it is very difficult to say that just this amount comes from salt that you added at the table, this amount from salt that you added while cooking, and that amount was added by the food processor, or put there by nature. I think the rule of thumb, which is never truly accurate because of this tremendous variation in our dietary patterns and the very nature of our food supply, is that about a third of it may come from the food itself in the natural state, a third of it is there because the food processor introduced it, and a third was added in the home or restaurant. The processor puts it into the food for a whole series of reasons, some of which are absolutely essential to the production and preservation of the food.

**Q.** *Why is sodium such an issue now? Haven't we known these things for years?*

**A.** We have known an awful lot about this for a long time. The history of a connection between sodium and high blood pressure goes back 50 years or more. However, when dealing with a disease that is multi-factorial, it takes a very long time for the scientific community to unravel enough of the facts to begin to develop clear concepts of what ought to be done. During roughly the last 10 years there has been a gradual development of what I would call a national medical and biological consensus about the association between sodium and blood pressure. That consensus has become national only in the last year or two. I am speaking of groups like the Food and Nutrition Board of the National Academy of Sciences and American Medical Association. Both only recently concluded that the sodium intake of the general public should be moderated. The FDA had a very detailed review of this matter undertaken by another of the most prestigious scientific organizations, the Federation of American Societies for Experimental Biology (FASEB), which reviewed the matter in great detail and came to the same conclusion. Then two major departments of the government—Health, Education and Welfare (now Health and Human Services) and Agriculture—stated in their published *"Dietary Guidelines for Americans"* that it would be reasonable to moderate the sodium intake in the American diet. The National Heart, Lung and Blood Institute (NHLBI) advisory groups on hypertension in the last several years have been saying the same thing. So that's what I call a gradual development of a national consensus.

**Q.** *Now that we have this national consensus, what is FDA planning to do?*

**A.** The basic thrust of FDA's program is aimed at moderation of the sodium content of processed and packaged foods plus a number of initiatives that pertain to labeling to inform consumers on how much sodium is present in the food they buy. We have a five-point initiative concerning sodium. The first is working directly with industry, particularly on moderation of the over-all sodium content of foods as they are processed, packaged and sold to consumers. Second is our labeling proposal. We're asking for comments from everyone—industry, professionals, consumers, the entire population—on having nutrition labels bear sodium labeling in terms of milligrams of sodium per serving. We will also permit, as we have in the past, sodium labeling to appear on a food label by itself without full nutrition labeling so that a manufacturer who decides not to do full nutrition

"The thing we don't know with precision that we wish we did know is if there is an 'ideal level' of sodium intake for a population."

labeling may still put the sodium content in milligrams per serving on the label.

Q. *That won't trigger nutrition labeling?*

A. That's right. It won't trigger full nutrition labeling. Another aspect of the labeling proposal is a series of definitions, so that there is uniformity in the marketplace that makes medical and scientific sense. We're offering definitions of "low sodium," "moderately low sodium," and "reduced sodium." The third part of the FDA program would be new legislation. For example, at the present time, the FD&C (Food, Drug, and Cosmetic) Act does not provide us with clear authority to require sodium labeling on all packaged foods (except under unusual circumstances which do not exist today). We have made a detailed analysis of this and have a whole series of options for consideration. However, for the moment, we wish very much to continue down a voluntary road because we anticipate serious interest by a very large part of the American food processing industry. Step four in our program will be tracking, with considerable care, what is actually happening to the sodium content of the food supply, as well as what is happening to labeling. How much "low sodium" food appears in the marketplace, how much "moderately low sodium" food appears, and how much "reduced sodium" food is getting into the supermarkets. We will watch those things with great care and have a pretty sophisticated system for tracking it. The final part of the FDA initiative is to work with industry, other government agencies and interested non-governmental organizations to help consumers make effective use of the new labeling and to raise consumer awareness of the effects of sodium on health.

Q. *Why three definitions of sodium on labels?*

A. "Low sodium" on the label would mean foods which could be eaten without great concern by individuals who are on sodium-restricted diets. The labeling of "moderately low sodium" foods would help individuals who either are already on a low sodium diet or a sodium-reduced diet. Foods labeled that way could be eaten quite extensively by a person on a low-sodium diet without getting into trouble. It would be very helpful in terms of maintaining good diversity in a sodium-restricted diet. "Reduced sodium" foods would be very useful to consumers who are simply attempting to reduce the sodium in their diets.

Q. *Will FDA's program tie in with what the National Heart, Lung and Blood Institute (NHLBI) is already doing on hypertension?*

A. I want to stress again that the primary focus of FDA in the sodium and hypertension issue pertains to moderation of sodium in the food supply and to labeling. The NHLBI, on the other hand, has a much broader role to play in the picture. The NHLBI has a large scale professional and consumer information and education program already in place. In fact, it has been going on for the better part of a decade. It originally focused on diagnosis because there were so many people in this country who had hypertension but did not know it. That has improved very substantially, so the Institute refocused on proper treatment by drug therapy. The NHLBI program is changing again now to further emphasize the relationship of dietary sodium to hypertension as well as weight control. We're working with the people at the Institute in getting the sodium message across.

Q. *How important is nutrition labeling? How much of the food supply is involved in nutrition labeling?*

A. Excluding fresh produce and fresh meat and poultry, about 40 percent of the food supply has nutrition labeling.

Q. *How much food has sodium labeling?*

A. We have just finished an analysis on that part of the food supply that we call the "core" of the supermarket. It is the part in most supermarkets that is in the mid-

*"Part of the problem is that there are some people who are particularly sodium sensitive and some who are not. It is very difficult to determine in advance who is and who isn't."*

dle of the store—foods that are in packages and cans distinct from the fresh produce and refrigerated produce. In this "core" in 1977, a little over 7 percent of the consumer's food dollar went for sodium-labeled foods. By 1979 the figure was 13.4 percent.

**Q.** *Of the 60 million hypertensives in this country, how many of them can control their hypertension strictly through diet?*

**A.** I don't think I can answer that question in terms of percentages. I am not sure that anybody knows the answer, but let me try to address it this way. There are a great many individuals with mild hypertension who can be treated initially by non-drug treatment including weight reduction, if necessary, and/or a low sodium diet. There are a great many hypertensives who can readily return their blood pressure to normal by simple drug treatment. Then there is a group of hypertensives who require much more vigorous drug treatment with much more potent drugs. But there is another point I think we ought to mention. Research on hypertension has made it crystal clear that the mild hypertensive should not be ignored. It has been clearly demonstrated that proper dieting and/or medical management of the mild hypertensive markedly reduces the prevalence and occurrence of stroke and high blood pressure-related heart attacks. Therefore, the medical community today is paying a great deal more attention to the mild hypertensive than was true even 5 years ago.

**Q.** *What has the food processing industry done thus far to make the public more aware of the sodium problem?*

**A.** The very existence of some sodium labeling out there in the marketplace shows there are major segments of the food industry that are paying particular attention already. Some examples are the ready-to-eat breakfast cereal industry, the baby food industry which voluntarily has greatly reduced the sodium content of baby foods over the past decade, and the shortening and oils part of the industry which has put sodium labeling on a fair number of their products; this also applies to baking mixes. A major manufacturer of soups has a large research program in a number of areas concerning sodium. One is the marketability of low-sodium foods, particularly soups. Another area they're looking into is sodium-reduced products with the best possible retention of flavor and other characteristics.

**Q.** *Doesn't the public want highly salted foods? Doesn't the public want sodium—like Al Capone said during Prohibition, the public is a guy who wants his beer?*

**A.** Fair enough. Let me emphasize that it is not the intent of the FDA to revolutionize the nature of our food supply. We are more interested in stimulating readily available choices to the consumer, particularly those individuals who have a medical need to moderate the amount of sodium that is taken in. I absolutely will not pontificate over the lowering of salt content in sauerkraut and dill pickles. I just won't do that. These foods are the way they are and I assume they should stay that way—readily recognizable as being salty. And needless to say, we haven't the slightest intention of castigating the existence of salt in a salt box in the supermarket; that would be patently ridiculous. So our thrust is aimed at choice and moderation where it is reasonable, feasible and marketable to do. But there is another facet to the thing from a medical point of view which has always fascinated me. When a patient is put on a sodium-reduced diet and stays on it for a moderate period of time, an interesting thing often happens. And it's frequently spontaneously offered to the physician as information. The threshold for the perception of saltiness goes down. After some months on a sodium-moderated diet, many patients will say to their doctors—and I have had this experience myself—"That food that I used to really like now tastes so salty that I can't eat it. But the food that I have been eating recently seems to be just about as salty as I really like." So my belief is that there is a high probability that if there was a reasonable moderation of sodium content in our food supply gradually and over time, that the American public will adapt to that quite readily.

# EVERYTHING DOESN'T CAUSE CANCER

**But how can we tell which chemicals cause cancer and which ones don't?**

U.S. DEPARTMENT OF HEALTH, EDUCATION, AND WELFARE
Public Health Service
National Institutes of Health

**Question:** What causes cancer?

**Answer:** Cancer is actually many different diseases with many causes. Most human cancers probably are caused in part by environmental factors. Cancer-causing agents, called carcinogens, include certain man-made and natural chemicals which may be found in small quantities in air, water, food and the work place. Cancer-causing agents also include x-rays, sunlight, and certain viruses. Contact with cancer-causing agents may result from individual actions, such as smoking or dietary habits. This brochure deals mainly with chemical carcinogens. It is not true that everything causes cancer, or that the problem is hopeless. Relatively few substances cause cancer. Most chemicals, even most toxic or other dangerous chemicals are not carcinogenic. Susceptibility to carcinogens may vary among individuals.

*Certain cancer-causing agents (carcinogens) are found in air, water, food and the work place.*

**Question:** Can cancer be prevented?

**Answer:** Yes, many cancers could be prevented by reducing our exposure to carcinogens. Some carcinogens can be avoided by personal choice, government regulation, corporate decisions or other societal actions. Reducing human cancer rates by reducing or avoiding exposure to cancer-causing agents is an achievable goal.

*Some carcinogens can be avoided by individual choice.*

**Question:** How soon after exposure to a carcinogen does the cancer appear?

**Answer:** Cancers develop slowly in man, usually appearing 5 to 40 years after exposure to a cancer-causing agent. Cancers of the liver, lung or bladder, for example, may not appear until 30 years after exposure to vinyl chloride, asbestos, or benzidine. This long latent period is one reason why it is so difficult to identify the causes of human cancer.

*Human cancers usually don't appear until 5 to 40 years after exposure.*

**Question:** How can we identify agents that cause cancer in people?

**Answer:** It is hard to do so directly because suspected carcinogens are tested on laboratory animals, not people. But direct human exposure to cancer-causing substances has often occurred, nevertheless, and we can study the exposed populations. For example, people who have been exposed to tobacco smoke or asbestos develop after many years a higher frequency of cancer of the lung and other organs than unexposed people.

From such population studies we have identified about 30 agents as human carcinogens.

*Mice or rats are used for laboratory tests because generally they are similar to humans in their response to carcinogens.*

**Question:** Aren't there other ways to identify carcinogens, without humans first getting cancer?

**Answer:** Yes, tests on laboratory animals can identify substances that are likely to be human carcinogens. Mice or rats are most commonly used for such tests because they are small, easily handled, more economical than larger animals, and similar to humans in their response to carcinogens, at least in a general way. Most major forms of human cancer have been reproduced in such animals through exposure to chemical carcinogens. Since their natural lifetime is two to three years, rodents provide information about the cancer-causing potential of test materials more quickly than do longer-lived animals, such as dogs or monkeys. Special strains of mice and rats have been developed to be particularly suitable for carcinogenicity testing.

**Question:** How well do laboratory animal tests predict whether or not a substance can cause cancer in humans?

**Answer:** There are two ways to answer that question. Of the approximately 30 agents known to cause cancer in humans, almost all cause cancer in laboratory animals.

Of the several hundred other chemicals that cause cancer in laboratory animals, however, it is not known how many are also human carcinogens. Nevertheless, materials that cause cancer in one mammalian species are usually found to cause cancer in other species. Furthermore, in some instances the risk predicted in advance by tests on rodents were later confirmed by the occurrence of cancer in exposed humans. Chemicals such as diethylstilbestrol (DES), vinyl chloride, and bis(chloromethyl)ether were shown to cause cancer in mice and rats before it was known that people exposed to those chemicals also had increased cancer rates.

We should assume, therefore, that agents that cause cancer in animals are likely to cause cancer in humans. To prevent cancer, we cannot afford to wait for absolute proof of carcinogenicity in humans. Instead, we must heed the warnings provided by laboratory animal experiments and reduce or eliminate human exposure to probable cancer-causing agents.

**Question:** How are these laboratory animal tests performed?

**Answer:** In brief, groups of about 50 mice or rats of each sex are exposed to the test substance at different dosages for about two years. Other groups, known as controls, are treated identically, but are not exposed to the test substance. At the end of the experiment, the animals are carefully dissected and examined by pathologists (doctors who interpret the changes in body tissue caused by diseases), and the frequency of tumors in the test groups is compared with that of the controls. Carcinogens produce a tumor frequency higher in the exposed animals than in the unexposed control animals. Noncarcinogens, by contrast, do not produce tumors.

**Question:** We often read about mice or rats being fed dosages much higher than those to which humans would normally be exposed. Are high doses really used, and if so, why?

**Answer:** Yes, high doses are often used to increase the ability of the tests to detect tumor-causing potential. The public often misunderstands the reasons for high dosage testing, and misinterprets the results.

In the human population, large numbers of people are exposed to low doses of chemicals, but the total impact may not be small at all. For example, a carcinogen might cause one tumor in every 10,000 people exposed to it, which may seem like a needle in a haystack. But exposure of 220 million Americans would result in 22,000 cancers—a public health disaster. We therefore need sensitive tests to detect those agents with the potential to cause only low cancer rates.

We obviously could not identify a carcinogen that causes one cancer in every 10,000 exposed mice by running the test on only 50 mice. To detect such a low cancer rate, we would need tens of thousands of mice, which would cost many millions of dollars per test. Testing more than a few chemicals in such an unwieldy fashion would be prohibitively expensive and time-consuming.

With high dosages, any potential carcinogenic effects are more likely to be detected in small groups of rodents because the cancer rate among the test animals is increased correspondingly. In the above example, a dose 5,000 times higher might cause cancer in 5,000 of every 10,000 mice, or 50 percent of the animals. If 20 or 30 of our test group of 50 mice developed cancers under such conditions, while the control group had only a few cancers, we could properly conclude that the chemical was capable of causing cancer. When high doses do not cause cancer, we also have greater assurance that the chemical is not a carcinogen.

**Question:** Won't the animals get sick and die if given such high doses?

**Answer:** No. National Cancer Institute guidelines for testing restrict the doses to levels that will *not* cause significant toxicity or unduly shorten the animals' lives. Because the animals must live long enough for tumors to develop, doses that kill the animals prematurely are not used in tests for carcinogenicity.

**Question:** Doesn't everything cause cancer if the dose is high enough?

**Answer:** No. High doses of many chemicals are toxic, but they will not cause tumors. Other forms of toxicity, such as loss of hair or weight, various organ malfunctions, or even death, should not be confused with carcinogenesis.

In one study, 120 pesticides and industrial chemicals were tested at the highest doses the mice could tolerate and survive. These chemicals were not randomly selected, but were chosen because they were suspected of carcinogenicity. Nevertheless, after two years of such treatment, only 11 of these chemicals caused cancer in the test animals.

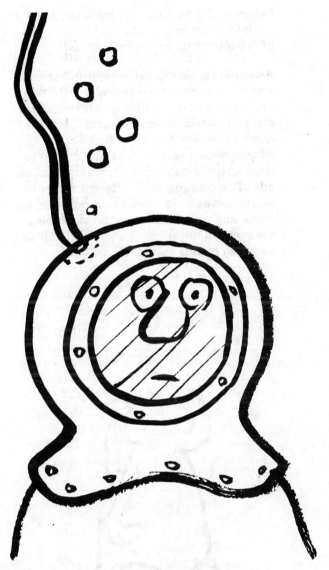

*High doses of many chemicals may be toxic, but not all of them will cause tumors.*

**Question:** Will a carcinogen cause cancer in the same organ when tested in different animal species?

**Answer:** Often, but not always. Different organs may become cancerous in different species. Thus a chemical that causes cancer of the liver in mice, for example, might cause cancer of the breast in rats and cancer of the bladder in humans. A carcinogen also can cause tumors in several organs in the same species, or even in the same individual.

**Question:** Why are test animals sometimes exposed to the test chemical in a way different from the way in which humans would be exposed? Hair dyes were *fed* to rats, for example.

**Answer:** The method of exposure need not be identical, although that is usually preferred. The tests seek to determine the *potential* to cause tumors, and distribution throughout the animal's body is the important part of the test. Hair dyes can enter the circulatory system through either the scalp or digestive tract.

**Question:** Aren't many of the animal tumors not cancerous?

**Answer:** Yes, but these tumors also indicate human cancer risk. Noncancerous (benign) tumors often become cancerous. No chemical is known to cause only benign tumors.

**Question:** How should we interpret the appearance of tumors in one species, such as rats, and the absence of tumors in another species, such as monkeys?

**Answer:** Positive evidence of tumors in one test is not cancelled by negative evidence, or the absence of tumors, in another test. A substance that is carcinogenic in one strain or species occasionally has little or no effect in another strain or species. Furthermore, not all tests are equally sensitive, and a negative result may mean merely that the effect was missed. Too few animals, too low a dosage, or too short a test period, for example, can lead to a false negative result. A 5-year test on 10 monkeys could fail to detect tumors that would have appeared after 10 to 15 years, or that would have been detectable in 50 monkeys. Any adequately performed positive test in laboratory animals indicates a cancer hazard for man.

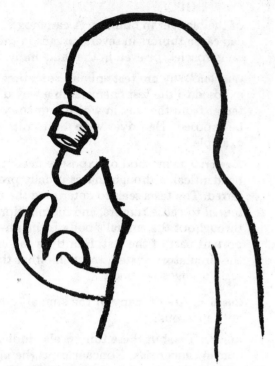

*We can't be sure there is any safe level of exposure to carcinogens.*

**Question:** Are there safe levels for human exposure to carcinogens; that is, are there threshold dosages below which we can be sure that no cancer will occur?

**Answer:** There is no adequate evidence for the existence of a safe threshold for any carcinogen. As far as we know, the frequency of tumor formation declines as the dosage declines, but the risk of carcinogenesis may not disappear until the dosage reaches zero. Although high doses often are used in the tests, we must not assume that only high doses cause cancer. On the contrary, we must assume that low doses will also cause cancer, but at proportionally lower rates.

Human cancers have occurred following very low level exposure. Asbestos brought home on the clothing of asbestos workers, for example, has caused fatal cancers in members of the workers' families.

**Question:** Can we estimate the magnitude of the human cancer risk from the results of laboratory animal experiments?

**Answer:** Yes, we can estimate it in various ways, but the estimates usually are crude and we can't always depend on them. Animal tests tell us there is a hazard, but they don't tell us the extent of that hazard. Attempts to quantify the risk can lead to large errors, resulting from different methods of calculating risk, different susceptibilities between laboratory animals and humans, and potential interactions between the agent in question and other chemicals to which man is exposed.

*The risk of cancer may be increased when people are exposed to several carcinogens at the same time.*

**Question:** What happens when people are exposed to several carcinogens at the same time?

**Answer:** The resulting cancer rate may be higher than would be predicted by adding the risks from each carcinogen alone. Cigarette smoking and asbestos exposure, for example, are both carcinogenic, but asbestos workers who smoke are subject to a cancer risk that is far higher than would be expected by adding the risk from smoking to the risk from asbestos.

**Question:** What is the best way to prevent cancer?

**Answer:** Individually and collectively, we must make every effort to reduce or eliminate human exposure to carcinogens. This effort applies to agents known to cause cancer in humans, especially tobacco smoke. It also applies to agents that, on the basis of evidence gathered in animal or other experimental studies, are suspected of causing cancer in humans. Because everything does not cause cancer, such an approach to cancer prevention is reasonable and workable.

*Many cancers can be prevented by reducing exposure to carcinogens.*

## For further information

Cancer Information Services (CIS) can provide toll-free telephone answers to your questions about cancer cause, prevention, diagnosis and treatment.

ALASKA: 1-800-638-6070
CALIFORNIA:
　FROM AREA CODES (213),
　(714) and (805):
　1-800-252-9066
COLORADO: 1-800-332-1850
CONNECTICUT: 1-800-922-0824
DELAWARE: 1-800-523-3586
DISTRICT OF COLUMBIA (METRO
　WASHINGTON): (202) 636-5700
FLORIDA: 1-800-432-5953
GEORGIA: 1-800-327-7332
HAWAII:
　OAHU: 524-1234
　NEIGHBOR ISLANDS: ASK
　OPERATOR FOR ENTERPRISE 6702
ILLINOIS: 800-972-0586
KENTUCKY: 800-432-9321
MAINE: 1-800-225-7034
MARYLAND: 800-492-1444
MASSACHUSETTS: 1-800-952-7420
MINNESOTA: 1-800-582-5262
MONTANA: 1-800-525-0231
NEW HAMPSHIRE: 1-800-225-7034
NEW JERSEY: 800-523-3586
NEW MEXICO: 1-800-525-0231
NEW YORK STATE: 1-800-462-7255
NEW YORK CITY: (212) 794-7982
NORTH CAROLINA: 1-800-672-0943
NORTH DAKOTA: 1-800-328-5188
OHIO: 800-282-6522
PENNSYLVANIA: 1-800-822-3963
SOUTH DAKOTA: 1-800-328-5188
TEXAS: 1-800-392-2040
VERMONT: 1-800-225-7034
WASHINGTON: 1-800-552-7212
WISCONSIN: 1-800-362-8038
WYOMING: 1-800-525-0231

　ALL OTHER STATES: 800-638-6694

# Should the Delaney clause be changed?

## A debate on food additive safety, animal tests, and cancer

In an attempt to clarify these complex issues, C&EN has invited prominent figures representing four distinct vantage points to present their contrasting views—a biochemist specializing in chemical carcinogenesis, a toxicologist, a public interest advocate with a medical background, and a member of the U.S. House of Representatives trained in chemistry. **Dr. William Lijinsky of the Frederick Cancer Research Center** supports retaining the Delaney clause unchanged and describes scientific evidence and reasoning behind this view, in particular the key issue of use of animal testing for carcinogenicity. **Dr. Frederick Coulston of Albany Medical College** favors changing the clause and criticizes current practices in testing and regulation of possible chemical carcinogens. **Dr. Sidney M. Wolfe of the Health Research Group** supports retention of the clause as is, and focuses in particular on public interest and regulatory aspects of ensuring food additive safety. And **Rep. James G. Martin (R.-N.C.)** discusses his bill to modify the Delaney clause, and questions of science and public policy involved in regulation of food additives.

Each participant submitted his written contribution, and then was sent copies of his three colleagues' papers and given an opportunity to write a rebuttal. Their contributions and rebuttals, appearing in the following pages, were completed before last week's epidemiological report.

The Delaney clause covers only materials deliberately added to food, including food packaging—and not such contaminants as pesticide residues or natural carcinogens (for example, aflatoxin), for which tolerance levels are set under other FDA regulations, and a few other exclusions. And despite its great fame (or notoriety), the clause has rarely been invoked.

However, stresses one FDA official, although the agency can and has banned food additives suspected of carcinogenicity under its "general safety" powers, the clause makes FDA "keep a closer eye on things." And "it's much easier to propose a saccharin ban when the Delaney clause is there. It'd be much tougher to deal with the pressures otherwise."

Indeed, there are charges by Delaney clause supporters that public emotion against a saccharin ban has been deliberately whipped up. For example, a trade association of 60 dietary food and beverage firms, the Calorie Control Council, is accused by Wolfe of using saccharin as a lever to change the clause and prevent banning of further food additives.

Robert C. Gelardi, staff director of the council, denies this. The council has not taken a position on revising the clause, he tells C&EN, although individual members have. But the council urges "more intelligent and rational interpretation of existing laws and regulations." And it strongly opposes a saccharin ban as "politics, not based on science," and "precipitous and arbitrary." It is "very important that this kind of experiment not become a precedent for studies on other food additives," he stresses. Or else "we may wind up without most of the substances on the market." And with 50 million people in the U.S. wanting diet foods and beverages, he says, he questions whether "consumer activists" really represent consumers.

Reprinted from CHEMICAL & ENGINEERING NEWS, Vol. 55, June 27, 1977, pp. 24-46
Copyright 1977 by the American Chemical Society and reprinted by permission of the copyright owner

*"Animal tests are predictive of carcinogenicity in man"*

Dr. William Lijinsky
Frederick Cancer Research Center

# U.S. health will be jeopardized if Delaney clause is abandoned

The Delaney clause, forbidding addition to food of any substance found to induce cancer in man or animals, removes discretion from the regulatory agencies. These agencies are subject to great pressure, both direct and indirect, from those with an economic stake in the matter, to set tolerances for carcinogens. The current argument about the widely used food additive saccharin is whether we should trust manufacturers, or anyone else, to decide whether the additive is safe to use, even though it is a carcinogen. As then Secretary of Health, Education & Welfare Arthur Flemming said in 1960 at the time of an argument similar to that over saccharin, no one knows how to set a safe tolerance for a carcinogen and, therefore, no cancer-causing substance can be legally added to food. That is the reason for the Delaney clause, which I strongly support.

It is not unexpected, nor unreasonable, that manufacturers that will be hurt economically by such a ban will use all of their economic clout to oppose it. However, neither can they claim to be objective in their view of the matter, nor to have the interest of the public as their guide. So let us examine the scientific evidence supporting the Delaney clause, and why the future health of Americans will be jeopardized if it is abandoned.

In this paper, I will first discuss the basic principles and methods of toxicological testing in animals for carcinogenicity of chemicals, and some widely held misunderstandings of these principles and practices. I will then discuss the saccharin case and the Delaney clause as applications of these principles, and explain why I believe that the Delaney clause should be preserved intact.

First, a few observations on chemical carcinogenesis in man. It is a sobering fact that every day 1000 Americans die of cancer. The probability that a majority of these cancers are caused by chemicals shows why there is such concern over the presence of carcinogens in the environment.

If, as this implies, cancer is a preventable disease, a major effort should be devoted to identifying cancer-causing substances to which people are exposed and to removing them, or at least to reducing exposure to them as much as possible. Although it is certain that nobody is in favor of carcinogens, past and present arguments over how to go about eliminating these substances have centered on their identification, and on whether a little of them is acceptable, even when a large amount is not.

The association of cancer with chemical substances in the environment goes back more than 200 years, when the surgeon Percival Pott related the scrotal cancer of chimney sweeps to their exposure to soot. But widespread interest in this association has burgeoned only in the past 20 years or so. The causes of death that were most common in the 19th century have been greatly reduced through better sanitation and better care, and cancer and circulatory diseases have now assumed the place of infectious diseases as the most common cause of death.

There is growing evidence that cancer and circulatory diseases are both related to our life style, rather than to genetics, and therefore are inherently preventable. Few would deny that cigarette smoking is a major factor in the large, and growing, incidence of lung cancer in most western countries. Epidemiological studies of cancer rates in Japanese people strikingly show the major influence of environment and life style in the induction of cancer. Japanese living in Japan have a low incidence of colon cancer and of lung cancer, but a relatively high incidence of stomach cancer. First- or second-generation Japanese living in Hawaii have a very similar cancer pattern to those living in Japan. However, Japanese of third and subsequent generations in Hawaii or the continental U.S.—who have to a large extent adopted an American life style—show the same pattern of cancer as the bulk of the American population, namely high incidence of colon and lung cancer and relatively low incidence of stomach cancer.

So it seems that the environment is very much involved in cancer. But where and how do we look for the carcinogenic agents? In the beginning, industrial cancers were the most obvious ones seen. For example, bladder cancer of dye workers in the Ruhr was related by Ludwig Rehn in 1895 to exposure to aromatic amines. Skin cancer of creosote workers, wax pressmen, and workers in the shale oil and coal tar industries was related to polynuclear hydrocarbons in materials to which these people were exposed. More recently, rare cancer of the liver found in some people who worked in plastics plants was related to exposure to vinyl chloride. To these must be added cancer of the vagina in young women whose mothers were given diethylstilbestrol during pregnancy. All of these associations were made because the cancers observed by alert physicians were rare in the rest of the population.

But how could we make such associations if the chemicals

happened to cause cancer of types commonly found in the rest of the population? And what might be causing these common types of human cancer, anyway? Such common cancers are lung, breast, colon, pancreas, stomach, bladder, esophagus, and leukemia—each causing the death of more than 10,000 Americans every year.

### History of carcinogen testing

In an effort to identify the causes of these cancers we have to turn increasingly to the testing of chemicals for carcinogenic activity. It is these tests which are currently the subject of such intense debate. As do all tests, they suffer from imperfections

> *"Regulatory agencies are subject to great pressure to set tolerances for carcinogens"*

and lack of precision. But they are certainly better than no tests at all, and when they do raise the warning flag of potential cancer-causing activity by a chemical, they cannot be ignored.

Animals have been used for centuries as surrogates for people in assessing the effects of therapeutic and other agents which might later be used in humans. It has not been common to carry out experiments in man without preliminary studies in animals. Were the results of tests of chemicals in animals not useful as predictors of their effects in man, few pharmaceutical compounds would be developed and introduced these days, and mankind would, no doubt, suffer as a consequence. This is true whether the tests in animals are to demonstrate a desired biological effect or the absence of an unwanted effect—for example, toxicity.

The induction of cancer is one such biological effect that has been observed in experimental animals during the 70 years since the first studies of experimental carcinogenesis. These studies were in 1906 and involved treatment of rabbits with a dye, scarlet red, proposed for use as an antiseptic. The dye was painted on their ears, and the rabbits developed skin cancer at the site of application.

In 1915 two Japanese scientists, Katsusaburo Yamagiwa and Koichi Ichikawa, painted coal tar on rabbits' ears—and again skin cancers developed there. Exposure to coal tar was known to be related to the common skin cancers in workers in that industry. This demonstration was the first example of a known carcinogenic agent producing similar results in laboratory animals.

This finding was followed up during the 1920's and early 1930's by Dr. Ernest Kennaway and his coworkers in London who, after a series of revolutionary experiments combining chemical analysis and biological testing, isolated a cancer-causing chemical, benzo[a]pyrene, in pure form from an environmental source, coal tar pitch. The isolation was made easier, though hardly simple, because of the intense fluorescence in ultraviolet light of this compound. It was possible to follow the compound through the fractionation procedure, because the carcinogenic activity in rabbit and mouse skin seemed to parallel the degree of fluorescence of the fractions. Nevertheless, it took almost 10 years to obtain 5 grams of benzo[a]pyrene from 2 tons of coal tar pitch.

Work on polynuclear hydrocarbons and their carcinogenic effects is still going on, and attention still focuses on benzo[a]pyrene in tobacco smoke, air pollution, petroleum products, solvents, and cooked food. A natural outgrowth was study of the mechanism of action of this and similar chemical carcinogens, work which continues to this day.

### Parallel carcinogenicity in man and animals

Thus was begun the ever-expanding investigation of chemical carcinogenesis, which can nowadays be called, more comprehensively, environmental carcinogenesis. This investigation has developed along two lines:

• Continuation of the pioneering medical studies of Pott, Rehn, and others into the relation between cancer incidence in man and exposure to environmental agents—which is categorized as cancer epidemiology.

• Investigation of the chemical nature of carcinogenic agents and their mechanism of action. This involves testing a great variety of chemical substances for carcinogenic activity in experimental animals, with the aim of revealing the types of chemicals that present a cancer hazard to people, as well as the way in which they do this.

So far, the relation of these two lines of investigation has been close. Epidemiological studies in man have directed us to the identification of certain substances which are carcinogenic in man, and these substances have been found equally and simultaneously to produce cancerous tumors in experimental animals.

In fact, the parallel is so close that almost all substances known to be carcinogenic in man have had the same effect in some suitable animal model. Polynuclear compounds show this parallel effect in man and animals, for example, especially as mixtures. Aromatic amines—2-naphthylamine and benzidine

> *"There is no way of establishing a safe threshold dose of a carcinogen"*

among them—are associated with bladder cancer in dye workers and also induce bladder tumors, as well as tumors of other organs, in dogs and hamsters. Several metal oxides and sulfides (such as nickel, cobalt, and cadmium) and other inorganic compounds (for example, beryllium salts, chromates, and asbestos) have been related to cancer in industrial workers and also have given rise to a variety of tumors in experimental animals.

In parallel fashion, several compounds that were tested for a variety of reasons were first shown to be carcinogenic in experimental animals, and these results later turned out to be accurately predictive of the compounds' carcinogenic effects in people exposed to them. A good example is vinyl chloride. It was found to induce liver cancer when inhaled by rats. A few years later, the same type of liver cancer—which is rare in humans—was diagnosed in factory workers who had been exposed to relatively high concentrations of vinyl chloride vapor.

Similarly, diethylstilbestrol (DES) was shown to cause cancer in mice more than 20 years ago. During the past few years, a rare cancer of the vagina developed in several young women who had been exposed in utero to DES (their mothers took it in quite small doses while pregnant).

The DES case illustrates several factors which are important in using experimental animals to test for carcinogenicity and then trying to use the results to predict possible carcinogenic effects in man. First, animal tests are predictive of carcinogenicity in man, who is not an exceptional species in this regard. Secondly, there is a dose-response effect: Larger doses of carcinogen given to experimental rodents make tumors appear within two years (untreated rodents normally live only a little longer than that), whereas the comparatively small doses to which people are exposed make tumors appear in them only after a much longer time. Thirdly, not all of the exposed people developed the cancer, suggesting a considerable variation in susceptibility to the carcinogen (which might have something to do with genetics), just as we find in experimental animals.

Even if we accept the probability that a majority of human cancer is caused by chemical factors, we are faced with a dilemma, since exposure to chemicals in the workplace seems to account for only a small proportion of chemically induced cancers. The dilemma is that we cannot prevent exposure to all foreign chemicals without sacrifice of what we call civilization, and yet we do not have adequate methods of determining which chemicals are carcinogenic and which are "safe."

It is quite likely that humans are exposed to a large number of chemical carcinogens in quite small amounts, rather than to large doses of a single carcinogen such as we administer to animals in the course of a test. Accumulation of the effects of these small doses—sometimes supplemented by relatively large doses of unusual carcinogens, as in the workplace—leads to cancer. That is why we cannot say that such and such a carcinogen *causes* cancer in man.

Instead we say that exposure to a carcinogen increases the risk of developing cancer—the higher the exposure, the greater the risk. Therefore, it is unwise for anyone to predict how many cases of cancer will result from exposure of people to a particular amount of carcinogen.

Both human data and the results of animal experiments tell us that individuals vary greatly in their susceptibility to the effects of carcinogens, but we do not know how to identify those most at risk. Not everyone exposed to materials containing polynuclear hydrocarbons develops skin cancer. Not everyone who smokes a lot of cigarettes develops lung cancer (although in this case the problem might be much more complex than simply exposure of the lungs to identifiable carcinogens in the smoke). Not everyone who worked with 2-naphthylamine or benzidine developed bladder cancer, although 20 to 25% did. In animal experiments, a fixed dose of a carcinogen, for example nitrosomorpholine, caused the death from liver cancer of one of a group of rats within six months. Other rats died of the same cancer at intervals over the subsequent 18 months, and a few rats died late in the experiment without tumors. Here again, the experience with humans and experimental animals is very similar.

**Animal tests for carcinogenicity**

Since most of our information about carcinogenesis has been obtained from experiments in animals, it is appropriate to describe such studies and to catalog what they have told us about carcinogenesis. Carcinogens elicit the maximum tumor response, dose for dose, when they are administered as a large number of small doses, rather than in a small number of large doses. This is almost universally true of all classes of carcinogen. Usually a single dose, however large, of most carcinogens does not give rise to tumors.

The response of animals to carcinogens depends upon the site of application. Some classes of carcinogen act mainly at one site and not at all at other sites. For example, polynuclear hydrocarbons act mainly on the skin and to a lesser extent on the lungs.

Other carcinogens act at distant sites whatever the route of administration, a type of action shown by aromatic amines (which cause bladder tumors even when fed), and by aflatoxins and nitrosamines. Nitrosamines thus cause tumors in a great variety of organs (depending on the chemical structure of the nitrosamine), whether inhaled, injected, or given by mouth. Nitrosamines seem to be most effective when given orally, an important consideration since the most common human exposure to them is through their formation in the stomach from amines and nitrite.

There are significant differences in response of different species to a given carcinogen. This difference can be tumors versus no tumors, or it can be one type of cancer in one species and another type of cancer in a second species.

> *"Even though one might agree that there must be a no-effect level, we do not know how to determine what that is"*

For example, polynuclear hydrocarbons applied to the skin of mice or hamsters induce skin tumors, but the response of rats is very much lower and guinea pigs do not respond at all. Man, of course, responds with skin tumors. Aflatoxins are very potent liver carcinogens in rats, but mice are almost totally resistant to their action. Aromatic amines give rise to bladder cancer in man, dogs, and hamsters, but they are much less effective carcinogens in rats and mice, including mainly liver tumors. And nitrosamines have not been related so far to any particular human cancer, but they induce tumors in all species of animal tested, and show very different tumor responses from one species to another. For example, 2,6-dimethylnitrosomorpholine induces cancer of the pancreas in hamsters but not in rats, in which instead it causes esophageal tumors.

This suggests that there is no way in which we can predict from animal experiments which type of tumor could be induced in man by any particular compound. It might be the same target as in the animal, or it might be a different target. However, it cannot be assumed that a compound that is carcinogenic in animals would not have a carcinogenic effect in people. Indeed, because of the refractoriness, largely unpredictable, of some species to the action of some carcinogens, it cannot even be assumed that a substance shown to be nontumorigenic in animals is necessarily without effect in man. But we have to draw the line somewhere, and to assume the worst would be to ban all foreign chemicals on the ground that they *might* have a carcinogenic effect in some susceptible person. Only the most radical would espouse such a view.

**Difficulties in animal testing**

The most misunderstood aspect of toxicity testing is the use of very large doses in experimental animal studies, particularly in rodents. There is a wide spectrum of susceptibility to the effect of a toxic substance, and statistics must be applied to arrive at an $E_{50}$ dose (the dose at which one expects half of the animals to show the effect). When the effect is death, the number is called $LD_{50}$, or the lethal dose that will lead to death of 50% of the animals. A similar number can be arrived at for induction of a tumor of a particular type, or the death of the animals from tumors. These numbers are, of course, estimates, since the size of the groups of animals is small.

In the specific case of testing compounds for carcinogenic effects, the problem is more complicated than in looking at a definite end point, such as death. However carefully we do the animal experiments, there are large variations, biological and chemical, which render the tests imprecise. As discussed previously, carcinogens are most effective when given as many doses over an extended period, often the life span of the animals. The animals increase in weight during the course of the treatment, and grow at different rates, and it is impossible to deliver exactly the same dose to each animal each time. Unless the tests are to be enormously expensive, the animals are housed in groups, and the consumption of food or water in which the test compound is incorporated will be an average, rather than a specified amount for each animal.

Assuming that the chemical administered is a carcinogen, a certain proportion of animals in the group will develop tumors, which may or may not lead to death. The animals will be ap-

> *"The saccharin furor is based more on emotion than on interpretation of scientific evidence"*

proximately the same age when the treatment starts, usually six to eight weeks. One end point of the test, then, is to kill all the animals at a fixed time, usually two years after the beginning, and to count the number of animals with tumors at that time. The other approach is to let the animals die naturally and to use the average or median age at death as an index of the carcinogenic effectiveness of the treatment, assuming that the tumors kill the animals.

Using the first approach, a larger number of animals with tumors would indicate greater carcinogenic effectiveness. By the second approach, the carcinogenic treatment that caused death from tumors at an earlier average or median age would be considered more effective.

In either case, it is the pathological findings that are the essence of the tests. All such tests have to include a comparison with a control group of untreated animals. It is not simply that the animals treated with a carcinogen develop tumors and the controls do not. All types of experimental animals develop tumors, including the rodents commonly used in carcinogenesis tests. The tumors appearing in untreated animals we call, perhaps unjustly, "spontaneous tumors"—mainly tumors of the breast and endocrine organs, most of which are "benign." However, control animals almost always develop a few tumors of the liver, kidney, brain, and other organs, but at a very low incidence, perhaps one in 100. We do not know and cannot speculate about the cause of these "odd" tumors.

If only a very small group of animals were used in a test, it would be impossible to demonstrate a significant carcinogenic effect. Therefore, minimal sizes of groups have been established for such tests. The number is usually 50 males and 50 females, a total of 100 animals.

A single test of one chemical usually consists of three dose groups of this size and preferably two species. Such a test on the 600 animals involved over a period of two years usually is estimated to cost about $150,000, setting economic limits on the maximum numbers of animals used.

If the group of animals to which the test substance is given develops a number of tumors that are not seen in untreated controls, then we say that the substance is a carcinogen. But how many tumors of, for example, the liver, lung, esophagus, or bladder are needed before we conclude that there is a carcinogenic effect? Obviously not one or two, since we might expect this number by chance. The number usually accepted as sta-

tistically significant is five. This is not to say that a test of a substance which induces four kidney tumors in a group of animals demonstrates the absence of a carcinogenic effect, but only that within the limits of the test a carcinogenic action is not proved.

In such a test, then, we would assume that diagnosis of an induced (meaning nonspontaneous) tumor of a certain type in four animals would not suggest a carcinogenic effect of the chemical, even though 4% of the animal population under test developed that tumor. Translated to the U.S. population, that would be equivalent to 8 million people. On the other hand, a finding of this tumor in five of the 100 animals would be considered statistically significant, and that would be equivalent to 10 million Americans.

Any substance which, when tested at the doses to which people are exposed, gave rise to an incidence of tumors greater than 5% in a group of experimental animals would, I think, be considered extremely hazardous. In practice, we never would find such a substance. Instead, we are looking for carcinogens to which man is exposed in very low doses, or for very weak carcinogens, and the continuing debate is about means of identifying them.

### Why high doses are used

A prime subject of controversy is use of high experimental doses on animals, doses which critics often charge are "unrealistic" compared to human consumption of food additives. The sensitivity of animal tests to weak carcinogens or to low doses of strong carcinogens is not very great. Thus, our tests have to be designed to compensate for the weakness in them.

The first problem is that experimental rodents live for only two years or so, and this must represent a possible human exposure of 50 to 60 years. So we increase the dose given to the animals above that reasonable for man, to compensate for the difference. It is most probable that tumors appear late in the life span of man, because of his exposure over many years to low doses of carcinogens. This is exactly the pattern observed in experimental animals—the lower the dose, the later in the life span the tumors appear.

Similarly, because of the statistics of the animal experiments, should we wish to detect a carcinogen that might cause cancer in 0.01% of the U.S. population (which would amount to 20,000 people), at the dose to which it is exposed, the dose given to the experimental animals must be greatly exaggerated. As has been mentioned, carcinogens evoke a dose response: The higher the dose, the earlier tumors appear, and the larger the number of tumors in any given group of test animals.

Of course, there is the alternative of using a group of 50,000 animals to test the substance, so that we could pick up the minimum of five animals at an incidence of 1 in 10,000. It is not necessary to dwell on such a ridiculous suggestion, which would mean devoting all our conceivable animal and manpower resources (especially pathologists, who are few in number) to the testing of a handful of compounds, to say nothing of the costs involved.

### Saccharin carcinogenicity

Let me turn now to the specific case of saccharin, about which a furor rages based more on emotion than on interpretation of scientific evidence. It is strange that many scientists, who are usually so logical, cannot accept the sound basis of toxicological testing when the results implicate as hazardous such a widely used substance as saccharin.

There have been many tests of saccharin for chronic toxicity during the past 20 or 30 years. Several of these were "suspicious," that is showing tumor incidences on the borderline of significance; others were negative. It was thought that some of the equivocal results might be due to the presence in saccharin of a common impurity, o-toluenesulfonamide (a starting material in the synthesis).

The recent Canadian tests—which were excellently conducted—used both saccharin and o-toluenesulfonamide (OTS) to settle the point. For reasons already explained, each compound was given separately to rats mixed in food at quite high doses, up to 5% of their diet, during their lifetime. In one part of the test, saccharin was also given to pregnant rats and then to the offspring for their lifetime.

In both types of saccharin tests, a significant number of urinary bladder tumors occurred in treated rats. As would be expected, there were more tumors in animals treated transplacentally. (Fetuses are often much more sensitive to carcinogens and teratogens than are children or adults.) However, there was no significant incidence of tumors induced in rats given OTS only. It was concluded that saccharin is carcinogenic and that OTS is not. Because this was the outcome, a host of authorities has questioned the validity of the tests. Had these shown that OTS was carcinogenic and saccharin not, I am sure that few would have questioned the validity of the tests, and the only pressure would have been to purify the saccharin.

What do the results of this test—and tests of other compounds that result in a significant incidence of tumors—mean? They do not mean that everyone who takes saccharin will develop bladder cancer, nor necessarily any other type of cancer. We do not know which organ of man might be the target of saccharin. Were the target organ the colon, for example, we should never know that some cancer is caused by saccharin, since the incidence of colon cancer is so high from other causes (equally unknown, incidentally).

Nor do the Canadian test results mean that saccharin has the same effect in all people who take it. They do mean, however, that among substances to which we are exposed, saccharin presents a substantially higher carcinogenic hazard than compounds that are not carcinogenic in animal tests—o-toluenesulfonamide, for example. Because of the special sensitivity of the fetus and small children to carcinogens, it is most important to avoid exposing children and pregnant women to saccharin.

The most controversial aspect of the saccharin problem, and that of all food additives found to be carcinogenic, is application of the Delaney clause of the Food, Drug & Cosmetic Act of 1958. This states, in essence, that no substance can be allowed as a food additive that induces cancer in man or animals. Use of the substance is not permitted at any level; there is no tolerance. The law takes this position because there is no way of establishing a safe threshold dose of a carcinogen.

Even though one might agree that there must be a no-effect level, we do not know how to determine what that is. In an animal test of reasonable size, it would not be possible to establish a dose level that would guarantee the safety of even 95% of the population. As discussed earlier, to determine a dose that would protect the most sensitive 0.01% of the population from any carcinogenic effect of the compound would require that 50,000 test animals would show no significant incidence of tumors compared with 50,000 untreated controls. And even that would not be good enough, since 20,000 people might still be at risk. Also, in most cases we do not know whether man is more sensitive or less sensitive than the experimental animals to a particular chemical carcinogen.

In trying to set a threshold, still other problems arise. A safety margin would have to be included, as in all toxicity evaluations. This would probably reduce the permitted level of use of a compound below that at which it is useful, as has happened with several moderately toxic (but noncarcinogenic) substances.

The problem would be simpler if saccharin and other food additives were used occasionally—as are several drugs prescribed in dire circumstances even though they are carcinogenic. But the additives are not used occasionally. Instead, they are consumed in as large quantities as people can be persuaded to buy them. Since the effects of carcinogens are cumulative, the more that is taken, the larger the risk of tumor induction.

Moreover, there is growing evidence that carcinogens can act synergistically. That is, the effect of one carcinogen can potentiate the action of another, so that the total effect is greater than addition of their effects. In this respect, our animal tests do not serve as perfect models of the exposure of people to a carcinogenic substance. In animal tests we administer a single substance, while in man the effect of that substance is added to the effects of many other carcinogens to which he is exposed—those in tobacco smoke, for example.

The studies of Dr. Irving J. Selikoff of Mt. Sinai School of Medicine, New York City, are relevant here. They show that workers exposed to asbestos have a manifold higher incidence of lung cancer if they smoke cigarettes than either asbestos workers who don't smoke, or heavy smokers who have not been exposed to asbestos; asbestos increases the risk. Several experiments with nitrosamines have also shown this type of synergism in carcinogenesis.

Because of interaction of the effect of one carcinogen with that of another—even if synergism were not involved, but simple additive effects—acceptance of a threshold for the large variety of food additives that might now, or in the future, prove to be carcinogenic would lead to a dangerous outcome. Assuming that 100 substances were so approved, each could be used at its threshold level. Their combined effects, added to those of possibly unavoidable carcinogens in the environment (unavoidable because we have not yet identified them), would provide a carcinogenic exposure that could be overwhelming. This is what must be avoided.

There are solutions to the dilemma, even for saccharin. Noncarcinogenic substitutes undoubtedly can be found, since most substances are not carcinogenic. In fact, carcinogenicity is a rarity—contrary to popular opinion that "anything will cause cancer if you feed an animal enough of it." For example, a survey of compounds tested for carcinogenic activity (U.S. Public Health Service publication 149) shows that less than 20% of the compounds tested are carcinogenic in animals. Since the compounds were especially selected for testing because of strong suspicion of their carcinogenicity, one would expect a far lower percentage of carcinogens among chemicals in general.

Furthermore, it might be possible to do without the particular property, sweetness, imparted by this food additive. After all, mankind survived before 1879, when the sweetening effect of saccharin was discovered. It even survived before refined sugar became an article of commerce. We might be better off if we could do without either.

And, considering the personal disaster that cancer represents to those who develop it, as well as to their families, we can do without any substance that is less than vital and for which there is any suggestion—let alone strong evidence—of cancer-causing effect. □

---

*Dr. William Lijinsky, 48, is director of the Chemical Carcinogenesis Program at the National Cancer Institute's Frederick (Md.) Cancer Research Center (Fort Detrick), which is run under contract by Litton Bionetics. Born in Dublin, Ireland, and a British citizen, he earned B.S. and Ph.D. degrees in biochemistry from the University of Liverpool, and did postdoctoral research at California Institute of Technology and McGill-Montreal General Hospital Research Institute. He served on the faculty at the Chicago Medical School in 1956–68, was at the Eppley Institute, University of Nebraska college of medicine, in 1968–71, and at Oak Ridge National Laboratory in 1971–76. Lijinsky has published some 160 papers on chemical carcinogenesis. His major research interests for more than 20 years have been environmental carcinogenesis and mechanisms of action of chemical carcinogens, with special emphasis on polynuclear hydrocarbons and N-nitroso compounds. A particular contribution has been his pioneering research on formation of nitrosamines in the environment and the body, their mechanisms of physiological action, and their significance in human cancer.*

## "I would like to see the Delaney clause repealed entirely"

**Dr. Frederick Coulston,** Albany Medical College

# Tolerance levels can be set for chemical carcinogens

On April 14, 1977, the new commissioner of the Food & Drug Administration, Dr. Donald Kennedy, reaffirmed FDA's decision to ban, or, if you wish, delist saccharin as a food additive. However, he stated that this action was taken not necessarily because of the Delaney clause; it would have been done anyway, even if the clause did not exist. In addition, he said that FDA will propose to continue use of saccharin as an over-the-counter item for sale in drugstores and supermarkets, just as aspirin or cough medicine is now available to the consumer, provided that there is medical evidence to prove that saccharin can be considered a drug.

The statement by the FDA commissioner that he did not need the Delaney clause in order to make the decision about saccharin is a key point, and must be considered by all thinking people. If the Delaney clause was not necessary in this decision making, then what is its value in protecting consumers from unwanted food additives that may be imminent carcinogenic hazards to man?

In many interviews with the media before Kennedy's April 14 announcement, I expressed my astonishment that FDA had taken the action that it did—claiming that it was banning saccharin on the grounds that the Delaney clause forbids its continued use. And I said that it would be far better and more traditional for the agency to simply delist saccharin as a GRAS (generally recognized as safe) substance or food additive, considering it instead as a possible drug in over-the-counter use and sales—an action FDA now proposes to take.

Certainly, no thinking person, either at the Canadian Ministry of Health or at FDA, considers saccharin to be an imminent carcinogenic hazard to man. If they did, remaining supplies of saccharin and diet foods on the shelves would be banned and destroyed immediately. It is indeed noteworthy that the new commissioner now has accepted the approach that should have been followed at the beginning of the saccharin controversy. The fact that the Delaney clause did not have to be invoked to ban saccharin emphasizes the need to re-examine the intent of the clause and its importance for the consumer and FDA.

However, before the Delaney clause can be discussed in any detail, it is important to understand the role of the toxicologist in these difficult situations. His responsibility is as simple to state as it is difficult to achieve: The toxicologist must assure the safety of man as he is increasingly exposed to chemical agents.

The human race has been exposed for hundreds of years to such chemicals as arsenic, lead, fluorine, copper, pyrethrum, natural flavors, and spices. Proof of safety was relatively simple in the past, compared with that needed for chemicals of the present day. The proof of safety simply consisted of the experience of man himself. If man got sick from too much arsenic or lead, the signs and symptoms were very readily recognizable and were well documented as early as the era of the ancient Greeks and Romans. Man himself, rather than animals, was the species of choice for the actual determination of toxicology.

In modern times, man has been exposed to hundreds and hundreds of new synthetic chemicals that are used as feed and food additives, pesticides, and medicines, and with which he has had no previous experience. Furthermore, chemical intermediates and metabolites, apart from their original purpose, manage to find their way into man and affect the normal physiological state. No one can deny the importance of these chemicals to general human welfare, but we must always assure man's safety as he is increasingly exposed to them.

Traditionally, toxicology is a study of poisons as they affect man, plants, and animals. The modern concept of toxicology applies a multidisciplinary approach, combining pharmacology, biochemistry, and pathology. Toxicology thus includes safety evaluation of the injurious effects of chemicals, drugs, natural products, and physical agents on the physiological states and pathological conditions of cells and tissues in plants, animals, and man.

We enjoy unmatched nutritional abundance and a superior level of health in very great part because of the inventiveness of chemistry in reshaping our food supply and our environment. Indeed, man can and must control his total environment in order to survive. Our interest today is not only in the pace of chemical synthesis, but in the character of it. To consideration of acute toxicity of chemicals, we must now add concern with chronic and cumulative reactions. Another very important problem is that of one or more chemicals acting in a way that interferes with or augments the effects of another chemical. Only recently have we begun to understand the chemical-chemical interactions that contribute to metabolic toxicities.

### Extrapolation from animals to man

Improvement of testing procedures is needed for food additives, whether they are intentionally placed in the food supply or occur inadvertently as chemical residues, such as pesticides or aflatoxin. Many eminent authorities agree that animal toxicity studies leave much to be desired in providing adequate margins of safety for man. The question "How can one be sure that extrapolation of animal data to man is accurate?" is perhaps more pertinent in assessment of direct and indirect food additives than in evaluation of new drugs. This is because proving the safety of a new drug implies studies in man, whereas for food additives, most of the decisions are based upon animal laboratory experiments.

Our knowledge of how chemicals are handled by the mammalian body has advanced so rapidly in the past 20 years that it now seems feasible to perform safety evaluation of a chemical in an animal within a rather limited period of time. The multidisciplinary approach has made it possible for the toxicologist to discover quite rapidly the extent to which a chemical is absorbed, metabolized, and excreted, what its distribution in the animal body is, and what organs are likely to be injured. It has become possible in some instances to pinpoint the precise enzyme or structural component that is affected by the chemical. However, in spite of our ability to understand the cell and to study its function on the molecular level, species differences soon return us to the question of relevance for man.

Absorption, distribution, metabolism, and elimination by the human organism of such substances as food additives must be known. This information may greatly affect decisions about the safety of any of these substances. Observations made in a species that deals with a chemical in a manner similar to that of man, if possible, are more likely to have predictive value for effects in man than observations made in a species by which the chemical is metabolized quite differently. In recent years, use of nonhuman primates has helped solve this problem of the toxicologist.

In most cases, the toxicologist can determine with great accuracy the target organ and physiological system that will be affected by a given chemical. Unfortunately, with cancer, it is very difficult to determine which animal species best predicts safety for man. It is well established now that nonhuman primates and man metabolize many chemicals differently than do rodents, cats, and dogs. This is particularly true for mice and cats, where the pathways of metabolism are often quite different than in any other species.

Unfortunately, we are now in an era where toxicologists are not making decisions concerning development of cancer from chemicals, but these decisions are rather being left to oncologists, pathologists, lawyers, and consumer advocates. Decisions are being made which brand chemicals as carcinogens, without consideration of basic principles of modern toxicology. The aim is to produce cancer, if possible, in the most sensitive strain of

---

## *"Regulatory agencies change basic concepts of pathology and toxicology to suit their regulatory needs"*

---

a species—at whatever dose, by whatever route of administration—assuming that this kind of data will predict safety to man. Nonsense!

Each experiment should be done in more than one species of animal, including if possible nonhuman primates, and certainly should be done at no less than three dose levels. And adequate controls should be used to redefine the physiological states that exist in the experimental groups of animals.

### Saccharin studies

The Canadian studies of saccharin, however, consisted of only one group of rats, which was fed a high dose of saccharin—so high that it interfered with the normal physiological state of the animals, resulting in a loss of electrolyte in the urine. It also resulted in loss of greater than 10% body weight, which is usually accepted by most toxicologists to indicate a chemical dose that is too high.

Furthermore, there were no adequate controls for the saccharin-fed animals. Extrapolation of these data to predict cancer in man is as farfetched as saying that the moon is made of "green cheese" even after the astronauts landed there.

For regulatory agencies to wave a "red flag" based on this kind of data is understandable. Indeed, if I had been in their shoes, perhaps I would have acted on saccharin—but in another way. Frankly, the Canadian study would never be acceptable to a regulatory agency if presented to it by an outside group.

Certainly, these studies should be done over, particularly in light of experiments by several colleagues and myself with rhesus monkeys. No cancer or other physiological or pathological change was produced in the monkeys when they were fed saccharin in relatively high doses (as high as 500 mg per kg) for more than six and a half years.

Current research at the National Cancer Institute with rhesus monkeys indicates that they are suitable for chemical carcinogenesis studies. However, the routine carcinogenic studies at NCI on rodents and hamsters, where the maximum tolerated dose of a chemical is given to one group of animals and half that dose to a second group, disregard completely a cardinal rule of toxicology and pharmacology: the dose response in terms of time.

There is a time-dose response for every chemical, and every chemical also has a no-effect level—even cyanide. If this were not true, most of the readers of this article would long since have departed from this earth. Chemists create and use chemicals and are always exposed to small amounts without undergoing any untoward long-term personal harm that we are aware of.

With respect to the no-effect level, I have never known a chemical that did not have some level that produces no effect. If I give a certain level of a chemical to an animal, such as a rodent, for its life span and find no carcinogenesis, central nervous system disturbance, or other pathological condition related to the chemical, I must assume as a toxicologist that this is an absolutely safe level for the life span of this animal.

Whether this chemical should be used for some purpose by man is another matter. If the second or even third dose in the experiment leaves enough of a safety factor between the no-effect level and the dose which produces an effect, then this chemical could be used safely for its intended purpose.

The question of benefit-risk, therefore, is of the utmost importance in any consideration of the safety of a chemical, be it a so-called carcinogen, teratogen, mutagen, or general physiological poison. Expert committees of the World Health Organization (WHO), Food & Agriculture Organization (FAO), and International Atomic Energy Agency (IAEA) have followed these concepts for years in setting maximum residue limits and acceptable daily intakes.

### Tolerances for carcinogens

Turning again to the Delaney clause, one can state unequivocally that there is nothing wrong with its intent and philosophy. No one wants to put a hazardous chemical that may be a carcinogen into our food.

But the way that the Delaney clause is formulated leaves no option for judgment based on some of the considerations discussed above. In fact, the clause discourages any reasonable interpretation of data by regulatory agencies. Nor does it allow any changes based upon new data and concepts developed over the years. It is inconceivable to me that in this day and age such a law could exist any longer.

Therefore, in spite of the implied risk, I advocate that even a potential chemical carcinogen could be used in food—allowing for a safety factor in its dose level—if on balance the substance is beneficial to man. Occasionally, the beneficial nature of some chemicals, such as nitrite, overrides any other consideration. Indeed, such decisions are made by FDA now—for example, allowing peanut butter and other peanut products to be used by setting a maximum residue limit for aflatoxin in them. Actually, there is more reason to ban peanut butter containing aflatoxin than there is to ban saccharin.

Indeed, tolerated doses have been established in most countries for potential chemical carcinogens—for instance 1 ppb of aflatoxin in smoked foods, and 20 to 50 ppb in animal feed. In fact, the joint meeting in Geneva in 1975 of the FAO Working Party of Experts on Pesticide Residues and the WHO Expert Committee on Pesticide Residues, which I chaired, reiterated the fact that it is possible to set tolerances for so-called chemical carcinogens—or for that matter, teratogens or mutagens—when a no-effect level has been established and the safety factor is great enough for the intended use of the chemical. And market basket surveys have clearly demonstrated that maximum residue limits and acceptable daily intakes are almost never exceeded in food samples tested in the U.S.

Toxicologists understand this kind of decision making. And in recent years, more and more oncologists and radiation experts have come to the conclusion that tolerances can be set for carcinogens, especially where malignant tumors form only in certain strains of mice, and in male but not female mice. One can understand, therefore, the decision of Canadian regulatory agencies to restrict but not ban saccharin, and FDA's proposal to ban saccharin as a food additive, but possibly allow its use as an over-the-counter drug.

More and more scientists recognize that the one-molecule cancer concept (one carcinogenic molecule hitting the nucleus of a target cell and initiating a process leading to cancer) is not applicable to modern day science or to ensuring human safety for most chemical food additives. This, as I have said, is the attitude of WHO, FAO, and IAEA.

Indeed, I would go so far as to say that if tumors are only demonstrated in a mouse with a tendency toward malignancy, but with no metastases or signs of invasiveness, the chemical should not be classified as a carcinogen, but perhaps should be given another name to reflect a whole new concept of the meaning of the results. Certainly, I would not classify DDT, saccharin, or cyclamate as a potential carcinogen for man.

### An acceptable concept

In the past five years, regulatory agencies, notably the Environmental Protection Agency, have tried to change the views of toxicologists and pathologists as to what is an acceptable concept for chemical carcinogenesis. There are some who would say that benign tumors caused by chemicals, or even single cells undergoing changes reminiscent of carcinogenicity (hypertrophy or hyperplasia), should be classified as definite carcinogenic hazards to man. The attempt by some people, even at the highest level of the regulatory agencies, to change basic concepts of pathology and toxicology to suit their regulatory needs is to be abhorred.

It is difficult to understand this attempt, time and time again, to prove their point—backed up by a few pathologists selected by the regulators. More and more toxicologists and pathologists recognize that there can be a no-effect level for chemical carcinogens in animals, particularly the mouse, and that benign tumors should not be called cancer unless there is definite and observable invasion of tissue by tumor cells and metastases to some other part of the animal's body. The regulators are coming close to saying that any inflammatory process or lesion is a cancer. In this case, any black eye or bruise could be considered a precancerous lesion and should be removed by a surgeon!

The mouse has become a standard animal for this kind of study, but many eminent pathologists and toxicologists realize that information obtained from a positive mouse liver tumor model should be accepted with great caution before prediction of cancer in man is established. Working toxicologists certainly look askance at the results of studies which indicate that only certain strains of one species develop cancer, while no other species in the same study have any kind of tumor, benign or malignant. Comparative studies in pathology, biochemistry, pharmacology, and toxicology have demonstrated time and time again that mice are often a poor experimental choice, since strains differ from each other and from higher animal species in their reaction to chemicals.

In rodent studies there are also very significant differences between the action on males and on females of chemicals such as saccharin or organophosphorus pesticides. Indeed, in the Canadian saccharin studies, only the male rats developed bladder tumors. There was no statistically significant effect in female rats fed a diet containing 5% saccharin. Thus, a no-effect level for females was indicated. Indeed, perhaps it would be appropriate for FDA to require, in the proposed warning label on the over-the-counter saccharin bottles, the following statement: "Warning—only to be used by females. Males should avoid this product!"

Because only male rats developed bladder tumors, one should immediately suspect that perhaps it was not saccharin that caused this effect, but rather physical changes in the rats produced by the high level of saccharin used, altering only the males' response to the chemical. Perhaps, for example, male rats retained urine longer than females and this caused tumor growth.

### No epidemiological evidence

Another important point is the lack of epidemiological evidence that saccharin causes cancer in man. When epidemiological data over many years make clear that a chemical does not cause cancer in man, then human experience must take precedence over any animal data, regardless of its nature. This is particularly true when the chemical only causes a tumor in rodents.

Saccharin has been used extensively during and since World War II. Yet, there is no evidence epidemiologically or from autopsies that saccharin has caused an increase in any kind of

> *"Canadian studies of saccharin fed rats a dose so high it interfered with their normal physiological state"*

cancer. Indeed, it would be very desirable to do extensive epidemiology on saccharin users before banning the chemical. And use of epidemiological data should be written into any revision of the Delaney clause.

Let me now turn to the statistics of carcinogenic risk. The kind of statistics used today to declare a carcinogenic risk—predicting how many thousands of people will die of cancer after consuming a given chemical—is only tenable if one indeed knows that the rodent studies are predictable to man. This kind of "numbers game" is used to determine from rodent studies how many deaths will occur for the entire U.S. population. This unscientific evaluation is unworthy of thinking people, and leads to emotional persuasion of the public that regulators are protecting them from likely harm, when the hazard is actually remote and unlikely.

Fortunately, in the case of saccharin, the large unorganized mass of consumers has at last been prodded to question this kind of reasoning. This public reaction is the basis on which everyone concerned must reconsider the meaning of the Delaney clause.

Anyway, who says that there should be no risk whatsoever?

If this were true, we would immediately ban or restrict use of the automobile and x-ray equipment used by radiologists or dentists. Accepted risk, based upon careful consideration of the benefits of a product, is the concept that should be followed by all regulatory agencies.

### Changing the Delaney clause

Obviously, the Delaney clause prohibits almost all of the types of approaches I have discussed. I therefore advocate modifying the clause, because it is redundant, too rigid, and not needed by FDA. The clause needs to be modified to fit updated

> *"There can be a no-effect level for chemical carcinogens in animals, and benign tumors should not be called cancer"*

toxicological concepts. And it is redundant because other regulations give FDA all the authority it needs to protect U.S. food safety.

Actually, I would like to see the Delaney clause repealed entirely. But since in my view this is not politically feasible, I support modification. With modification, it would be possible to arrive at an acceptable daily intake for such chemicals as saccharin and cyclamate, based on the concept of a no-effect level plus a safety factor. Under the current Delaney clause, this is impossible.

Strict interpretation of the Delaney clause, as it exists today, will set back science and prohibit use of foodstuffs in this critical period, when food shortages already exist. As food shortages grow in the world, man will need to use more and more chemicals to augment the food supply, including pesticides, fertilizers, and food additives. Inevitably, these chemicals will get into the food we eat as residues, metabolites, and parent compounds. The Delaney clause as it exists today thus will hinder development of future food supplies needed by all of us.

If the clause is not changed to allow balanced judgment by regulators, it will be more and more strictly imposed. And as analytical procedures become increasingly sensitive—they are already down to parts per trillion—more and more substances will have to be banned. Certainly, the presence of aflatoxin would raise the question of whether peanut sale and consumption should be allowed.

### How other nations cope

I hope that the U.S. will follow the route of other nations, including Canada, which decide whether to ban a food additive or chemical, or restrict its use, according to reasonable interpretation of whether it is an imminent carcinogenic hazard to man or not. For example, Canada did not ban cyclamate and will not ban saccharin, but instead allows over-the-counter sale of both. Canada reserves the right to use these chemicals wisely, for the consumer's benefit.

Moreover, FDA bans Red Dye No. 2, but Canada does not. (FDA does permit use of Red Dye No. 40, although Canada does not.) And FDA and EPA do not allow nitrilotriacetic acid as a phosphate substitute in laundry products, but Canada has allowed its use without any difficulty for years.

One could ask why decisions in the U.S. are so different from those of our neighbor to the north, or for that matter any country in Europe? Are we reaching so far to protect the consumer that reasonable interpretation of scientific data is distorted? Have we become overzealous in our pursuit of happiness, to the point where regulatory officials make decisions that rightfully belong to the public? Or is the philosophy of the Delaney clause responsible for this kind of action by our regulatory agencies?

The tendency to ban substances outright by FDA, EPA, and the Occupational Health & Safety Administration is wrong, in my view. Good chemicals that can be used correctly for their intended purposes should be retained in our society at all costs, with the consumer being advised of any potential risk that may be determined. Certainly, if a chemical would cause cancer in every animal species, including nonhuman primates, I would be the first to say "Ban that chemical—and never consider its use in any way, even though it may be very beneficial." But when one strain of a species—and only the males—is positive, and all other species are negative, how can a judgment be made whether saccharin is safe or not for man?

Thus, benefit-risk relationships, socio-economic costs, and acceptable risk levels for food additives must all be part of reconsideration of the Delaney clause. Above all, administrators of regulatory agencies should be given the right—based upon adequate scientific data as presented by experts—to accept a reasonable risk if it is in the public's interest.

The idea that no risk whatsoever can be taken with any chemical, where cancer is concerned, is a concept that can never be realized. This idea is usually based on statistics derived from rodent studies and really can scare anyone, until it is realized that the basic premise of the data is open to question and, although the statistics look good, they are flawed by the very data they are based on.

I have always given great credit to the role of FDA in developing concepts of safety and modern toxicology. Men like Dr. Arnold Lehman, Dr. Oscar Fitzhugh, and Dr. Arthur Nelson developed these concepts, which were accepted worldwide by other scientists. We must never forget their contribution, and we would do well to re-examine how they served and protected the public.

The trouble is that other agencies do not have the background of FDA, and they are using the Delaney clause's philosophy in their decision making. They are furthering the erroneous concept of "Why take a chance—better to ban the chemical." Most of these decisions are not scientific, but are often based on the ideas of a few people in an agency—often lawyers—who decide what is good or bad for the public. Let us go back to the days when good men made rational decisions based upon scientific facts and the needs of the consumer! The public needs to be protected, but it also needs to be served. ◻

---

*Dr. Frederick Coulston, 62, holds professorships in toxicology, pharmacology, and pathology and has been director since 1963 of the Institute of Comparative & Human Toxicology and the Animal Research Facilities at Albany (N.Y.) Medical College of Union University. He received B.A., M.A., and Ph.D. degrees from Syracuse University, and has held posts at the University of Chicago, Du Pont, Christ Hospital in Cincinnati, Sterling-Winthrop Research Institute, and other institutions. He has been president of the Society of Toxicology, was founding coeditor of its* Journal of Toxicology & Applied Pharmacology, *and was given the society's Merit Award in 1975 as "a pioneer in advancing the frontiers of toxicology." He is currently editor of four journals and a member of numerous international and U.S. advisory groups. His institute has a staff of 280 and annual budget of $3 million—55% from FDA, NIH, and other federal agencies, and 45% from industry. He emphasizes that all grants have "no strings attached." He has published more than 200 papers, and has pioneered use of monkeys in drug safety evaluation, including research on saccharin and other possible carcinogens.*

## "Food additives are largely of benefit to industry"

**Dr. Sidney M. Wolfe,** Health Research Group

# Delaney clause should be strengthened, not weakened

In 1958, and again in 1962, Congress amended the Food, Drug & Cosmetic Act with respect to the benefits of regulated products. Until 1962, a drug manufacturer was required only to submit evidence of safety to gain approval to market a new drug. With increasing questions being raised about the effectiveness of many marketed drugs, and with the increasing evidence of risk from drugs, Congress said—in passing the 1962 Humphrey-Durrum amendments—that "implied" benefit was not enough. There had to be scientific evidence that a drug worked.

It makes good sense for a category of products—all of which are supposed to improve health and which have varying degrees of toxicity—that approval be based on balancing benefits and risks. Had this efficacy amendment been around in the 1950's, for example, millions of women would have been spared exposure to diethylstilbestrol (DES) when they were pregnant, exposure predicated on the invalid "assumption" that the drug prevents miscarriages. Instead, the drug was pushed and hundreds of DES daughters have cancer, and thousands of sons have reproductive tract birth defects.

If the only purpose of the class of chemicals called drugs is to improve health, FDA needs evidence that a drug really works to improve health, and that this benefit is not counteracted by safety risks that worsen health.

In 1958, Congress enacted the Delaney clause, which also deals with the benefits of chemicals. Although the explicit language of the clause speaks to safety and prohibits use of any food additive demonstrated to cause cancer, implicit is a strong statement about benefits.

Food additives are largely of benefit to industry—hustling the sale of billions of artificial products that appear to taste, smell, feel, and look like something they are not. In passing the Delaney clause, Congress said, in essence, that no benefit to consumers of any food additive can be so great that it outweighs the risk, however small, of cancer to the large proportion of the population using food additives.

In the 19 years since its passage, the Delaney clause or its principle has been invoked only nine times, including saccharin. In none of these prior instances—including two packaging adhesives, oil of calamus, Violet Dye No. 1, safrole, trichloroethylene, DES as an animal feed additive, and diethylpyrocarbonate—was there an outcry by the public against the ban.

Only with saccharin has there been a reaction—and there, entirely because of "perceived" benefits. Yet, the only published study on use of artificial sweeteners for weight reduction fails to show any difference in weight loss between obese people dieting with or without the aid of artificial sweeteners. It seems that although—for a moment—diet soda provides less calories than sugar-containing soda, the moment is short and people compensate for this by eating other kinds of calories later.

If saccharin, or any other food additive, actually did have such an extraordinary benefit on health as attainment or retention of reduced weight, it would by definition be a drug and therefore subject, as over-the-counter saccharin will be, to the drug efficacy laws.

### Relevance of animal studies to man

Another objection to the Delaney clause by industry and some of its academic friends is the so-called irrelevance of large-dose animal studies to man. Three of the major occupational or environmental chemicals found to cause cancer in humans since 1970 were all originally determined to be carcinogens in animal experiments:

- Estrogens (similar to those now used for menopause and for birth control) were originally found to be carcinogens in large-dose animal experiments in the 1930's. Now, they have been found carcinogenic in humans, too.
- Bis-chloromethyl ether was found to cause lung cancer in animals in the late 1960's (after suspicion of human cancer). Now, it's been found carcinogenic in humans, too.
- Vinyl chloride—at 5000 to 10,000 ppm—caused liver cancer in animals in 1969–71. Now, it's known to be carcinogenic in humans, too.

In each instance, definitive nationwide action to eliminate worker or consumer exposure awaited human evidence of cancer, years after animal experiments were positive and decades after human exposure first began.

A third category of objections to the Delaney clause has to do with so-called "trace" amounts of chemicals, which were previously undetectable and which now "exist" only because of advances in instrumentation. There are two major reasons why this objection—like the "benefit" and "animal irrelevance" objections—is itself irrelevant.

The overwhelming majority of food additives are not present in trace amounts. Indeed, flavors, colors, emulsifiers, and other additives often reach levels of hundreds of parts per million or more.

For the largest category of food additives not present in these large concentrations—packaging materials—the argument against applying the Delaney clause is simply a thinly disguised version of the claim that it is possible to determine "safe" exposure levels of carcinogens. Although it is theoretically possible that there is a dose or exposure level of a carcinogenic food additive below which none of the 200 million Americans who use it regularly will get cancer, in practice there is no way to determine what this "safe" threshold is.

If everyone drank 800 cans of saccharin-sweetened soda a day and were as sensitive to its cancer-causing properties as the Canadian male rats, there would be about 40 million cases of bladder cancer. If everyone drank only one can of such soda a day over a lifetime, this could produce as many as 1200 cases of bladder cancer a year, according to FDA.

If a lot of carcinogen causes a lot of cancer—more than one fourth of some workers exposed to large doses of $\beta$-naphthylamine or benzidine got bladder cancer—less carcinogen will cause less cancer, but not no cancer. Moreover, as appears to be the case with these dye intermediates, humans may be more sensitive than animals.

### Natural carcinogens and pesticide residues

A fourth objection to the Delaney clause goes something like this: It isn't "fair" to focus only on food additives that enter the food supply at the point of manufacture. What about chemicals in food that are not deliberate food additives, but which are carcinogens?

Carcinogenic food additives are much easier to control, in that a manufacturer merely stops using them. This is not the case with environmental contaminants and natural carcinogens such as aflatoxin, and the Delaney clause does not apply to them. These substances are regulated under a section that permits FDA to set tolerances for so-called unavoidable dangers in foods, and by action levels, which are informal agreements by FDA not to seize products.

Aflatoxin is a very potent carcinogen, produced by mold on peanuts and other products. It is controlled mainly by proper care of peanuts in the field—drying them immediately and storing them properly. Many peanut products, including most peanut butter samples, are entirely free of aflatoxin. Nevertheless, FDA considers aflatoxin to be an "unavoidable" contaminant of peanuts. It currently allows up to 20 ppb aflatoxin in peanut products and has proposed to lower the tolerance to 15 ppb.

However, aflatoxin causes 100% incidence of cancer at 15 ppb in the diet of rats [Wogan, G., *Cancer Res.*, **27**, 2370 (1967)]. Furthermore, in experiments in trout it has caused cancer even at 0.45 ppb.

Equally disturbing is FDA's "action level" on DDT. Again, environmental contamination by carcinogens is not as simple a situation as are carcinogenic food additives. Because of decades of profligate use of DDT by farmers, almost all fish carry some DDT residues, and setting a zero tolerance level would jeopardize a major source of protein. However, current tolerances allowed by FDA are 5 ppm in fish and 7 ppm in apples—levels that are clear hazards since DDT has induced cancer at 2 ppm in the diets of mice.

Thus, the answer to this contention about environmental carcinogens in food is not to weaken the Delaney clause for food additives but rather to strengthen FDA's regulation of these other food carcinogens.

### Extrapolation from animals to man

Using animal evidence of carcinogenicity to ban human food additives underestimates the problem. As mentioned previously, humans may well be more sensitive to a carcinogen than animals. Equally important, however, is that humans are exposed to many carcinogens rather than just one carcinogen.

Unlike the rat that is exposed to a single carcinogen, a human may get drugs, air, water, and occupational exposure laced with carcinogens, to say nothing of other food additives that may not yet have been tested to see if they cause cancer. A little bit of this plus a little bit of that seems to be, at the least, additive and, at worst, synergistic.

The Delaney clause has served us well. Now, as opposed to 19 years ago when it was passed, we know much more about the chemical origins of cancer and, unfortunately, the rate of cancer has increased since then. Rather than thinking about weakening this important law, unleashing even more cancer-causing chemicals into the nation's food supply, serious attention needs to be given to strengthening it.

"Rigid," "inflexible," "no discretion" are the weapons Delaney opponents use to attack the law. There is, in fact, quite a lot of discretion given to FDA in using this law. FDA can and has rejected animal experiments purporting to show the carcinogenicity of a chemical if there were too few animals used, the experimental animals did not get an appreciably larger number of tumors than control animals, or other experimental deficiencies were present. This is a proper kind of discretion that will continue.

There is another kind of discretion that, I believe can be argued, FDA has abused. That is the discretion to reject the validity of well-conducted animal experiments that do show carcinogenicity.

Although there is much furor now about use of the Delaney clause to ban saccharin—despite the fact that FDA says it could have banned the chemical without the clause—it is the case of saccharin that best exemplifies this abuse of discretion. By 1973, there were two studies which, like the 1977 Canadian study, showed a statistically significant excess of bladder tumors in male rats that were fed saccharin over their lifetimes, and that were previously exposed in utero by feeding saccharin to their pregnant mothers.

An excuse now given for not banning saccharin on the basis of those earlier studies is that the saccharin used in those studies (and ingested by millions of Americans) contained an impurity, $o$-toluenesulfonamide (OTS), which some thought might be the culprit. As far as science was concerned, it was of interest to find out which was the carcinogen. As far as public policy was concerned, however, people were ingesting both, and the cancer risk was present.

Had saccharin been banned four years ago on the basis of these studies, we would likely now have an artificial sweetener on the market—for those who like a sweet taste, even though it doesn't help lose weight. It even would likely be one that does not cause cancer.

---

*"No benefit to consumers of any food additive can be so great it outweighs the risk, however small, of cancer"*

---

Two other food additives—both artificial colors—are also still being eaten by millions of people in this country, even though they are both clearly carcinogens. Citrus Red Dye No. 2—used to make greenish Florida oranges appear orange—was examined by the World Health Organization, which said in 1969 that because of its carcinogenicity it "should not be used as a food additive." A 1975 monograph on food dyes by the International Agency for Research on Cancer (funded by the National Cancer Institute) states that Citrus Red Dye No. 2 "is carcinogenic in mice and rats."

Furthermore, Red Dye No. 40, the main substitute for now-banned Red Dye No. 2, has caused lymphomas in animals and has been described as clearly a carcinogen by an FDA toxicologist, Dr. Adrian Gross. Nevertheless, 1.5 million lb of Red Dye No. 40 were certified for use by FDA in 1976.

If there is too much discretion that allows FDA—in apparent violation of the Delaney clause—to permit marketing of known carcinogens such as saccharin, Citrus Red Dye No. 2, and Red Dye No. 40 long after there is evidence of their carcinogenicity, what can be done?

Public Citizen's Health Research Group currently is discussing with members of Congress and the Carter Administration the need for a strengthening amendment to the Delaney clause. According to such an amendment, FDA officials who fail to invoke the Delaney clause where there is sufficient scientific evidence to support the banning of a carcinogenic food additive would be subject to punishment for failing to uphold the law.

Congress has moved steadily forward in giving FDA more power to protect the public from unsafe food additives, drugs, medical devices, and other consumer items. Such an amendment would be yet one more step forward. It is not likely that when all arguments are out on the table, Congress will go backwards to weaken or destroy the Delaney clause, as proposed in several bills now pending, including that sponsored by Rep. James Martin.

---

*Dr. Sidney M. Wolfe, 40, founded in 1972 and continues to direct the Health Research Group (HRG) of Public Citizen, a public interest organization established by consumer activist Ralph Nader. Wolfe has a B.S. in chemistry and an M.D. from Western Reserve University. He did clinical research for several years at the National Institute of Arthritis & Metabolic Diseases of NIH, including study of the alcohol withdrawal syndrome and of blood clotting. Particular focuses of HRG interest have been possible health hazards of food, drugs, cosmetics, pesticides, and unsafe consumer products; occupational health and safety; and accountability of the U.S. health care delivery system. As HRG director, Wolfe frequently has testified before Congressional committees and regulatory agencies on these and other topics, including recent hearings on a proposed saccharin ban and earlier hearings on toxic substances control legislation. HRG also has filed a number of court suits to stop worker and consumer exposure to suspected or known carcinogens and other possibly hazardous chemicals and drugs, including a recent petition to FDA to ban use of six food dyes.*

---

## "The clause is an inconsistent anachronism"

**Rep. James G. Martin, (R.-N.C.)**

# The Delaney clause must be amended and modernized

Saccharin, the last of the approved artificial sweeteners, is about to be banned.

That will do more harm than good. That will not serve the public health interest, but will aggravate it. There's little anyone can do about it, however, until Congress amends the law governing food additives or specifically exempts saccharin.

The Food & Drug Administration is required by law to ban saccharin or any other food additive if massive daily overdoses of it cause cancer in test animals. That is a single, absolute test with zero tolerance. FDA must disregard the substantial benefits that 50 million consumers will lose when saccharin is banned. FDA must disregard the fact that for all the studies on saccharin, there is no evidence that anyone has ever gotten one tumor from normal use of saccharin. Regardless of such factors, saccharin must be banned "at the drop of a rat."

The Delaney clause provides that "no additive shall be deemed to be safe if it is found to induce cancer when ingested by man or animal." This requires saccharin to be banned solely on the indisputable evidence that a significant increase in bladder cancer results from drastic overexposure of test rats to ridiculously extreme concentrations of saccharin prior to birth (in utero), followed by daily massive overdoses thereafter. How absurd! How absolute!

Clearly, but cautiously, the Delaney clause must be amended and modernized.

The Delaney clause already had fallen into scientific disrepute, as analytical chemistry steadily lowered the threshold of trace detection to a few parts per billion. The clause is an inconsistent anachronism, because it will ban a food additive with only a very remote risk of carcinogenicity, while permitting the consumption of a host of "natural" foods containing traces of far more potent "natural" carcinogens. It is about to become a bizarre "hazard to your health," because without a noncaloric, noncarbohydrate sweetener, millions of Americans will cheat on their otherwise bland diet, gain weight, and increase their risk of cancer (colon and breast), cardiovascular disease, diabetes, and hypertension.

These preventive medicine benefits of saccharin in diet control are enormous.

Therefore, I have introduced with 201 cosponsors (list available on request) a bill, H.R. 5166, to allow an exception to the Delaney Absolute if saccharin or any other suspected food additive is found to have public benefits outweighing the risk attributed to it. Priority would be given to health and nutritional benefits to the general public. Benefits to producers and investors would not be counted. This measure is a cautious balancing of consumers' interests. It is widely supported by consumer groups—representing real consumers who actually use the stuff, their parents, their children, and their doctors.

Ironically, the principal advocate of the saccharin ban is the Ralph Nader-connected Public Citizen's Health Research Group. Their "consumer" position is that the Delaney clause needs to be extended to cover all exposure to carcinogens (in which case we would starve).

Their view is that saccharin has no benefits, but is merely a nonessential convenience. Some convenience!

---

*"Banning saccharin will do more harm than good; it will not serve public health interest, but aggravate it"*

---

It is true, of course, that thousands can do without any sweetener in their food. On the other hand, there are tens of millions who lack that elite, ascetic self-discipline; they cannot marshal the iron will that will be required of them. They will resent those who take away another of the free choices they are still allowed to exercise in a world swarming with risks.

There is clear evidence that in the absence of diet drinks, those who have become accustomed to them will just shift to sugar-sweetened colas. When cyclamate was banned in 1969, the annual consumption of diet colas decreased 71 million cases (from 235 million to 164 million). This was accompanied by an increase in sugar-sweetened drinks of 159 million cases (from 1445 million to 1604 million). The trend of increasing consumption of all soft drinks hardly showed a dent when diet drinks were in short supply.

Furthermore, on the basis of 16,000 interviews each year, the number of Americans drinking a low-calorie diet drink on an "average day" dropped from 21 million in 1968 to 11 million in 1970. The number taking a sugar-sweetened (only) soft drink increased from 82 million in 1968 to 96 million in 1970. It is obvious that most of the decline in diet cola consumers simply shifted over to sugar-sweetened cola. The risk of this occurring again if saccharin is banned is extremely high, and the consequences are frightening.

### Risk of bladder cancer

What is the risk of bladder cancer if saccharin is permitted to continue as a food additive? It is remote at worst, as can be seen from the following calculation. Let us assume that:

• There is a rectilinear (proportional) dose-response relationship.
• Humans are as sensitive at normal use as the test rats in the Canadian studies were at very high overdoses.
• The sensitivity of human females (there being no significant incidence of bladder cancer in second-generation rat females) is one third that of human males (one third being the existing statistical ratio of current incidence).
• The amount of saccharin actually consumed by humans in the U.S. annually is 6.0 million lb.
• The proportion of pregnant women using saccharin is approximately the same as for nonpregnant women.
• The percentage of second-generation male test rats developing cancer is the percentage actually observed (that is, 24%, not the figure of 40% used by FDA in its risk calculations, based on its use of the maximum experimental margin of error plus rounding off of the numbers). Then the only valid conclusion is that no more than 26 additional bladder tumors might be expected annually, resulting in no more than eight deaths (see table on this page).

Although those numbers are regrettably large, they are nowhere near the inflated calculation of 1200 tumors fobbed off on us by FDA. Furthermore, there is no evidence and little probability that anyone has ever gotten one tumor from normal use of saccharin. So the rational risk is somewhere in the range of zero to 26.

There is, moreover, no evidence that test rats get cancer from any conditions less severe than the dual protocol of high concentration prenatally plus near pathological overdosage every day thereafter. Testimony by the Health Research Group has claimed that a 1973 Canadian study showed cancers in test animals fed 0.2% and 1.6% levels of saccharin in their diet. That misuse of data conveniently overlooks the reported fact that animals at intermediate and even higher doses had no tumors, and test animals at all exposure levels and control animals at zero exposure had precisely the same percentage of cancer—0.9%.

It is customary and accepted practice to overdose test animals because of the small number used and the short life span. To be sure, a hypothetically weak carcinogen could not be detected in a hundred animals without resorting to a massive overdose.

One irony of this is that some toxic substances—such as hydrogen cyanide, formic acid, and trichloroethylene—can be approved as food additives at lower than lethal concentrations without any way of knowing whether they are carcinogenic. Test animals would not survive even a modest overdose, let alone a massive one. Consequently, a poisonous substance that may be moderately carcinogenic has a better chance of being approved than saccharin does, under the existing law.

What is the probable mechanism of bladder tumor formation in second-generation rats? In all probability, the observed effect of saccharin was that of a mechanical irritant or abrasive, occurring only under the most extreme conditions. Through unrelenting abuse of bladder tissues, it produced higher sensitivity (or lower resistance) to a carcinogenic effect of some other chemical present. The general presence of microcrystals in the rats' urinary tracts cannot be ignored.

Consider the fact that first-generation rats show no incidence

### Calculation estimates remote cancer risk from saccharin

1. **Daily U.S. consumption of saccharin:**
   6.0 million lb per year × 454 grams per lb ÷ 365 days per year = 7.46 million grams per day

2. **Maximum number of consumers taking 150 mg per day, on average:**
   7.46 million grams per day ÷ 0.150 gram per person per day = 49.7 million persons (23% of total population)

3. **Lifetime tumor expectancy (maximum) for 25 million males using 0.1% of the dose causing cancer in 24% of the second-generation rats:**
   25,000,000 × 0.24 × 0.001 = 6000
   **Lifetime tumor expectancy for 25 million females:**
   6000 × 0.333 = 2000

4. **Annual tumor incidence (maximum) for these 8000 men and women during 72-year lifetimes**
   8000 ÷ 72 = 111

5. **Since only 23% of the people in steps 2 through 4 would be exposed to the critical double jeopardy of both prenatal (in utero) and postnatal lifetime exposure to saccharin, annual tumor incidence (upper limit) drops to:**
   111 × 0.23 = 26 tumors

6. **Extra bladder tumor fatalities, if any (maximum), based on the usual 30% mortality rate for bladder cancer:**
   26 × 0.30 = 8

of saccharin-induced cancer, and second-generation rats show significantly higher incidence of bladder cancer only at the highest doses employed. There is thus no evidence of a proportional, linear relationship between dose and response—

> *"If benefits are substantial and risks remote, put a warning label but don't ban. Let people make their own choice"*

suggesting that there is a noneffect threshold in these experiments and that it is just a little bit lower than the massive exposure of the fetal rats in utero.

Consider also the reported fact that the 1977 Canadian study found the crucial second-generation test animals to be 20% underweight at birth! Clearly, they had survived, but on the verge of a pathological overdose.

### Table top compromise

What about the "table top compromise," announced by FDA on April 14? I commend the agency and wish it well on that. It may be the most that it can do legally to ease the public hazard of its ban, until Congress changes the absolute law.

If FDA succeeds in approving saccharin for over-the-counter sale as a single-ingredient sweetener (under the drug section of the law), it will be a clear demonstration that the benefits are held to outweigh the risks. FDA is authorized to consider that balance in the law on drugs, but not on food additives. If it succeeds, it will reaffirm my belief that the same balance of public interests ought to be weighed in the food additives section of the law, as well—thus adding impetus to my bill.

Unfortunately, there are two major problems with this "compromise." First, there is great doubt that saccharin can be reclassified as an efficacious drug. In the Food, Drug & Cosmetics Act there is too sharp a statutory demarcation. Thus, there won't be a chance to weigh benefits against risks unless a way can first be found to include "this last of the approved noncarbohydrate, noncaloric, artificial sweeteners" within the statutory definition of a drug, according to a law proposed by Sen. Estes Kefauver in 1962.

> *"There is no evidence that anyone has ever gotten one tumor from the normal use of saccharin"*

In the second place, even if saccharin is continued as a "table top" sweetener (sold over the counter as tablets, powder, and liquid concentrate)—for which several million adult diabetics who use it only in their coffee and tea will be grateful—what about several million others who are accustomed to diet drinks? What about 2 million juvenile diabetics, facing enormous peer-group pressures at school and social events?

Then what about many obese millions who will lose control of their low-calorie diets, shifting irresistibly to sugar-sweetened desserts and beverages? Won't they be left with nothing but the old, proverbial "fat chance" of dietary management?

No, the "compromise" is inadequate. The law must be changed to allow all the evidence to be weighed, including the epidemiological evidence of public health statistics. A growing series of such statistical studies fails to detect any causal association of normal use of saccharin with cancer.

For example, Dr. Bruce Armstrong of Oxford University in the U.K. and coworkers found in 1976 that 5971 diabetics had a lower than expected incidence of bladder tumors. Omitting all kinds of cancers suspected of being due to smoking, they found that instead of the predicted occurrence of 213 other cancers, only 189 were found. That is not an increase, but a 12% decrease overall [Armstrong, B., et al., *Brit. J. Prev. Soc. Med.*, **30,** 151 (1976)].

### Let the public decide

The time has come to amend the Delaney clause and any other part of the food additives law (such as the general safety provision) that would ban saccharin on such flimsy grounds. If the benefits to the public are substantial and the risks remote, then put a warning label on it, but don't ban it. Let the people make their own choice, just as they do with alcohol, cigarettes, riding in automobiles and airplanes, swimming, hiking, and eating grilled steaks and various other wholesome but suspect foods.

According to a recent poll, the public opposes any saccharin ban by a margin of 5 to 1. An estimated million Americans have written to their government or their representatives to oppose the ban. The American Diabetes Association, the American Cancer Society, the present and past director of the National Cancer Institute, four consecutive former FDA commissioners, and thousands of private physicians have argued for a cautious modernization of the Delaney clause.

What can you do?

You can add your voice to that of others, expressing your own judgment on the merits of this issue. From thoughtful consideration of the points of view presented from all sides in this News Forum, you can reach your own conclusion as to the scientific validity of the evidence and arguments.

Then, as citizens, you can sort out the priorities of public policy like anybody else. Having done so, I suggest that you carry out your obligation to speak out on this subject—to your Congressman, your Senator, your local paper—and ask your neighbors to do the same. I think I know what the majority of you will conclude. But either way, you need to speak up for yourself, before it's too late. □

---

*Rep. James G. Martin, 41, a Republican, since January 1973 has represented North Carolina's ninth Congressional district (centered around Charlotte) in the U.S. House of Representatives. He earned a B.S. in chemistry from Davidson College and a Ph.D. in chemistry from Princeton University. He was a member of the chemistry faculty at Davidson from 1960 until his election to Congress, specializing in organic chemistry—including stereochemistry and conformational analysis of organic compounds and machine computation of distorted ring systems. He has been a member of the American Chemical Society since 1959. Exploiting his scientific background, he is currently a member of the important House Ways & Means Committee, including its Subcommittee on Health, and of the House Select Committee on Energy. He also chairs the House Republican Task Force on Health. Martin has been active in opposing FDA's proposed ban on saccharin. He is the author of a House bill, with 201 cosponsors, that would modify the Delaney clause to allow continued use of saccharin or any other food additive suspected of carcinogenicity if the public benefits can be shown to outweigh the risks.*

# REBUTTALS

## Rep. James G. Martin

### High doses can cause cancer by irritation, metabolic disruption

Although the Lijinsky paper is a remarkably elegant survey of the general subject of carcinogens and their evaluation, it is not really focused on the subjects at hand: the Delaney clause and the saccharin issue. Furthermore, it neglects the emerging evidence of "no effect" or "threshold" levels for no fewer than six carcinogens: nitrilotriacetic acid (NTA), calcium salts, selenium, 2-acetylaminofluorene (2-AAF), aflatoxin, and saccharin.

It is becoming increasingly recognized that some substances show carcinogenic effects only when test animals are exposed at such high levels that prolonged physical torture of the tissues lowers resistance to other carcinogens, natural detoxification mechanisms are overwhelmed, or secondary effects of the overdose displace other chemical substances from their detoxifying refuge, allowing them to cause cancer. Any of these pathways would operate only at high dose. Although animal experiments certainly justify the use of exaggerated overdoses, they do not justify ridiculously exaggerated overdoses.

Four additional comments apply to the Lijinsky article. In comparison with the fact that 1000 Americans die of cancer every day, the outside risk of cancer from saccharin would not exceed one more death per month. Putting that another way, Lijinsky could have said that regardless of whether we ban saccharin, 1000 Americans will die of cancer every day anyway.

As do many other commentators, Lijinsky emphasizes "environment" as being a major factor with cancer. He should have explained that this means one's "personal environment," including primarily excessive smoking, drinking, and general eating habits.

As to nitrosamines being formed from amines and sodium nitrite, those who postulate this as a basis for banning sodium nitrite as a meat preservative should be prepared to tell us not only what to do with the resulting increase in botulism, but also what to do with our saliva, which also contains sodium nitrite. Must we diligently spit it out?

Finally, his inclusion of the hazards of vinyl chloride must take into account two new circumstances: Vinyl chloride vapor is allowed in factory environments if it is reduced to a low, but detectable, level; and unreacted monomer in polyvinyl chloride packaging is currently substantially reduced from what it was in the earlier experimental plastic samples, from which it could be detectably extracted.

Turning to the essay by Dr. Wolfe, I find it to be a marvelous collection of inconsistencies and non sequiturs. Self-restraint limits this rejoinder to six points.

Wolfe maintains that there is no evidence that saccharin is associated with diet control—a point that was anticipated and proven, contrary to his view, in my reference to data on soft drink consumption. Moreover, it is one of those classic ironies that Wolfe seeks to discredit epidemiological statistics based on samples of 5000 to 20,000 subjects on the one hand. Then on the other hand he serves up a clearly nonrepresentative sample of 145 patients (whose problem was so acute that they had been to an obesity clinic) as "evidence" to support his absurd notion that saccharin is of no value for dietary discipline. Tens of millions of saccharin users would find his attitude cavalier.

Next, he proceeds to exemplify the basis for my concern regarding the inflated mathematical calculation of 1200 additional cases of saccharin-induced cancers by quoting that estimate as though it were valid and defensible. To put that baloney back in the freezer where it belongs, I included in my paper the arithmetic of my own calculation. This shows that if there be any additional bladder cancers at all, they will not exceed a predicted maximum of 26, based on the Canadian rat tests.

Perhaps the most bizarre element of his article is the meaning attached to the allowed exposure to aflatoxin, the most potent carcinogen known. If reason were to prevail, there would be far more attention to reducing exposure to aflatoxin from natural sources (nuts, coconut, rice, yams, cabbage, tomatoes, fish, and other sources) and corresponding less flatulence about saccharin, which is relatively innocuous. The real significance of aflatoxin, DDT, selenium, phenobarbital, and estrogens is that these and other carcinogens are allowed at levels where there is no significant risk to the population. The same standard should reasonably apply to saccharin.

Wolfe asserts that the rate of cancer has increased in recent years. To the contrary, what has increased is the proportion of citizens in upper age brackets—who are more likely to suffer cancer. On an age-adjusted basis, the incidence of cancer for any narrow age group has declined.

Most intriguing was Wolfe's conjecture that if saccharin had been banned, we would now have an artificial sweetener "that does not cause cancer." This is an incredible irony! Without approval of saccharin or cyclamate, a manufacturer probably could best win FDA approval only by proposing a noncaloric sweetener that also happened to be a poison. Its toxicity would preclude testing for carcinogenic response at massive overdoses, and so it would be approved as a food additive at levels that weren't toxic. This is true of formic acid and hydrogen cyanide. If only they were sweeteners!

Finally, I wish to take for my own the Wolfe statement that we know more now than in past years about the chemical origins of cancer. Among the things we know is the frequent observation that physical irritants can cause precancerous inflammation of tissues.

Moreover, it has been demonstrated that excessive doses of one substance can cause functional discontinuities of metabolic effects involving it or other substances, leading to tumors only when a large overdose is employed. Apparently, one or the other of these effects accounts for the fact that significant incidence of bladder tumors is found only when male rats are dosed in utero at extremely high concentrations of saccharin. Below that level, saccharin is quickly and harmlessly excreted. □

## Dr. Sidney M. Wolfe

### Industry using saccharin as lever to undermine Delaney clause

I agree with Dr. Lijinsky's thoughtful discussion of the Delaney clause and saccharin, especially with his conclusion that, "we can do without any substance which is less than vital and for which there is any suggestion—let alone strong evidence—of cancer-causing effect." On the other hand, most of the statements by Rep. Martin and Dr. Coulston are labored attempts either to make the use of saccharin (or other food additives) appear "vital" or to trivialize the evidence for the carcinogenicity of saccharin.

First, I wish to emphasize that the risks of saccharin have been established. The just completed Congressional Office of Technology Assessment report, "Cancer Testing Technology and Saccharin," concludes that "laboratory evidence demonstrates that saccharin is a carcinogen," and it is "a potential cause of cancer in humans."

I concur with this conclusion and with the finding, about which the Health Research Group testified earlier, that all three second-generation rat studies on saccharin done to date—including the recent Canadian study and two previous ones—show statistically significant increases in bladder tumors in males. After the first two such studies, completed in 1973, cancer pathologist Melvin Reuber analyzed all other studies on saccharin carcinogenicity, in particular experiments where saccharin-fed animals had more tumors than control animals. Although very few of these other studies showed statistically significant increases of tumors, in many subgroups saccharin-fed animals did have more tumors than control animals. It was these findings that the Health Research Group referred to in its Congressional testimony in March on the saccharin ban, stating that the findings were consistent with saccharin's being a carcinogen (testimony which is questioned in Rep. Martin's paper).

Furthermore, I wish to emphasize that there is no evidence for benefits of saccharin. Rep. Martin, assuming that the health benefits of saccharin are established, raises the red flag of health hazards related to increased obesity, which he believes will occur if saccharin is banned.

In a review of the published scientific literature on the "benefits" of using artificial sweeteners, we have been unable to find any study demonstrating that people using these chemicals can achieve or sustain weight loss any better than people who do not use them. The only two published studies that even explore this issue conclude that there is no benefit in using artificial sweeteners for obese people (1) or female adult-onset diabetics (2). In these studies, weight loss or adherence to a carbohydrate-restricted diet is no different with or without the use of artificial sweeteners.

Curiously, Martin himself must sense that the starting premise of his classic Aristotelian syllogism is false, for he states that "there is great doubt that saccharin can be reclassified as an efficacious drug." In other words, faced with FDA's requirement to show medical benefits or effectiveness, Martin does not appear to think saccharin will meet the challenge. If there is no evidence that saccharin helps to lose weight, why all the furor to keep it around?

The Calorie Control Council—a trade association for the diet drink industry and diet food manufacturers—says it had already spent almost $1 million by mid-June of this year to fight the saccharin ban. Expenditures have included two-page advertisements in dozens of major U.S. daily newspapers—public relations work to draw attention to voices, such as Martin's, which oppose the ban—and strong, inflammatory messages urging consumers to write to members of Congress to urge opposition to the ban.

Indeed, in an article by Julian Armstrong on why there is much less reaction in Canada to the proposed saccharin ban than in this country, the director of corporate relations for Coca-Cola is quoted as saying that one reason is the lack of "response" from the food industry in Canada, "such as the advertising campaign of the U.S. Calorie Control Council" (3).

Indeed, the council appears to be using the saccharin dispute as a lever against the Delaney clause. By undermining the saccharin ban with its advertisements ridiculing use of animal studies to predict human cancer, the council is also undermining the very basis upon which the Delaney clause rests. Since dozens of other food additives used in diet foods and beverages, such as food colors, might also in the future come under the gun of the Delaney clause, the Calorie Control Council understandably has an interest in the broader issue of modifying the Delaney clause.

Aside from distorting the benefits and risks of saccharin, Coulston and Martin also attack the Delaney clause, urging its modification. Coulston says that its strict application would threaten the world food supply. Does he know of yet other carcinogenic food additives that are still being ingested because of FDA neglect? If Coulston does, he should be specific about these additives.

As for modification of the Delaney clause, Rep. Martin cites only part of a Harris poll—the part that says a majority of people oppose the saccharin ban—but neglects to cite the most important finding from that poll, touching on the broader issue of the Delaney clause. Of those surveyed, 61% agreed "that we don't know much about what causes cancer and the law gives us strict protection against food additives that have been shown to have any connection with cancer."

Even though FDA has stated that it could have banned saccharin without the Delaney clause, by using discretion granted by other parts of the food law to remove unsafe additives, the Delaney clause is essential. If, even after conclusive evidence of carcinogenicity, FDA still has the option to leave a widely used food additive on the market, industry will push this discretion to its advantage and to the public detriment by lobbying for the chemical's life.

The public, I believe, wants to retain the Delaney clause as it is or stronger. And it will not look kindly on those Congressmen who want to weaken it "on their behalf."

### References

1. McCann, M. B., Trulson, M. F., and Stulb, S. C., *J. Amer. Diet. Assoc.,* **32,** 327 (1956).
2. Farkas, C. S., and Forbes, C. E., *J. Amer. Diet. Assoc.,* **46,** 482 (1965).
3. Armstrong, J., *Montreal Star,* April 2, 1977, page A-11.

## Dr. Frederick Coulston

# Odds are poor that mouse tests are good predictor of human cancer

I have read with great interest the papers of Dr. Wolfe, Dr. Lijinsky, and Rep. Martin. I can discuss the first two papers together since they both, more or less, state the same half-truths—namely that chemically induced cancer in rodents is related directly to cancer in man. Since there are now more than 1600 chemicals that produce cancer in mice, and only about 15 are known to cause cancer in man, the odds are poor that the mouse is a good predictor of cancer to man.

Lijinsky and Wolfe, to be consistent, should demand that all 1600 chemicals be banned immediately! Why pick on saccharin over any other chemical on this list? If these chemicals were banned, an economic disaster would occur, not only in the U.S. but worldwide. Indeed, we would be plunged deeper and deeper into the "dark ages of the 1970's."

Obviously, neither of these authors understands the basics of modern toxicology. They say that the mouse or rat predicts to man, but none of the 15 known chemical carcinogens affects only one animal species. For example, the often cited $\beta$-naphthylamine causes bladder tumors in more than one species, including dogs and rhesus monkeys as well as man. Estrogens also induce tumors and often cancers in many species of animals, again including dogs, monkeys, and man.

To my knowledge, all chemicals that induce cancer in man also cause cancer in many animal species. The argument that the nonhuman primate is no good for this type of study is fallacious, since we now know that cancer can be produced by chemical carcinogens in a relatively short time in monkeys, usually in less than one fourth of the monkey's life span.

On the other hand, DDT, dieldrin, mirex, and phenobarbital cause hepatic tumors only in certain strains of mice, and rarely in any other species—including rats and monkeys. Certainly there have been no instances of cancer formation by these chemicals in man.

As is stated in the Delaney clause, Lijinsky and Wolfe have reiterated again and again that if a chemical causes a tumor in any species or strain, even if it's benign, this is sufficient grounds for banning the chemical, even without evidence that more than one species is affected. In fact, they completely ignore the basic rules of toxicology on no-effect levels and dose responses to any chemicals given an animal.

Most pathologists and toxicologists would not agree with the concept that any morphological alteration in cells that could possibly lead to even benign tumors should be considered precancerous, or, for regulatory reasons, cancerous. A tumor can only be called a tumor if the cells have lost control and the morphology of the cells has changed to a hyperplastic condition with many mitoses; if invasion by these cells of surrounding tissues is evident; if metastases are observed; and if the cells can be transplanted and grown.

Lijinsky and Wolfe should stop attempting to change the teaching and disciplinary nature of toxicology and pathology to suit their personal thoughts and wishes. If their viewpoints were carried to their logical conclusion, the U.S. chemical and food industries would be set back so drastically that the food supply could be cut in half. The need for not only our country but the rest of the world to use safe and effective food additives goes without saying, if we are to avoid a calamity where people will eat anything, regulated or not, to survive.

All that is needed is a reasonable, scientific approach to the problems at hand, which necessitates changing the Delaney clause. Rep. Martin addresses himself very well to this proposition, and in general I agree with his statements. The recommendations made by him, Sen. Kennedy, and Rep. Rogers and Rep. Foley all suggest that re-evaluation of the Delaney clause is necessary at this time. The reasons presented by Martin for this re-evaluation are correct and proper, in my view.

I shall not be concerned with Martin's calculation concerning the number of human bladder cancers that could occur in the U.S. from saccharin—extrapolating from rats. But I will re-emphasize his point concerning physical irritancy and inflammation caused by large overdoses of any chemical, particularly where the bladder is concerned. I repeat what I said in my paper: When a high dose changes the normal physiological conditions of cells or tissues, a pathological state must result, provided that the chemical is given continuously over long periods of time, including life span studies.

Any toxicologist realizes the difficulty of extrapolating from animals to man. All toxicologists recognize that to do this demands data on a comparative basis from many species of animals. This is done all the time when a chemical is called a drug—it is necessary for FDA approval for new or old drugs.

Food additives must be considered as if they were drugs, and the same principles must apply to them. Even chemicals or drugs that are known to be the cause of cancer in man can still be used, provided the benefit to the consumer exceeds the risk. The intended use of the chemical is the key to the interpretation of safety and risk relationships, and the socio-economic relationships must also be considered. □

## Dr. William Lijinsky

## Idea that irritation causes cancer has long been discarded

The comments of Rep. Martin are rather unscientific. He claims to know that saccharin is safe at the doses used because no one has developed cancer from using saccharin. How does he know? The relevant epidemiological studies have not been done. Indeed, they are very difficult to do, especially if the target for the carcinogenic action of saccharin in man is not the bladder, as it is in rats, but some other organ.

If animal tests showing the cancer-inducing action of chemicals are no indicator of carcinogenesis by these same chemicals in man, then what is responsible for the enormous incidence of cancer in the U.S. population? And what do we do about it—nothing?

When animal tests indicate that a chemical causes cancer, we must take some action to reduce exposure of people to that chemical, especially when it is a substance that is added to our food. The problem is more complicated when the carcinogen is a naturally produced substance, such as aflatoxin, although here, as well, we must reduce people's exposure to it (as is done by setting maximum tolerance levels in food).

Especially must we reduce exposure of young children and yet-unborn children to such carcinogens, since the risk to them is greater than to adults. And risk is what we are talking about.

It is not useful to say that a material causes cancer when what we mean is that the material increases the risk of cancer, and the greater the exposure, the higher the risk. This is shown very well in the case of cigarettes. Not everyone who smokes even a large number of cigarettes develops lung cancer. Yet undoubtedly the smoker of many cigarettes is at greater risk than someone who smokes few. Women who smoke cigarettes seem to be at lesser risk than male cigarette smokers.

It is the matter of risk assessment which makes it impossible, and perhaps foolish, to estimate how many people who have taken saccharin will develop bladder cancer. It is wiser to say that people who take saccharin are at a greater risk of developing cancer than those who do not take saccharin, because the results of tests in rats show that such a risk exists.

Furthermore, it is baseless to complain that the tests of saccharin in rats were invalid because the high doses given led to other toxic effects in the bladder, especially irritation. The idea that irritation causes cancer was believed by many a century ago, but has long since been discarded by everyone who understands carcinogenesis (whether they believe in the Delaney clause or not). Many carcinogens have other toxic effects besides those leading to cancer, but their ability to induce tumors is not ignored because of this.

A great deal is made by both Rep. Martin and Dr. Coulston

of the absence of a dose-response in the saccharin tests in rats. Such a result is not uncommon and is the reason for testing a chemical at several doses, of which only the highest might be effective. Similarly, because there is sometimes a major difference in response between males and females, both sexes are usually tested. In testing a substance for carcinogenic activity, both males and females, more than one species of test animal, and several doses of the substance are used to be sure of circumventing all the biological variations and yet obtain a tumor response in one small group of animals. We can be fairly confident of the safety of any substance that passes such a battery of tests.

The faulty reasoning on the part of Martin and Coulston—and the trouble this might cause—is illustrated by tests in animals of 2-naphthylamine, which is a known bladder carcinogen in man. This compound was tested in dogs, and induced bladder tumors after administration of very large doses for five years or more. In Syrian golden hamsters, a significant incidence of bladder tumors was seen at a level of 1% in the diet for two years; no bladder tumors appeared at 0.1%. In rats, the compound did not induce bladder tumors at all; instead it gave rise to liver tumors.

These results could have been ignored, since the exposure of man, particularly people who worked with 2-naphthylamine in factories, was so much lower than that of the hamsters. In spite of that sincere logic, a quarter of all those who worked with 2-naphthylamine developed bladder cancer, even if exposed for only five years.

The suggestion that we follow the example of other countries in formulating and applying laws regulating drugs, food additives, and other substances amazes me. The U.S. probably has the best such laws, and they would be even better if they were always fully enforced. We should recommend that other countries follow us. The testing required of synthetic chemicals to which people are exposed is our best guarantee of protection from ignorant merchants. It was this protection that prevented the tragedy of thalidomide babies in the U.S., while there are many of these unfortunates in countries with less rigid laws.

If there is a decision to allow an expert, or a committee of experts, to evaluate the information about saccharin, to assess the risks and benefits, and to set a safe level for its commercial use, who is wise enough to do it? If the committee makes a mistake, no one will know, since the effects may not show up for decades, and even then might stay undiscovered. Yet we would have failed to take an action that would prevent exposure of millions of unsuspecting people to still another carcinogenic risk.

Surely the argument that people should be allowed to make their own decisions about using possibly harmful materials is not serious, particularly when those subject to the greatest risk are children? Even those who argue most zealously for individual rights (and I am one of them) would not suggest that everyone be given access to heroin, cocaine, or marijuana, or that children be allowed to buy and use cigarettes or alcoholic drinks. Unfortunately, we have to depend upon the government to protect us from ourselves, whether our instinct is to steal, murder, smoke cigarettes, or merely to ingest a food additive, such as saccharin, which might help us develop cancer. □

# CANCER PREVENTION and the DELANEY CLAUSE

**HEALTH RESEARCH GROUP**
2000 P ST. NW
WASHINGTON, D.C.

*FUNDED BY PUBLIC CITIZEN, INC.

# CANCER FROM THE ENVIRONMENT

One in four Americans contract cancer during their lifetime. Two-thirds of these--about 1,000 every day--die from the disease.[1] At least one cancer victim will probably be someone you know.

The World Health Organization has concluded that more than three-quarters of human cancers are influenced by factors in the environment.[2] A leading cancer specialist, Dr. E. Boyland, estimates that

> "not more than 5 percent of human cancer is due to viruses and less than 5 percent to radiation. Some 90 percent of cancer in man is therefore due to chemicals, but we do not know how much is due to endogenous chemicals and how much to environmental factors."[3]

Many human cancers have already been traced to chemical exposures.[4] Groups of workers and people in the neighborhood surrounding their plants are exposed in large doses to carcinogens such as asbestos, which induces lung cancer.[5] Chromate workers are far more likely to get lung cancer than the general population,[6] as are uranium miners and workers in nickel refineries.[7] Workers who handle certain textile dyes contract cancer of the bladder,[8] as do laboratory workers who perform medical tests using benzidine.[9] Workers who make polyvinyl chloride plastics contract cancers of the liver, lung, and central nervous system.[10]

The fact that urbanites get far more lung cancer than rural dwellers do, testifies to the carcinogenic properties of air pollutants.[11] Exhaust from cars and diesel engines, which contains carcinogenic polycyclic aromatic and aliphatic hydrocarbons, nickel "anti-knock" chemicals, aromatic amino compounds, aniline dyes from gas coloring and asbestos dust from break linings, can cause cancer. Factory smokestacks release arsenicals, asbestos, aromatic amino and nitro-compounds, nickel and chromium particles, and beryllium. Burning of coal in furnaces, in coke ovens, and in refuse piles near mines, pumps 1,300 tons annually of carcinogenic benzopyrenes into the air.[12] It is not surprising that 95 percent of lung cancer is of environmental origin.[13]

Household products such as rubber and waxes, cleaning fluids, pesticides, and cosmetics sometimes contain carcinogens.[14]

Other examples are medical treatments such as radioactive phosphorus given for blood disorders, immunosuppressive drugs used in transplants, and alkylating agents used in treating cancer. Coal tars and creosote used by dermatologists and estrogens used in contraception and face creams are also carcinogens.[15]

Thus, many cases of cancer which were thought to be "spontaneous" or "accidental" are explainable by chemicals in the environment. Cancer has reached epidemic proportions in recent decades[16] and is increasing yearly as industrialization and exposure to great amounts of synthetic chemicals has increased.[17]

The good news about environmentally caused cancer is that the majority of cancers are preventable.[18] Known carcinogens should be removed from the environment wherever possible, or their use severely restricted. All widespread chemicals should be tested for their long-range effects on health.

> "Cancer is a social disease, largely caused by external agents which are derived from our technology, conditioned by our societal lifestyle, and whose control is dependent on societal actions and policies."[19]

The search for cancer _cures_ is a humanitarian task. This search has not, in spite of the billions of public dollars devoted to it, been very successful.[20] It is a stopgap measure until cancers are _prevented_. Disproportionately large amounts of money should not be spent on treatments and drugs for cancer, but more resources should be focused on preventing and eliminating carcinogens from our environment.[21]

## FOOD ADDITIVES

One important source of consumer exposure to man-made chemicals is through chemicals deliberately added to the food supply. Close to a billion pounds of food additives are produced annually. Consumers eat an average of five pounds of them a year,[22] and the amount is expected to greatly increase.[23]

The benefits of food additives are generally very slight. They are put into food products to improve their taste, appearance, texture, smell, longevity and, occasionally, nutritive value. They are used most in processed and synthetic foods such as "convenience" foods, and snack foods such as soda pop and candy. Convenience foods

> "are prepared under more severe conditions of temperature, pressure or agitation. Therefore they may require special flavorings, flavor enhancers, colors, and other additives to make up for the partial loss of flavor, color, texture and other properties caused by processing."[24]

Most additives are purely cosmetic. Many involve outright deceit. For example, red dyes are added to strawberry ice cream to make the consumer think it contains more strawberries than it does.

Additives allow manufacturers to simulate natural products. A typical example is Cool Whip, a fake whipped cream, which contains: water, hydrogenated coconut and palm kernel oils, sugar, vanillin, sodium caseinate, dextrose, polysorbate 60, sorbitan monostearate, carrageenan, guar gum, and artificial coloring and flavoring. Synthetic foods generally are less nutritious than natural ones.

The food industry gets its largest profits from marketing synthetic and processed foods.[25] It is cheaper to add one of the 1,610 artificial flavors to a product than to sell high quality, naturally flavorful ingredients; or to add an artificial filler and texturizer rather than the full share of ordinary ingredients. It is cheaper for the food industry to add a preservative which retards the growth of mold, than to use fresh, nutritious ingredients and remove old products from the shelves.

Should individual additives show any chance of harm to health, that risk far outweighs the limited benefits they offer. The food industry's argument that "calculated risks" to life and health are necessary for "progress" in "food innovation" are pleas for private profit, nothing more. This is particularly true when the risk is cancer, a very serious, irreversible disease, which develops covertly over many years.

In the rare case where a risky additive provides a benefit to consumers--as opposed to producers--the benefit can be provided by another additive which is safe.

## THE DELANEY CLAUSE

> "No additive shall be deemed to be safe if it is found to induce cancer when ingested by man or animals, or if it is found, after tests which are appropriate for the evaluation of the safety of food additives, to induce cancer in man or animal...."[26]

The Delaney Clause prohibits the use of <u>any</u> amount of a food additive which causes cancer. If an additive causes <u>other</u> kinds of harm, such as birth defects, liver damage, etc., the Food and Drug Administration (FDA) may allow low amounts, or "tolerances," of the additive if it determines that only high amounts cause such damage. The law prohibits the FDA from determining that any amount of a carcinogen is "safe." It prohibits FDA from permitting acknowledged hazards on the basis of "overriding benefits."[27]

The clause is named after Congressman James J. Delaney (D-N.Y.), who was chairman of a House of Representatives special committee to investigate the safety of food chemicals in the early '50s. Delaney was chief spokesman for the clause, which passed in 1958. He was influenced by the Committee on Causative Factors of Cancer and Cancer Prevention of the International Union Against Cancer, which in 1956 issued a list of food additives which had cancer-producing characteristics, and recommended that:

> "...as a basis for active cancer prevention, the...various countries promulgate and enact adequate regulations prohibiting the addition to food of substances having potential carcinogenicity. Any substance which causes cancer in man or which, when tested...is shown conclusively to be a carcinogen at any dosage level, for any species of animal, following administration by any route, should not be considered innocuous for human consumption."[28]

This recommendation, combined with the ominous fact that FDA had previously allowed residues of a carcinogenic pesticide, Aramite, on food, convinced Congressman Delaney that "zero tolerance" for carcinogenic additives was necessary to protect consumer health.

## "IF IT IS FOUND TO INDUCE CANCER...IN MAN"

Most evidence in man of cancer causation is found by survey-type studies called epidemiology. Such studies scan hospital and other records for a correlation between a greater incidence of human cancer and exposure to a given substance. Diethylstilbestrol (DES) was identified as a human carcinogen in this way. A great rise was detected in cases of vaginal cancer, a very rare form, in women 14-22. Scientists found that the mother of each victim had been given DES while pregnant.[29] In India, it was discovered that most cancers of the oral cavity and salivary glands were due to chewing betel nuts.[30] Epidemiology is most likely to discover carcinogens like these, where the cancer is an unusual type, where there are few or no other causes of that type, and where the exposed population is limited in size and can be studied.

Cancer experts can determine what chemicals one job involves which other jobs without high cancer rates do not involve. For example, when a number of insulation workers contracted mesothelioma, a fatal form of cancer which rarely occurs in the general population, the cause was identified as asbestos, an insulation material.

## LIMITATIONS OF EPIDEMIOLOGY

Epidemiological studies cannot detect carcinogens when there are not clear differences between groups in amount of exposure to a substance. If everyone smoked cigarettes, it is unlikely that lung cancer would have been traced to smoking. A rise in lung cancer over the years would have been spotted, but the culprit would not have been isolated from the other chemicals in the environment. According to Dr. Samuel Epstein, a well-known cancer expert:

> "Epidemiological techniques are unlikely to detect weak carcinogens and other toxic agents unless there are sharp differentials in exposure of the population, as with cigarette smoking. For widely dispersed agents, such as intentional or accidental food additives, to which the population at large is exposed, human experience is unlikely to provide any meaningful indication of safety or hazard."[31]

Collection of human evidence is also limited because

certain kinds of cancer may be caused by several substances. Human bladder cancer may be induced by, among other things, beta-napthylamine, benzidine, and 4-amino-diphenyl.[32]

Epidemiologists are also unlikely to spot carcinogens if they cause cancer by a single exposure, since cancer patients are unlikely to be aware of such an exposure many years back. Animals can get cancer from single doses of dimethyl nitrosamine, which can be formed in food or in the gut by reaction between naturally occurring amines and the preservative sodium nitrite,[33] from single doses of aflatoxin, a product of fungi on lentils,[34] and from dibenzanthracene, a combustion product.[35]

The long latency period of human cancer makes the detective work much more difficult than for diseases which appear quickly. Once affected by a carcinogen, body tissues do not return to their normal state.[36] In the presence of conditions favorable to cancer growth, tumors appear long after the causative agent has been in contact and disappeared. The latency period ranges from under five to over 40 years.[37] The latency period for cancer from beta-napthylamine, for example, averages 18 years. New cancer cases this year reflect exposures perhaps decades ago.

Human studies can occur only after there are cancer victims. Epidemiology is an after-the-fact technique. Because of the long latency period, the carcinogen will have been in the environment many years and many people exposed, before it is identified. Victims will continue to appear years after the substance is removed from the environment.[38]

While human evidence may be available for very potent carcinogens such as beta-napthylamine, which produces dramatic statistics, epidemiology usually cannot spot the less potent carcinogens--the ones which may cause cancer in a few thousand people rather than many thousand. Yet, "it is assuredly the less potent carcinogens that seem to be more important in human cancer, and it is these that provide the real problem for evaluation."[39]

Thus, human evidence of cancer causation is likely to be available only when a scientist has studied a distinct population whose exposure to a chemical dates back several decades and only if the chemical causes a high percentage of cancer in those exposed.

## "IF IT IS FOUND TO INDUCE CANCER...IN ANIMALS"

Because the collection of human data has so many limitations, scientists have traditionally relied on the results of animal tests to determine whether substances entail a risk of cancer for humans.

Scientists also use animal tests to confirm human data. If they can induce cancer in animals by administering the suspect agent, they are sure that their epidemiological conclusion was correct. With the possible exception of arsenic, still under experimental study, all chemicals and other agents found to be carcinogenic in man have also caused animal cancer.[40] Some carcinogens first identified in man and then confirmed by animal experimentation are coal tars and soot, creosote, residues of petroleums, and tobacco.

Some substances first shown to cause cancer in animal tests and only later in humans are vinyl chloride and estrogens.[41] In these cases, humans have been exposed to the carcinogen long after the animal evidence was in.

There is no doubt that extrapolation from animal data to humans carries some difficulties, since animals and humans sometimes metabolize substances in different ways and have varying defenses to disease. However, cancer scientists believe that while the fact that a substance causes cancer in animals does not make it <u>certain</u> that it has the same effect in humans, with a well designed experiment, it is <u>highly probable</u> that a similar effect would occur in humans. It is "almost incredible that our species could be exempt from such causes of cancer when other species are susceptible," says one cancer scientist.[42]

Extrapolation from animal data is sometimes complicated by the fact that different animal species do vary in their sensitivity to a carcinogen and in the kind of cancer contracted. Benzidine, a dye intermediate and test-laboratory material, for example, causes cancer of the liver and large intestine in rats, and bladder cancer in dogs and man.[43] It is highly unusual, however, for one species to contract cancer from a substance and another species to contract it not at all.[44]

Animal tests yield knowledge about cancer hazards without human cancer victims being involved, so that human exposure to carcinogens can be prevented. Animal tests also yield knowledge about "weak carcinogens" when human evidence is unlikely ever to be obtained.

Because we cannot conduct tightly controlled experiments on humans--exposing them to suspected poisons, keeping them in a clean environment for many years to exclude other causes of disease, and dissecting their organs for detection--and because the survey-type human study has so many limitations, animal evidence is usually the most reliable information we have.

## WHY HIGH DOSES ARE USED

Some carcinogens, such as aflatoxin,[45] cause cancer in animals at very low doses. Others cause cancer in animal tests at higher doses but not lower ones. Opponents of the Delaney Clause say that the clause is "unscientific" because it prohibits adding any amount of a carcinogen to food, even though animals in a test show no cancer at lower levels. When FDA announced a ban on saccharin in March 1977 because saccharin caused cancer in a high-dose animal test, critics said, "You would have to drink 800 bottles of diet soda a day to get cancer."[46] This is misleading because it implies that fewer bottles a day are safe.

Test animals are fed high doses for three reasons: (1) to compensate for the short lifespan of animals as compared to humans;[47] (2) to compensate for the very fast metabolism and excretion of chemicals by animals as compared to humans;[48] and (3) to compensate for the small number of test animals compared to the number of humans exposed to the chemical.

High doses of a chemical increase the percentage of cancer caused, making the cancer rate high enough to be detected in the 50 or so animals used in the test. If one in 10 animals get cancer at high doses, the rate is high enough to see in a test which uses 50 animals. At lower doses, the risk of cancer is smaller. It may go down from 10 percent to .001 percent, a rate too low to detect in 50 animals but which would cause thousands of cases of cancer in humans. For example, if everyone drank one bottle of diet soda a day, FDA estimates saccharin would cause 1,200 cases of bladder cancer each year.[49] Although the percentage of victims is small, the actual number is large.

Animal test systems cannot show cancer causation from low doses if the rate caused is low.[50] Nevertheless, what data scientists do have indicates that carcinogens can be dangerous at very low doses. The ability of a minute amount of a carcinogen to sometimes cause cancer is believed to be related to its ability to penetrate the DNA (genetic material) of a cell and cause it to go haywire.[51] (The DNA regulates cell growth.) Also, unlike other toxic chemicals, a carcinogen's impact on a cell appears to be permanent and cumulative.[52]

Even at high doses, only a tiny fraction of chemicals tested cause cancer. Of known carcinogens, many have a similar chemical structure.[53] It is true that very high doses of almost any substance can quickly kill animals because of non-specific acute toxicity, but they will not cause cancer. Since the Delaney Clause was enacted in 1958, out of over 3,000 food additives in use, only about ten have been banned for causing cancer, including saffrole (root beer flavoring) and oil of calamus (vermouth flavoring).

## THERE IS NO SAFE DOSE OF A CARCINOGEN

> "Our advocacy of the anticancer proviso...is based on the simple fact that no one knows how to set a safe tolerance for substances in human foods when those substances are known to cause cancer when added to the diet of animals."[54]

Animal tests are useful to show whether a substance can cause cancer, but they cannot show how _much_ cancer a certain amount of the carcinogen would cause, or whether a certain amount would be safe.[55]

Man and animals can differ in their sensitivity to a carcinogen. Beta-napthylamine, for example, causes cancer at a high rate in man and dogs, but at a lower rate in mice, rats, and guinea pigs.[56]

Sometimes the target organ for man and a test animal is the same, but sometimes it is different. For example, 4-aminobiphenyl, an industrial chemical, causes bladder cancer in man, mouse, rabbit, and dog, but causes cancer of the liver, breast, and intestine in the newborn mouse and rat.[57]

The development of tumors is influenced by many factors. Non-carcinogenic substances can promote cancer after a carcinogen has initiated it. For example, in one experiment, skin cancer in mice increased 1,000-fold when the mice were also exposed to a non-carcinogenic substance, n-dodecane, found in petroleum waxes.[58] The food additive safrole, now banned, causes many more animal cancers when fed with phenobarbitol, a sleeping pill.[59] Liver tumors have been greatly increased by exposure to piperonyl butoxide, a pesticide.[60] Diet can affect the potency of a carcinogen. For example, aflatoxin produces more cancer in malnourished than well nourished people.[61]

Exposure to two carcinogens can potentiate the danger of each. For example, both cigarette smoking and asbestos are proven to cause human lung cancer. A cigarette smoker who is also an asbestos worker has an eight times greater risk of dying from lung cancer than do other cigarette smokers, and a 92 times greater risk than normal.[62] Thus, we must be concerned not only about exposure to individual carcinogens, but the total body burden of these chemicals.

Individual variations also affect susceptibility to tumors. Genetic heritage may affect the chances of contracting a few kinds of cancer, such as albinism and blondness which increase the chances of skin cancer.[63] Conditions such as Down's syndrome or schistosomiasis increase the risk of cancer developing. Age at exposure is a factor for some carcinogens; for example, urethan produces cancer in newborn but not adult animals.[64] The presence of bodily hormones affect the potency of a carcinogen, as do physical influences such as mechanical injury and even ultraviolet light.

The human population consists of people of differing genetic background, differing age and health, and differing chemical environment. By contrast, laboratory animals are genetically homogeneous, healthy, and raised in an environment clean of other environmental toxicants.

We do know that cancer in man and animals are strikingly similar in many ways.[65] However, we are still ignorant of the basic mechanisms by which cancer is caused.

Because of the complexity of the many factors influencing the cause of cancer in humans, including the wide variety of other chemicals to which humans are exposed, a "no-effect" dose in animal studies cannot predict a safe dose for humans.[66]

In 1970, a committee of eight scientists evaluated the risks of low levels of exposure to chemical carcinogens for the U.S. Surgeon General. The committee supported the Delaney Clause:

> "The principle of a zero tolerance for carcinogenic exposures should be retained in all

areas of legislation presently covered by
it and should be extended to cover other
exposures as well. Only...where contamination of an environmental source by a carcinogen has been proven to be unavoidable should
exceptions be made...[and then] only after
the most extraordinary justification is
presented."[67]

The recommendations of this committee are supported by leading cancer specialists in government and universities, including the Mrak Commission, the FDA Advisory Committee on Protocols for Safety Evaluation of Cancer, and a National Academy of Sciences Committee evaluating risks from pesticides, some of which are carcinogens.[68] A report on "Chemical Carcinogenesis" in the New England Journal of Medicine noted that

"the known carcinogens have not yet been
effectively banned from our environment,
and efforts must be made to educate populations and governments about their presence.
In this regard, it should be remembered that
weak carcinogenic exposures have irreversible
and additive effects and cannot be dismissed
lightly as standing 'below a threshold of
action.'"[69]

## NO NEW WAYS TO DETERMINE A SAFE DOSE

Since the Delaney Clause was first passed by Congress in 1958, knowledge in the field of carcinogens has progressed. However, new research has not aided in determination of safe doses for humans. If anything, new research has shown that determination of a safe dose would be more complex and difficult than appreciated before.

There have been improvements in chemical analytical methods for detecting carcinogens in food, as well as in water and air. For example, in 1954 FDA permitted DES to be added to cattle feed to stimulate growth, in the belief that none of the cancer-causing chemical ended up in table meat. Chemical analytical methods have improved over the years, and DES residues can now be detected in meat, with the result that FDA has proposed to ban DES in feed.

Advances in detection methodology provide a greater

awareness of the hazards, and a better means to determine total environmental exposures to known dangers.[70]

Another advance is a new method for identifying carcinogens. It is called the Ames test and is based on the newly realized fact that many carcinogens can be observed to mutate genes when they are added, together with liver metabolizing enzymes, to bacterial cells in a petri dish.[71] The cost of performing this test is about $250 per chemical, as compared to many thousands of dollars for ordinary, two-year animal studies. It may prevent newly synthesized carcinogens from being introduced into commercial products. This new research tool has not provided, so far, new means for determining safe doses of carcinogens.[72]

A June 1977 report by the U.S. Congress Office of Technology Assessment in connection with saccharin concluded: "The Delaney Clause reflects the present state of technology."[73]

## THE DELANEY CLAUSE ALLOWS SCIENTIFIC DISCRETION

A lobbyist for the National Canners Association has stated, "The Delaney Clause...categorically rules out the valid exercise of scientific reason and interpretation."[74] Other lobbyists state that the clause mandates banning an additive if one "bizarre" test yields a tumor in test animals.

This is false. The Delaney Clause gives FDA the scientific discretion to decide whether the food additive has in fact been shown to cause cancer. FDA can examine the test design, the way the additive was administered, the purity of the additive used, the reliability of the diagnosis of cancer, etc., to determine if the tests are methodologically acceptable to the scientific community.

The general principles and criteria for good carcinogenicity tests have been agreed upon by a number of expert committees, including the World Health Organization, the International Union Against Cancer, The European Committee on Toxicity, and the Food and Drug Administration Committee on Protocols for Safety Evaluation. Their recommendations are widely accepted by the scientific community.[75]

One of the disputes among scientists is whether producing cancer by injecting a substance under the animal's skin is a valid test for deciding whether a substance is carcinogenic. Substances occasionally cause cancer in these tests which have not yet been verified by other tests and experience.[76] There is no question that animal tests should be conducted under conditions that reflect human exposure as closely as possible. Injection tests have been widely used, however, because they allow scientists a greater degree of control and hence a greater ability to quantify their results.[77] Whatever the outcome of this controversy, the Delaney Clause gives FDA the power to decide for itself the appropriateness of the injection test or other test conditions.

The Delaney Clause "allows the exercise of all judgment that can be safely exercised on the basis of our present knowledge," the judgment to determine whether valid tests have shown that a substance causes cancer. The clause is grounded on the scientific fact of life that no one, at this time, can tell us how to establish for man a safe tolerance for a cancer-producing agent.[78]

## THE DELANEY CLAUSE IS A POLICY JUDGMENT

Congress determined that since cancer scientists cannot say at what level a carcinogen is "safe," the country should not suffer *any* risk of cancer from chemicals deliberately added to food. The judgment has so far not been extended to pesticides, carcinogens in the workplace or water supply, natural carcinogens in food such as aflatoxin, drugs, or other consumer products, presumably because the benefit-risk balance is more complex.

Certainly, a drug which caused cancer would be acceptable if proven to cure a disease which was life-threatening in the present. Drug manufacturers must submit solid evidence of benefit to FDA before marketing approval. In contrast, food additives are not required by law to have proven benefits; the only subject of scientific testing is their safety. Any food additive which presents a health risk must, whether under the Delaney Clause or the other parts of the food safety law, be banned. FDA cannot permit a hazardous additive on the basis of overriding benefit.

The judgment Congress has made is a social one. No additive is worth the risk of cancer, a covert, irreversible, and usually fatal disease. Any benefit should be provided by a safe alternative. Since food additives involve exposure of virtually the whole population, over a lifetime, the risks of allowing FDA to make a judgment of such social importance for a single food additive are too great.[79]

## PRESSURES AGAINST CONSUMER PROTECTION AT FDA

The FDA is over-influenced by the industries it is supposed to regulate. The food industry gives top priority to repealing the Delaney Clause, since it wants greater leeway in adding chemicals to its products. The food industry, with its trade associations and research foundations, is well financed and highly organized to pressure the FDA. Industry representatives visit, phone, and write FDA officials about every decision that could affect their financial position, constantly badgering, cajoling, presenting facts and figures, to persuade officials of their points of view. There is no comparable input from the consumer side.

Industry pressures gradually erode even the most vigilant FDA officials. In delicate situations, where human risk is suspect and not proved--as is almost always the case with cancer--industry presence makes the difference. If the FDA acts on the side of caution, and prohibits an additive, industry can quantify their losses in terms of a dollars-and-cents economic loss. Human loss, on the other hand, is never a definite sum of money.

If the agency permits a carcinogen, even if there results a two or three thousand increase in the number of cancer cases 15 years later, it will probably never be connected and will never be traced back to FDA's decision. The temptation to downplay the risk of industry products, and to emphasize the benefits, is overwhelming.[80]

FDA is subject to high-level political pressures against safety. In some cases, the White House acts on behalf of wealthy party supporters.[81] The chairman of the House of Representatives appropriations subcommittee which approves food agency budgets is Representative Jamie Whitten (D-Miss.), who is an avowed advocate of all pesticides and has a naive "show me the victims" attitude toward the hazards of pesticides and other chemicals in food. With its control over FDA money, his committee has far more influence than pro-consumer Congressional committees such as those of Representative L. H. Fountain (D-N.C.).

Individual congressmen exert extreme pressure on FDA on behalf of wealthy businessmen in their districts. Senator Everett Dirksen, for example, was a powerful force in FDA's five-year delay on action against cyclamates, on behalf of the powerful Abbott Laboratories. While companies such as Abbott have no trouble getting Senate advocates on their behalf, consumers as individuals or groups lack sufficient organization and money to do so.

Sometimes anti-consumer pressure comes from traditional toxicologists with little understanding of recently developed knowledge about long-range, as opposed to immediate, health hazards. Scientists such as William J. Darby and Julius Coon are active in downplaying the long-term risks of additives, apparently because they are unaware of the special properties of carcinogens.[82]

Because the main political pressures on FDA militate toward approval of carcinogens, FDA should not be given the power to allow these additives in food.

## CONSUMERS SHOULD SUPPORT THE DELANEY CLAUSE

The Delaney Clause does not protect us from all exposure to carcinogens. Some natural foods, such as the tropical cycad nut, contain carcinogens.[83] Contact with products of combustion is unavoidable. Water and air, household products, and drugs carry carcinogenic pollutants. Several thousand new chemicals are invented every year. Hundreds enter commercial channels.[84] Our environment is increasingly filled with synthetic chemicals, many not required by law to be tested for safety. Our bodies must fight a greater total carcinogenic burden.

Compulsory testing of all environmental chemicals is needed. Better FDA restriction of foods containing aflatoxin and DDT--not falling under the Delaney Clause--is needed. The Delaney Clause principle should be applied to other routes of exposure, such as pollution of drinking water sources by factories.

Food additives are chemicals of small benefit.[85] Once added to food, additives are widely distributed, making their effects on long range health impossible to trace, and of enormous potential danger. People are exposed to food additives throughout their lifetime, starting pre-birth. Exposure is easily preventable.

Prohibition of carcinogenic food additives is a sane, manageable approach to health preservation because it helps to prevent cancer. Defense of the Delaney Clause from the pleas of private industry is a top priority for all consumers.

## SUMMARY

The Delaney Clause of the Food, Drug and Cosmetic Act prohibits adding any amount of a cancer-causing chemical to foods. The food industry has recently intensified their campaign to cripple the clause by permitting so-called "safe" amounts of such chemicals in foods. This assault on a law designed to prevent cancer comes at a time when <u>no scientific evidence exists for what a "safe" amount of any cancer-causing chemical might be.</u>

The majority of the more than 600,000 new cases of cancer each year are caused by chemicals in the environment and, as such, are preventable. <u>Some human cancers have already been traced to cancer-causing chemicals</u>, in the workplace, in consumer products, and in drugs.

Animal tests provide much of the evidence for carcinogenic characteristics of substances. Studies which rely on <u>human</u> evidence for determining the causes of cancer are extremely difficult to conduct, because human cancer does not appear for many years--even decades--after initial contact with the carcinogen, and because humans are exposed to so many synthetic chemicals. <u>Animal tests can be conducted before there are human victims.</u>

Cancer scientists believe that if a substance causes cancer in animals (which is true of only a small percentage of chemicals which have been tested), it is highly probable that man will also be susceptible. Almost all substances known to cause cancer in humans are also carcinogenic in animals.

When test animals develop cancer from high doses of a chemical but do not contract cancer from low doses, it does not mean that humans are safe eating low doses. Experimental animals are fed high doses of substances in order to increase the sensitivity of the tests and compensate for the small number of animals used. A high dose of a chemical does not produce false results. It does, however, produce <u>more</u> cases of cancer, so that if the substance is inherently <u>carcinogenic</u>, it will be evident in the relatively small number of test animals used.

The results of properly conducted animal tests can thus be used confidently to determine whether a food chemical is carcinogenic. However, they do not tell us at what amounts, if any, a known carcinogen is safe.

Food additives offer minimal or, more often, non-existent benefits, and therefore consumers should be protected from <u>any risk</u> of cancer from them. The major use of food additives is as a cheap substitute for high quality natural ingredients. Food additives do not have the benefits of certain carcinogenic drugs which are worth the risk because they lessen an immediate threat to life when no other treatments are available.

Once chemicals are added to the food supply, virtually the whole population is exposed, children, adults and old people, healthy and sick.

THE DELANEY CLAUSE IS A SOCIAL JUDGMENT THAT ONCE A SUBSTANCE IS KNOWN TO CAUSE CANCER, ITS USE AS A FOOD ADDITIVE IS NOT WORTH ANY RISK TO THE MORE THAN 200 MILLION AMERICANS WHO MIGHT INGEST IT. The Health Research Group strongly supports the Delaney Clause and asks consumers and their Congressional representatives in Washington to defend it from any change.

## FURTHER READING

"Cancer and the Law," s. 38.45a - s. 38.65, <u>Lawyers' Medical Cyclopedia</u> (1972).

<u>Chemicals and the Future of Man</u>, Hearings before the Senate Government Operations Committee (1971).

<u>Color Additives</u>, Hearings before the House of Representatives Committee on Interstate and Foreign Commerce (1960).

"Evaluation of Environmental Carcinogens," Report to the Surgeon General, USPHS, April 22, 1970, Ad Hoc Committee on Evaluation of Low Levels of Environmental Chemical Carcinogens, National Cancer Institute. Reprints from Dr. John A. Cooper, Building 37, Room 3A-07, National Cancer Institute, Bethesda, Md. 20014. Reprinted also in <u>Chemicals and the Future of Man</u>.

Hueper, "Environmental Cancer Hazards," <u>Journal of Medicine</u> <u>14</u> (1972) 149-153.

Weisburger and Weisburger, "Food Additives and Chemical Carcinogens: On the Concept of Zero Tolerance," <u>Food and Cosmetics Toxicology</u> <u>6</u> (1968) 235-242.

## RECOMMENDED GENERAL ARTICLES

Samuel Epstein, "Environmental Determinants of Human Cancer," <u>Cancer Research</u> <u>34</u> (October 1974) 2425.

John Cairns, "The Cancer Problem," <u>Scientific American</u> <u>233</u>, 5 (1975) 64.

## FOOTNOTES

1. American Cancer Society, *Cancer Facts and Figures 1977*.

2. World Health Organization, *Technical Report #276*.

3. "The Correlation of Experimental Carcinogenesis and Cancer in Man," in Homberger and Karger, eds., *Experimental Tumor Research* (1969) 222-234.

4. Kotin and Falk, "Atmospheric Factors in Pathogenesis of Lung Cancer," *Cancer Res* 7 (1963) 475-515.

5. International Agency for Research on Cancer, Monographs on the Evaluation of Carcinogenic Risk of Chemicals to Man (hereinafter IARC), Vol. 14, *Asbestos* (1977).

6. Imprescia, *Amer Coll Chest Physicians* 22 (1952) 347-355.

7. Testimony of Dr. Samuel Epstein, *Chemicals and the Future of Man*, Hearings before Senate Government Operations Committee (1971) 46. Hereinafter called Epstein.

8. IARC, Vol. IV (1974).

9. IARC, Vol. I (1971).

10. Waxweiler et al., "Neoplastic Risk Among Workers Exposed to Vinyl Chloride," in *Occupational Carcinogenesis, Annals of the NY Acad of Sciences* 271 (1976) (hereinafter called *Occup Carcin*) 40.

11. National Academy of Sciences/National Research Council (NAS/NRC), *Particulate Polycyclic Organic Matter* (1972).

12. *Ibid.*

13. U.S. Atomic Energy Commission, *Inhalation Carcinogenesis* (1964).

14. Hueper and Conway, *Chemical Carcinogenesis* (1964).

15. See Hoover and Fraumeni, "Drugs," in Fraumeni, ed., *Persons at High Risk of Cancer: An Approach to Cancer Etiology and Control* (1975) 185; and Jablon, "Radiation," *ibid.*, 151.

16. Hueper, "Environmental Cancer Hazards," *J of Med* 14 (1972) 1949; Epstein, "Cancer and the Environment," *Bull of Atomic Sci* (March 1977) 23.

17. Kotin and Falk, op. cit.

18. Higginson, "Present Trends in Cancer Epidemiology," Canadian Cancer Conf 8 (1969) 43; WHO, ibid., 4.

19. Saffiotti, "Risk-Benefit Considerations in Public Policy on Environmental Carcinogenesis," Canadian Cancer Conf 11 (1976).

20. See Greenberg, "The Cancer Program: A New Burst of Criticism," New Eng J Med 293 (1975) 1379.

21. See Cairns, "The Cancer Problem," Scientific American 233, 5 (1975) 64.

22. Fortune (March 1972) 64.

23. Zwerdling, "The Pollution of Food," Intellectual Digest (October 1971) 38.

24. "Food Additives," Chemical and Engineering News 44 (October 10, 1966) 104.

25. Zwerdling, op. cit.

26. 21 U.S.C. 348(C)(3).

27. The general food additives law, apart from the Delaney Clause, also prohibits allowing hazardous additives because of their benefits.

28. Acta Union Internationalis contra Cancrum 13 (1957) 169.

29. The drug was given to prevent miscarriage and has subsequently been shown to be ineffective for this purpose. There are now several hundred victims. See Adam et al., Cancer Res 37 (1977) 1249.

30. Hirayama, "An Epidemiological Study of Oral and Pharyngeal Cancer in Central and Southeast Asia," Bulletin of the World Health Organization 34 (1966) 4-69.

31. Epstein, note 7, 53.

32. Hueper, Occupational and Environmental Cancers of the Urinary System (1969).

33. Magee and Barnes, Adv Cancer Res 10 (1967) 163.

34. Carnaghan, *Brit J Cancer* (1967) 118-914.

35. Weisburger, "Food Additives and Chemical Carcinogens: On the Concept of Zero Tolerance," *Food Cosmet Toxicol* 6 (1968) 235; Huggins et al., *Nature* 189 (1961) 204.

36. Ryser, "Chemical Carcinogenesis," *New Eng J Med* 285 (1971) 723.

37. Searle, "Chemical Carcinogens...," *Chem in Britain* 6 (1970) 7.

38. *Evaluation of Environmental Carcinogens*, Report to the Surgeon General, USPHS, April 22, 1970. Ad hoc committee on the Evaluation of Low Levels of Environmental Chemical Carcinogens, National Cancer Institute, Bethesda, Md. Reprinted in *Chemicals and the Future of Man*, 181.

39. Testimony of Dr. C. Burroughs Mider, Associate Director, National Cancer Institute, *Color Additives* (1960). Hearings before House of Representatives Committee on Interstate and Foreign Commerce. Hereinafter called Mider.

40. Testimony of Dr. Umberto Saffiotti, Associate Scientific Director for Carcinogenesis, National Cancer Institute, *Chemicals and the Future of Man*, 171. Hereinafter called Saffiotti.

41. Maltoni, "Predictive Value of Carcinogenesis Bioassays," *Occup Carcin*, 431.

42. J.A. Miller, "Carcinogenesis by Chemicals: An Overview," *Cancer Res* (March 1970) 560.

43. Hueper and Conway, 175.

44. When such results have occurred, they have been explained by poor experimental design, or by differences in the way of administering the substance in the two groups of animals (Shubik, "Survey of Compounds that have been treated for Carcinogenic Activity," Supp I (1957) as quoted in Mider). Animals may have natural defenses to the substance if injected, but not if eaten, and vice versa. In a few tests, one species gets cancerous tumors from a substance and another, benign tumors. However, no adequately tested chemical has been found to produce benign tumors only, and a substantial percentage of benign tumors in mice have eventually become malignant. (Report of the Secretary's Commission on Pesticides and their Relationship to Environmental Health (1969) 482.)

45. Aflatoxin has produced cancer in trout at .45 parts per billion (Sinnhuber et al., JNCI 41: 711-718) and 100 percent cancer in rats at 15 ppb (Wogan, "Naturally Occurring Carcinogens," The Physiopathology of Cancer, Vol. I (1974) 72).

46. Saccharin had caused cancer in other animal studies at doses as low as the equivalent of 1.6 cans of diet soda a day. See testimony of Health Research Group, Proposed Saccharin Ban: Oversight, Hearings before the House Committee on Interstate and Foreign Commerce (1977).

47. Since the latency period is thought to shorten as doses increase, large doses are used to make the cancer appear during the two- to three-year lifespan of the animal. Large doses simulate the effects of accumulation of the substance in humans over many years of exposure to small amounts. Druckrey, in Truhaut, ed., Potential Carcinogenic Hazards from Drugs, UICC Monograph Series, Vol. 7 (1967) 60; Albert and Altshuler, in Ballou et al., eds., Radionuclide Carcinogenesis, AEC Symposium Series (1973) 233.

48. Agriculture, Environmental and Consumer Protection Appropriations for 1975, Pt. 8: Food and Drug Administration "Study of the Delaney Clause and Other Anti-Cancer Clauses," Hearings of the House Committee on Appropriations (1974) 296. Hereinafter called FDA Delaney Clause Study.

49. If conditions are otherwise similar to laboratory conditions. 42 Fed Reg 73 (April 15, 1977) 20001.

50. At one time, some scientists planned to use tens of thousands of animals in one study--a so-called mega-mouse study--numbers that would theoretically make it possible to observe low incidence effects of low doses (or lack thereof), but these studies could not be conducted as a practical matter because of the huge financial and scientific resources required.

51. World Health Organization, Assessment of the Carcinogenicity and Mutagenicity of Chemicals, Technical Report #546 (1974) 9.

52. Weisburger, op. cit., 238; Druckrey, op. cit.; Crump et al., op. cit. The technical way of expressing this unique quality of carcinogens is that the dose-response curve is linear; that is, that some cases of cancer will be caused until the exposure level is zero. All the experimental

evidence that is available supports this dose-response curve, as does human evidence on radiation-induced cancer. See Saffiotti, "The Scientific Basis of the Delaney Clause," <u>Preventive Med</u> <u>2</u> (1973) 125; National Academy of Sciences, <u>Contemporary Pest Control Practices and Prospects</u> (1975) 82 ff; FDA Delaney Clause Study; Druckrey, <u>op. cit.</u>; Crump <u>et al.</u>, "Fundamental Carcinogenic Processes and Their Implications for Low Dose Risk Assessment," <u>Cancer Res</u> <u>36</u> (1976) 2973; <u>Environmental Health Perspectives</u> <u>21</u> (1977).

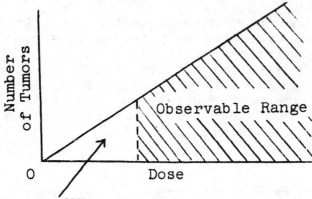

53. The carcinogens vinyl chloride, ethylene dibromide (an anti-knock gas additive), trichloroethylene (a solvent used to clean machinery and to decaffeinate coffee), and chloroform, for example, have a common structural element.

54. Testimony of HEW Secretary Arthur S. Flemming, <u>Color Additives</u>, 61. Hereinafter called Flemming.

55. Mider, 54.

56. Hueper, <u>op. cit</u>. Used in making dyes and as a stain in laboratory tests.

57. Tomatis, in <u>Occup Carcin</u>, 400.

58. Bingham <u>et al.</u>, <u>Arch Environ Health</u> <u>19</u> (1969) 779.

59. Wislocki, <u>Cancer Res</u> <u>37</u> (1977) 1883.

60. Weisburger, <u>op. cit</u>.

61. Oettle, "Epidemiology and Cancer," Canadian Cancer Conf 6 (1966) 411.

62. Selikoff, "Multiple Risk Factors in Environmental Cancer," in Fraumeni, op. cit., 468.

63. See Mulvihill et al., Genetics of Human Cancer (1977).

64. Steward, Lancet (1960) 1185-1188.

65. FDA Delaney Clause Study, 314.

66. "To postulate quantitative human significance to data derived from the use [of indexes of carcinogenic potency] in animals requires the assumption of a ludicrous situation in which carcinogenic agents act on humans, as it were, in a vacuum free of any exogenous and endogenous modifying factors." Mider, 54.

67. "Evaluation of Environmental Carcinogens," op. cit., reprinted in Chemicals and the Future of Man, 180-198.

68. NAS, Contemporary Pest Control Practices and Prospects (1975).

69. "Chemical Carcinogenesis," New Eng J Med 285 (1971) 18.

70. Saffiotti, 129.

71. McCann and Ames, "A Simple Method for Detecting Environmental Carcinogens as Mutagens," Occup Carcin, 5. Other quick cell tests are being developed. "Chemical Carcinogenesis," Science 183 (March 8, 1974) 940.

72. "For most mutagens linear dose-response curves are obtained." McCann, 5.

73. Office of Technology Assessment, Cancer Testing Technology and Saccharin (1977) 8.

74. Testimony of Ira Somers, Food Additives, Hearings before Senate Select Committee on Nutrition and Human Needs (September 1972).

75. Ad Hoc Committee on the Evaluation of Low Levels of Environmental Chemical Carcinogens, Evaluation of Environmental Carcinogens, Report to the Surgeon General, USPHS (April 22, 1970); World Health Organization, Prevention of Cancer, Technical Report Series 276 (1964); FDA Advisory Committee on Protocol for Safety Evaluation, Panel on Carcinogenesis,

"Report on Cancer Testing and the Safety of Food Additives and Pesticides," *Toxicol Appl Pharmacol* (1970); Committee on Causative Factors of Cancer and Committee on Cancer Prevention, "Report of Symposium on Potential Cancer Hazards from Chemical Additives and Contaminants to Foodstuffs," *Acta Un Int Cancer* 13 (1957) 179-193; Joint FAO/WHO Expert Committee on Food Additives, *Evaluation of Carcinogenic Hazards of Food Additives*, WHO Technical Report Series 220 (1961) 1-32; Food Protection Committee, Food and Nutrition Board, *Problems in the Evaluation of Carcinogenic Hazard from Use of Food Additives*, NAS/NRC Publ. No. 749 (1960); *Cancer Res* 21 (1961) 429-456; Berenblum, *Carcinogenicity Testing*, UICC Technical Report Series, Vol. 2 (1969); Technical Panel on Carcinogenesis, "Carcinogenicity of Pesticides," *Report of the Secretary's Commission on Pesticides and Their Relationship to Environmental Health* (1969) 459-506; Shubik and Sice, "Chemical Carcinogenesis as a Chronic Toxicity Test: A Review," *Cancer Res* 16 (1956) 728-742; Clayson, *Chemical Carcinogenesis* (1962); Hueper and Conway, *Chemical Carcinogenesis and Cancers* (1964); Weisburger and Weisburger, "Tests for Chemical Carcinogens," in Busch, ed., *Methods in Cancer Research*, Vol. I (1967) 307-387; Arcos, Argus, and Wolf, "Testing Procedures," in *Chemical Induction of Cancer*, Vol. I (1968) 340-463; Golberg, "Dose Selection and Administration," in *Carcinogenicity Testing of Chemicals* (1974); Proceedings of the NAS/NRC Drug Research Board Conference on Carcinogenesis Testing in the Development of New Drugs (1974); NAS, *Principles for Evaluating Chemicals in the Environment*, a report of the Committee for the Working Conference on Principles of Protocols for Evaluating Chemicals in the Environment (1975); Environmental Studies Board, NRC, *Contemporary Pest Control Practices and Prospects* (1975); National Cancer Institute, *Guidelines for Carcinogen Bioassay in Small Rodents*, NCI Technical Report #1 (1976).

76. Higginson, *Canadian Cancer Conf* 8 (1969) 63.

77. Mider, 51.

78. Flemming, 501.

79. See Turner, "The Delaney Anticancer Clause: A Model Environmental Protection Law," *Vanderbilt Law Rev* 24 (1971) 889.

80. Wellford, *Sowing the Wind: Food Safety and the Chemical Harvest* (1972), 184.

81. <u>Ibid.</u>, 183. See HEW Review Panel on New Drug Regulation, Investigation of Allegations...FDA (1977) 521 ff.

82. "There is not a shred of evidence or even a basis of reasonable suspicion that any such damaging effects have ever been caused by the additives or pesticides in food consumed in North America. Certainly some defects have been observed in test animals after they have been fed exceedingly large amounts of some additives. But it is a long, frequently too long, step from the observation of the effects of such provocative and bizarre experiments to those of man's daily diet." <u>Industrial Medicine</u> <u>39</u>, 10 (1970) 31.

83. NAS/NRC, <u>Toxicants Occurring Naturally in Foods</u> (1966) 30.

84. Council on Environmental Quality, <u>Toxic Substances</u> (1971) iv.

85. If a carcinogenic food additive is thought to be so crucial to civilization that it is worth possibly great cancer risk, Congress can be approached for an exemption for that individual additive.

July, 1981

# NUTRITION ACTION

# The Delaney Clause Draws Fire

### By Greg Moyer

Thirty years have passed since Rep. James Delaney, then Congressman from New York, opened hearings that sought to extend the protective umbrella of the landmark 1938 Food, Drug and Cosmetic (F,D&C) Act over newly discovered hazards in the nation's food supply. Since passage in 1958 of a controversial amendment outlawing *absolutely* most cancer-causing additives destined for America's processed foods, industry has resisted—and resented—the more aggressive oversight brought on by Delaney.

Until now, industry has not been in a position to challenge this state of affairs. But a confluence of developments in the spheres of science and politics offers food manufacturers their best chance in a generation to recast the intellectual foundations that undergird the federal food safety laws. What's at stake is best symbolized by the "Delaney Clause", a straightforward expression of legislative intent that continues to serve the public interest very well—despite charges to the contrary. It simply states:

> "... no additive shall be deemed to be safe if it is found to induce cancer when ingested by man or animal, or if it is found after tests which are appropriate for the evaluation of safety of food additives to induce cancer in man or animal."

In 1900, USDA's Harvey Wiley (seated center) tested additives on 12 human guinea pigs.

But these are rough times for this far-sighted food safety provision. During the past decade scientists devised ever more powerful tools for detecting small amounts of carcinogenic—or cancer-causing—substances in food. Delaney critics now claim that absolute zero is less than the Queens Congressman and his colleagues ever could have imagined; that detectable levels of a carcinogen are often so low as to be insignificant to human health.

### FDA Takes 'Black Eye'

Meanwhile, Americans watched as the Food and Drug Administration (FDA) confronted the food industry over the safety of two common additives: saccharin and nitrite. In both cases the agency lost the public relations battle,

*Also contributing to this story were BRUCE SILVERGLADE and MICHAEL JACOBSON.*

despite acting responsibly to fulfill its legal charge.

And finally, the era closed with the country experiencing an anti-regulatory backlash that helped Ronald Reagan become President of the United States. While Reagan did not take to the hustings bemoaning the inflexibility of America's food safety laws, the Delaney Clause did warrant a mention in the 1980 GOP platform. The Republicans went on record favoring "a legislative effort to revise and modernize our food safety laws, providing guidelines for risk assessment, peer review, and regulatory flexibility which are consistent with other government health and safety policies."

Judging from the shrillness of the debate, it's hard to believe that the Delaney Clause has only been formally invoked twice in its 23-year history:[1] once in 1967 to ban Flectol-H, and again in 1969 to ban chroanaline, both components of food packaging adhesives. But the outcome of these two decisions is not the issue. What industry disdains about Delaney is its clear and uncompromising standard that instructs an executive agency not to muddle around with something as serious as cancer.

The logic of Delaney is as irrefutable today as it was in 1958: since there are no known thresholds for inducing cancer, all food additives found to be

'A rigorous risk/benefit ratio is intrinsically impossible with our present knowledge of food additives.'

—**Sanford Miller**
**FDA Bureau of Foods**

carcinogenic in pre-market tests should be banned from the food supply. Though the FDA relied on the general safety provisions of the original F,D&C Act to ban chloroform, DES, and a handful of food dyes, the agency has cited the Delaney Clause four times in the last 10 years to buttress its no-nonsense decisions.[2] If the spirit of Delaney is scuttled in the rewriting of the food safety laws—as industry and their sympathetic legislators now propose—FDA rulemaking would be cut adrift in a sea of subjectivity where political and economic considerations are likely to count for more than good science.

Revise and modernize . . . risk assessment . . . peer review . . . regulatory flexibility . . . these are the code words used by the "Dump Delaney" forces to rally intellectual opinion around their cause. Peter Hutt, former Chief Counsel for the FDA, now partner in a D.C. law firm specializing in food and drug issues, and leader of a "Dump Delaney" movement, outlined his rationale in a recent article for *Legal Times of Washington*. Hutt is paraphrased below:

• As more and more food ingredients are tested for carcinogenicity, more and more are found to be carcinogenic in at least some of the test animals.

• As analytical chemistry has become more and more sensitive, more and more of the carcinogenic substances have been found in the food supply.

• As the number of epidemiological studies multiply, more and more common components in the American diet are implicated with cancer.

### Cancers Galore

"The result," concludes Hutt in apparent frustration, "is that it would be impossible today to survive on a diet consisting only of substances that have not been found to be carcinogenic by some type of scientific study."[3]

Of course, not everything causes cancer. A 1969 study by the National Cancer Institute tested 120 pesticides and industrial chemicals thought to be carcinogens. After feeding animals the maximum doses permitted by good laboratory protocol, only 11 of the compounds caused cancer. Thus, it is fair to assume that the number of carcinogens among 120 randomly selected substances would be even fewer still.[4]

But Hutt and his allies manipulate people's fears and dim memories of yesterday's headlines to sway the debate on an emotional level. A common tactic is to stand before lay audiences and recite long lists of foods where there is said to be some evidence of carcinogenicity. Here's the list delivered by Rep. William Wampler (R.-Va.), a Delaney foe, in a recent speech to the National Pork Producers Council: bracken fern in greens and salads, cadmium, caffeine, chloroform, carbon tetrachloride, egg yolk and egg white, ergot in rye, nickel, oil of calamus, aflatoxins, safrole in spices, selenium, tannic acid in coffee, tea, and cocoa, and vitamin D-2.

What's particularly deceptive about this approach is that possible "natural" carcinogens such as aflatoxins, ergot, black pepper, or safrole are not even covered by the Delaney Clause. While acknowledging the dangers inherent in natural carcinogens, the FDA has no authority to ban these food components using the food additive amendment. Delaney only addresses one category of food constituents: ingredients intentionally added—directly or indirectly—to fabricated foods.

### Ignorance Is Bliss

Ironically, Hutt, the friend of the food technologist, seems embarrassed by the fruits of our innovative science. Instead of taking comfort in the newfound techniques for detecting carcinogens in ever smaller amounts, and thereby better protecting the public's health, Hutt would have us return to less complicated times marked by scientific innocence. To compensate for our remarkable scientific advances, he wants us to dull our regulatory sensibilities, so as not to destroy the public's "confidence" in the American food supply. But Hutt and the other critics are smart enough to anticipate the outcry that would ensue were their ultimate goals—to weaken the mandate protecting the nation's food and health—presented in so naked a form. Thus, they have justified their designs in the language and logic of the pseudo-science currently in fashion: risk assessment.

Simply stated, risk assessment seeks to weigh the health consequences of exposing citizens to risk against the perceived benefits derived from taking the same action. Both saccharin and nitrite are examples commonly cited by

**If the spirit of Delaney is scuttled in the rewriting of the food safety laws, FDA rulemaking would be cut adrift in a sea of subjectivty.**

industry where the principles of risk assessment might have produced a policy decision controverting the Delaney Clause.

In 1977 the FDA acted to ban saccharin on the basis of a Canadian study that uncovered—for the umpteenth time—the carcinogenic potential of this artificial sweetener. An uproar fanned by the Calorie Contol Council, a trade association for makers of artificially sweetened products, led Congress to "bomb the ban", as Rep. James Martin (R.-N.C.) put it. The legislators prevented the FDA from limiting saccharin use on the grounds that the health benefits to the diabetic and obese outweighed the risk of possibly promoting several thousand cases of bladder cancers.

A similar argument unfolded over nitrite—though the most devastating charges against the chemical ultimately were judged to be groundless. When an animal study conducted by an MIT pathologist appeared to link nitrite to cancer, the FDA ordered the preservative phased out of cured meats. Industry balked, claiming that sodium nitrite is the most effective preservative known to inhibit the growth of deadly botulism spores. (Nitrite, as well, contributes to the pink color of cured meat and enhances the flavor.)

An independent examination of all the microscopic slides generated by the MIT experiment contended that the original analysis was wrong—nitrite did not appear to cause cancer directly. But the agency continued to express concern about nitrosamines, another potent cancer agent that is derived from nitrite in food. Throughout the controversy the meat industry framed the debate in terms of a trade-off: what is more feared, the chance of a few people developing cancer or an epidemic of botulism poisonings?

### Risks of Risk-Benefit

Disregarding the truthfulness of these industry conundrums—for some scientists doubt that saccharin actually helps take off pounds and others contend that meats can be safely preserved without sodium nitrite—the theoretical argument remains: can regulators be expected to impartially weigh the health risks against hard-to-measure benefits?

Sanford Miller, the current head of the FDA's Bureau of Foods and the person who would be given the responsibility for implementing risk assessment, has repeatedly cautioned against entrusting public health policy to this primitive methodology. Miller's thoughts are worth quoting at length:

> *"A rigorous risk/benefit ratio is intrinsically impossible with our present knowledge of food additives. The term calls for dividing a very uncertain probability of risk in an ill-defined series of controversially adverse physiological responses in undefined units, by a heterogeneous collection of socalled benefits, such as economic gains expressed in dollars, convenience expressed in unknown units, palatability expressed in yet-to-be-agreed-upon measures, aesthetic appeal not expressible in numbers of any kind and so on."*[5]

### Industry's Stand

Miller sees clearly that risk assessment is fraught with ambiguities on both sides of the equation: determining risks and determining benefits. But that has not impeded the Dump Delaney forces who suggest that this voodoo science would be an improvement on the current system. Rep. James Martin has submitted a bill, H.R. 2313, that would invalidate Delaney and reconceptualize the basic principles underlying the food safety laws. That its specific provisions will become law is far from certain—the National Soft Drink Association, the Grocery Manufacturers of America, and the American Meat Institute are all peddling their own versions of draft legislation—but the Martin bill is a concise statement of industry thinking.[6]

The legislation would establish a "rebuttable presumption" of carcinogenicity that allows industry to appeal an FDA ban before it becomes law. The grounds for rebuttal would be two-fold, either:

• industry must show that the benefit of the additive outweighs the risk; or

• industry must show that the risk, calculated by an accepted extrapolation technique, is so small as to be acceptable.

The Martin bill tells FDA just how it should evaluate the animal data upon which the presumption of carcinogenicity is based. If the dose to the animals is so high as to cause dysfunction of the animal's detoxification mechanisms, then the study must be discounted.

### New Criteria

H.R. 2313 also instructs the FDA to consider factors new to food safety rulemaking before the agency arrives at a final decision. These factors include:

• health risks and benefits;

• nutrition needs and benefits;

• nutritional value, cost, availability, and acceptability of food;

• environmental factors;

• the "interest of the general public."

And lastly, the Martin bill would permit the FDA to substitute a labeling no-

---

### Additive Alert

Here is a partial list of some questionable food additives you should avoid next visit to the supermarket. Clip this card and carry it in your wallet for handy reference. More complete listings are available on CSPI's "Chemical Cuisine" poster (see "Order" form on page 15).

• **Artificial colorings** (most foods of low nutritional value). Especially try to avoid Citrus Red #2, Red #40, and Yellow #5.
• **BHT** (cereals, chewing gums, potato chips, oils, etc.).
• **Brominated Vegetable Oil** (soft drinks).
• **Caffeine** (soft drinks; occurs naturally in coffee, tea, or cocoa).
• **Sodium Nitrite** (cured meats).
• **Saccharin** (diet products).

Two common food constituents, sugar and salt, become dangerous when consumed in excess. Cut back on both.

●CSPI, 1981

tice for an outright ban where it considered his alternative approach more appropriate.

These proposals go far beyond the legitimate need to square the food safety laws with current technology. They take a hatchet to a mission of legislative reform that could be better handled with a scalpel.

For the FDA to determine whether the risks of a cancer-causing additive are outweighed by the benefits, the agency needs to estimate the number of cancers the substance is likely to create. The first step is for scientists to use small-scale animal tests to develop hard data. From these findings, they use mathematical models to extrapolate the risks from a few animals to a human population. For the precise judgments of risk demanded by the Martin bill, both the animal tests and the mathematical models must be reliable. Unfortunately neither yields results conclusive beyond a reasonable doubt.

**Tests Lack Precision**

Look at the animal tests first: they are not sensitive enough to detect anything less than a five percent increase in cancer. Weak carcinogens slip through this loosely woven net. The only consolation is a troubling one: if an animal test shows positive results (i.e., the substance tested induces cancer), it surely means that the chemical is a serious health hazard.

Furthermore, the test animals are a pampered lot of genetically similar stock. They may be far more resistant to cancer than human beings, who reflect diverse genetic backgrounds, live in an array of habitats, and are exposed to a slew of environmental carcinogens in the course of leading normal lives. All these human susceptibilities have been eliminated from the animal experiments (*Nutrition Action*, June '81).

Now consider the mathematical models scientists use to translate the animal tests into an estimate of human cancers. Government scientists have adopted a model which presumes a "linear" or direct relationship between the amount of a carcinogen consumed and the likelihood of developing cancer. While this linear model best describes current experimental evidence, it also yields the most conservative and prudent recommendations for public health policy. Several other models predict fewer cancers per exposure, particularly at low dosages. Needless to say, industry is fond of quoting them.

Examples of the variation in predicting the potency of a carcinogen are

---

*Recommendations for Change*

## CSPI's Stand on Food Safety

Some changes in the food safety laws could benefit the consumer. As the debate opens on Capitol Hill, CSPI, in conjunction with a coalition of consumer groups, suggests that any rewrite of the Food, Drug, and Cosmetic Act include these provisions:

● *Establish new categories for chemicals in food to eliminate historical inconsistencies.* A new law should consider these substances the same way: food additives, color additives, substances now "generally recognized as safe", substances approved before the passage of Delaney in 1958 (prior-sanctioned), and essential nutrients consumed at high levels.

● *Review the safety of additives every ten years.* Deficiencies in testing, as judged by contemporary standards, should be corrected.

● *Broaden the testing regimen of food additives.* Additives should be tested for effects on behavior when deemed appropriate and together with common drugs and additives. This second test helps evaluate the possibility of dangerous interactions.

● *Alert the public, perhaps through labeling, when traditional foods pose a health risk.*

---

striking. In the report of the Food Safety Board of the National Academy of Sciences on saccharin and food policy, scientists used four different models to calculate the risk involved if one quarter of the population consumed one diet soft drink (0.12 grams of saccharin) a day. The results varied five millionfold, from an estimated 0.0007 to 3,640 additional cases of bladder cancer each year.[7]

With fluctuations of this magnitude, the choice of models is crucial. A taste of future industry strategy is already offered by Peter Hutt. He suggests that the FDA average the numbers from the linear model with the less conservative prognosticators to find a "more realistic" estimate of the actual health risk.

**Is Some Risk 'Acceptable'?**

The next component of Martin's bill discusses the establishment of an "acceptable" risk. Critics of Delaney noted that analytical techniques have so sharpened the scientist's vision, that it is now possible to track chemicals down to parts per billion or even parts er trillion. These methods are bet ween one thousand and one million times more sensitive than anything that was known when the Delaney Amendment passed in 1958. Subsequently, scientists have found very small quantities of suspicious substances in food. Often these substances are indirect additives, such as traces of chemicals used in packaging. A few of these substances are known to cause cancer in rats. But the human exposure is so low, that statistically, these carcinogens may account for only one cancer in a billion or more people, calculated over their lifetimes. Under the Delaney Clause, these substances must be banned from the food supply.

For lack of another word, we'll call these trace substances "contaminants." They pose a problem less serious than direct food additives, which Americans consume at the rate of about nine pounds per year.[8] The Dump Delaney forces are using the scientist's ability to spot trace contaminants as an argument for completely abandoning the Clause. Meanwhile, others are suggesting that contaminants should be shuttled off into another category of food hazards outside the purview of the Delaney Clause. This would lessen the pressure on Delaney, and allow it to continue to protect consumers from direct additives shown to be carcinogenic.

**New Variables**

The Martin legislation permits industry to rebut the FDA's presumption on the basis of benefits new to the rulemaking process. Rep. Martin suggests these criteria should include: nutrition needs,

> What lies beneath the rhetoric is a debate of first principles. To whom is industry accountable when it chooses to alter our foods?

cost, availability, environmental effects, and a vague notion of "the interest of the general public," among others. This invitation to rebuttable presumption is riddled with theoretical flaws. Again, Sanford Miller:

*"We all agree that health considerations should be included in such an exercise, but would we also agree that economic considerations belong in the rubric of such an evaluation? How shall we include aesthetic concerns . . . which in the end are the fundamental reason why people select one food over another? Shall we include the long range effect of a particular action on the fabric of society itself? What about societal impact—the psychological desire of an individual for a particular substance such as that for saccharin? . . . How do we evaluate these secondary, tertiary, and quarternary benefits.[9]*

There's a practical fallout to the introduction of benefit analysis as well. The decision to ban a carcinogen would revert to case-by-case rulemaking under the Martin scheme. More discretion would be vested in the FDA to determine when the risks posed by a chemical exceed some estimated threshold of benefits. At a minimum, this approach would severely strain the workload of the present regulatory apparatus. FDA decision-making, already arthritic, would virtually cease. And through it all, industry would be in a far stronger position to advocate its interests than would the general public, due to the disparity of resources each side could muster for the fight.

At the height of the saccharin moratorium controversy several years ago, Rep. Martin urged his colleagues to follow his lead in voting to scrap the Delaney Clause. He said: "We must be reasonable. If we were to ban every food and every item we contact that posed a risk comparable to saccharin, by the time the rat breeders and rat feeders finished their experiments, we would have to ban most of the natural food supply, most of the production jobs, and most of the universe. We would not allow you to go outside or to stay inside. But you would be safe, after a fashion."

Given that apocalyptic vision, one wonders how the Republic has managed under the Delaney Clause this long. The truth is that the "crisis" is nowhere near as serious as Delaney opponents suggest. The list of carcinogenic food additives is relatively short; with some exceptions, the dangerous substances are easily replaced with safer alternatives.

What lies beneath the rhetoric is a debate of first principles. To whom should industry be held accountable when it chooses to alter the contents of our foods? Do we regulate to protect the health needs of individuals, particularly those who may be unusually sensitive, or do we regulate to maximize some ill-defined benefits accruing to society-at-large? And how will FDA react when the risks of an additive are borne by consumers, while the benefits accrue mainly to producers?

During his Senate confirmation hearings, Secretary of Health and Human Services Richard Schweiker said that the Delaney Clause should be ditched. "We have to define the Clause in terms of a risk-benefit ratio," he said. "The regulation has to relate to present technology."

And Secretary of Agriculture John Block told a recent food policy conference that, "Less regulation can be more effective regulation . . . Perhaps, in this inflationary time, we should consider the wisdom behind an immediate ban on a questioned substance, the wisdom that does not assess the economic disruption that may result (from implementing the Delaney Clause.)."

Whether those concerned about preventing cancer will be able to expose the fraud of risk-benefit thinking could determine the fate of our current food safety law: the Food, Drug and Cosmetic Act. The juices are flowing for what looks like a noisy Congressional dogfight. Certainly consumer groups will come out swinging, for it may be another 30 years before the food safety laws again step into the political spotlight. ∎

---

### Speak Out on Delaney

Letters expressing concern for the changes in the food safety laws being discussed should be sent to Rep. Henry Waxman (D-Ca.) and Sen. Orrin Hatch (R.-Utah). As chairmen of the House Subcommittee on Health and the Environment and Senate Labor and Human Resources Committee they will be influential legislators when the debate starts in earnest. Let them hear from other than industry sources on this important issue, and send carbons to your own legislators.

Send to:
**Representative
House of Representatives
Washington, D.C. 20515**
 or
**Senator
U.S. Senate
Washington, D.C. 20510**

---

*For a copy of CSPI's* Does Everything Cause Cancer? *send $1.00 to:* CSPI, 1755 S St. NW, Washington, D.C. 20009.

---

[1] Anonymous, *HEW Working Group Report on the Delaney Clause*, Washington, D.C., 1977.
[2] *Ibid.*
[3] Hutt, Peter, *Legal Times of Washington*, Apr. 27, 1981.
[4] Brunin, Greta, and Jacobson, Michael, *Does Everything Cause Cancer?* CSPI Reprint.
[5] Miller, Sanford, *Journal of Food Technology* 32(2): 1978.
[6] Harris, Stephanie, CSPI Report on the Delaney Clause, 1981.
[7] Anonymous, "Food Safety Policy: Scientific and Societal Considerations", IOM,NRC/ Assembly of Life Sciences NAS, Washington, D.C., Mar. 1, 1979.
[8] Epstein, Samuel, *The Politics of Cancer*, Sierra Club Books, San Francisco, 1978.
[9] Miller, speech, *op. cit.*

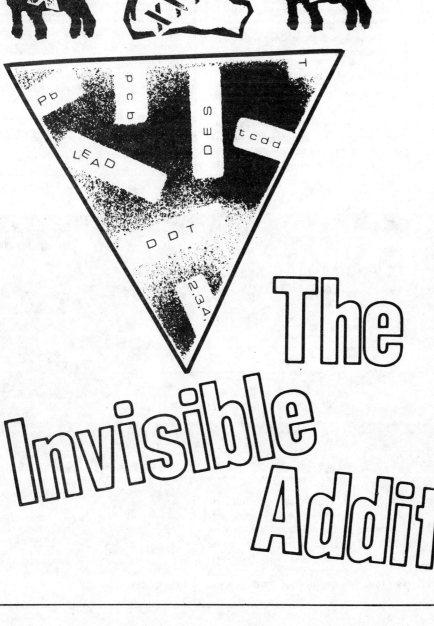

# The Invisible Additives

*By Linda Pim*

As long as you get out of bed in the morning, life is fraught with risks. From jaywalking through rush-hour traffic to living in the shadow of a nuclear power plant, potentially hazardous situations abound in everyday lives.

In an era when sweatsuit-clad business executives puff through their lunch hours on the jogging track, commuters exchange their transit tickets for 10-speed bicycles, and bran muffins encroach on the domain of the donut as a quick snack, public interest in chemical risks present in food is not surprising. All these shifts in lifestyle indicate a new appreciation of the responsibility each of us must assume for our own good health. In short—preventive medicine with a fresh face.

Several significant health threats are related to the food supply. They include: poor nutrition, bacterial contamination (food poisoning in the traditional sense), toxins naturally present in foods (aflatoxin in peanuts), and intentional food additives (colors, preservative, and flavors). But another serious and insidious source of food chemicals is in the form of *invisible* or unintentional additives.

These additives are contaminants in food—chemicals such as pesticides used in crop production, drugs used in livestock ranching, mycotoxins (poisonous molds), and air and water pollutants which become food pollutants. They are "invisible" in the sense that their names do not show up on food labels. They are substances which inadvertently get into food in the course of its passage from farm or factory to dinner plate. The appearance of these environmental additives in food is unintentional, inconsistent, and often difficult to monitor.

*LINDA PIM is a biologist and food chemicals researcher with the Pollution Probe Foundation, a public interest environmental organization based in Toronto, Canada.*

Controversy about the extent of the health risks of small amounts of environmental contaminants in the food supply rages unabated among scientists, regulatory agencies, the agriculture and food industries (collectively, "the experts"), and consumers. The following "catalog of environmental food risks" provides an overview of the potential hazards involved.

## Pesticides

By definition, a pesticide kills a living thing. For that reason alone, we must entertain the notion that pesticide residues in our food may not be the most benign addition to the menu.

While it is impossible to lump all agricultural pesticides together with regard to type and degree of toxicity, the health effects of some pesticides include the following:

- interference with the nervous system enzymes (for example, parathion);
- carcinogenicity (for example, ethylenethiourea, a by-product of some common fungicides);
- reduced sperm counts in some exposed males leading to infertility.

By far the largest proportion of pesticide exposure in the general population (that is, in people not occupationally exposed) comes from residues in food. Other sources—air, water, dust, cosmetics, clothing and aerosols—contribute only about 15 percent.

Pesticide residues in food may result from either spray application to the crop surface (topical residues) or soil treatment with possible uptake of the chemical into the inner tissues of the crop (systemic residues). Residues decrease over time, such that the level showing up on our dinner plates is much lower than the level on the crop immediately after application.

### Infants at Risk

Due to the phenomenon of bio-concentration or bio-accumulation—the increase of residue levels as we move up the ecological food chain—animal products (meat, eggs, dairy products) have contributed to the lion's share of pesticide residues in the typical North American diet. At the top of the food chain is human breast milk, which, at the height of DDT use in the 1960s and early 1970s, contained alarmingly high levels of the chemical.

In general, some pesticides residues in food are now considerably lower since certain chemicals were banned from agricultural use 10 or 20 years ago. This is largely due to the switch from persistent chemicals like DDT and the other organochlorine pesticides to substances that degrade much more quickly, such as the organophosphorus and

---

## The Pesticide Menace

- *New Mothers.* If you believe you have had undue exposure to pesticides by any route (food, air, water), you could have your breast milk tested. For the names of laboratories, check with your public health department. Private labs charge about $75 for the tests. One such firm in the Washington, D.C. area is Biospherics, Inc., 4928 Wyaconda Road, Rockville, MD 20852 (301-770-7700). They will supply information upon request and explain how to prepare samples for mailing.

- *The Rest of Us.* Congress will soon review the law designed to protect us from harmful pesticides. Already, manufacturers are lobbying hard to keep information on pesticide hazards confidential, and to weaken other aspects of the law. Consumers' voices need to be heard.

To air your views, write:
**Rep. George Brown
House Agriculture Committee
Washington, D.C. 20515
attn: Chuck Benbrooke**

For more information on the pesticide menace, contact Steve Kirk at the *National Association of Farmworker Organizations* (toll-free: 800-424-5100), or write the *National Coalition Against the Misuse of Pesticides* (NCAMP) at 1751 N St., NW, Washington, D.C. 20036.

---

**Airplanes and helicopters are commonly used to apply herbicides, pesticides, and fungicides.**

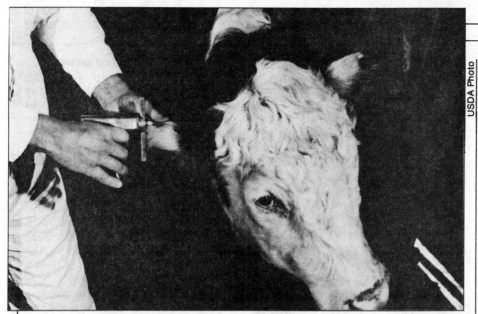

A capsule of the drug, stilbestrol, is implanted into the ear of this test animal at a USDA lab.

carbamate pesticides. Still, as recently as 1978, DDT was the most common pesticide residue in food on this continent.

Current pesticide residues are but a small fraction of the Acceptable Daily Intakes (ADI) established jointly by the Food and Agriculture Organization (FAO) and the World Health Organization (WHO). But the whole concept of ADIs as a reliable indicator of human toxicity has been questioned, especially considering possible interactions among substances in the chemical sea around us, and that there may be no "acceptable" level for carcinogens and mutagens.

**Exposure Data Questioned**

Under the U.S. Federal Food, Drug, and Cosmetic Act, maximum permitted residue levels, or tolerances, are given for many pesticides. These tolerances are derived from the ADIs, and from what are considered typical food patterns. But the Food and Drug Administration (FDA) has been criticized by the General Accounting Office (GAO) for its handling of the Market Basket Analyses which purport to indicate total consumer exposure to residues in food.

The GAO has complained that several pesticides with stated intolerances are not routinely checked, that sample sizes are too small, and that the practice of lumping similar foods together into composites obscures the kind and amounts of residues that specific foods contribute.

No simple solution exists to the puzzle of what is an "acceptable" pesticide level in our food. As more and more health problems are linked to chemicals in the environment, there must be increased emphasis on cutting down, even cutting out, the use of toxic substances in agriculture. One technique now being used successfully in this regard is Integrated Pest Management (*Nutrition Action*, Jan. '78). Going all the way to chemical-free "organic" crop production is possible for some crops in some areas, especially on small and medium-sized farms.

## Animal Drugs

If you thought drugs were only for people, just open the barn door. Animal medications are a big business. About 40 percent of all antibiotics produced in the United States each year is destined for livestock, not only to treat and prevent disease, but to boost growth as well.

Drugs commonly used in livestock production include penicillin, streptomycin, tetracycline, sulfonamides, nitrofurans, copper and arsenic compounds, and, to a lesser extent, hormonal preparations. (The infamous hormone diethylstilbestrol, DES, was banned for livestock use in 1979 in the U.S., and five years earlier in Canada.)

The health risks for people eating meat, milk, dairy products, and eggs from medicated animals are two-fold:

• Due to heavy use of antibiotics, in both animal and human therapy, pathogenic bacteria in our bodies may become resistant to the effects of antibiotics, such that these drugs become less effective in treating human illness (*Nutrition Action*, Feb. '80).

• Even low residues of antibiotics in food may pose a significant problem for individuals who have allergies to these drugs, according to the Office of Technology Assessment. In addition, some of the drugs—the nitrofurans, for example —are at least suspected of being carcinogenic.

While intensive livestock production is probably here to stay, drug use on "factory farms" can be moderated. Some key steps for reducing chemical risks to consumers include:

• better sanitation to prevent the spread of disease naturally;
• elimination of those drugs in animal feeds that are used in human treatments;
• use of growth promoters that do not leave tissue residues.

## Toxic Molds

Contrary to popular belief, natural is not always better. Among the most toxic environmental contaminants in our food are mycotoxins—poisons from molds. They inflict themselves on certain food crops in conditions of high moisture and humidity.

Some molds on grains, fruits, and vegetables simply detract from the foods' appearance or cause economic spoilage. But other fungi produce substances toxic to animals and humans. Mycotoxins have been known and feared for centuries—for example, ergot in rye, taken in high doses, can cause convulsions, gangrene, and hallucinations. Modern grain sorting and handling techniques have virtually eliminated any serious risks from ergot.

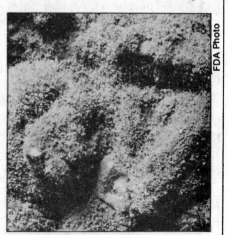

Despite improved storage techniques for peanuts and grains, aflatoxin-producing mold (pictured above) continues to be a serious health threat.

Other mycotoxins, however, have been recognized only in this century, and are a continuing health threat. Aflatoxin falls in this category.

Aflatoxin derives its name from the mold species that most commonly produces it, *Aspergillus flavus*. While almost any grain, fruit, or vegetable is susceptible to aflatoxin contamination, peanuts seem to be especially vulnerable. This mycotoxin is one of the most potent carcinogens known. It is *the* most active promoter of liver cancer, producing tumors in laboratory animals at the lowest analytically detectable level of one part per billion (ppb). The U.S. maximum permitted level of aflatoxin in food is 20 ppb.

Peanut butter is intermittently laced with small amounts of aflatoxin. In 1978, *Consumer Reports* found aflatoxin in at least one jar of every brand (including so-called "natural style" brands) and type of peanut butter tested. But aflatoxin contamination varies from year to year depending on the weather, which explains why similar testing by the Consumers Union in 1972 and 1973 detected aflatoxin in less than one-fifth of the samples.

Ideally, *no* food should have any cancer-causing aflatoxin. The question mark is the extent to which aflatoxin contamination of nuts and grains is avoidable. Since there's not much to be done about the weather, emphasis has been placed on proper drying of the crop at the farm level and proper sorting and storage at the processing level.

## Pollutants

When pesticides are applied directly to food crops or drugs are administered to livestock, one would expect residues of these toxic chemicals to appear in food products. After all, treating a crop or animal with a pesticide or antibiotic is deliberate and planned. But the accidental lacing of the food supply with the by-products of industrialization is food contamination in its most insidious form. Some of the chemicals we intend to contain in one place for one purpose are going to end up elsewhere.

Two broad categories of pollutants in food are synthetic organic chemicals—most often chlorinated hydrocarbons like PCBs and dioxins—and heavy metals, particularly lead, mercury, and cadmium.

PCBs, or polychlorinated biphenyls, have captured much attention and well

**One source of pollution is chemicals lost to the soil around city dumps. It is likely that the soil beneath these discarded tires contains high levels of cadmium, a heavy metal used in the tire manufacturing process.**

they should, since their suspected or proven human effects are legion. These include chloracne, cancer, liver disease, birth defects, and effects on disease immunity. The key source of PCBs in food is fish from polluted waterways. About 90 percent of any PCB that is eaten in food is retained in the body. It bio-concentrates in fat tissue and breast milk (*Nutrition Action*, Nov. '80).

Though new industrial uses for PCBs are blocked due to regulatory restrictions, the persistence of these chemicals in the environment means that they will continue to plague us for some time.

### Deadly Dioxins

The term "dioxin" actually covers several related substances. But the chemical usually referred to simply as dioxin—thought to be "the most poisonous substance known to man"—is highly toxic TCDD or 2,3,7,8-tetrachlorodibenzo-p-dioxin. It and other, less toxic dioxins, are formed as unwanted contaminants in the production of some pesticides, such as the herbicide 2,4,5-T.

TCDD at the level of less than one part per billion is a suspected carcinogen in laboratory animals. In man, overexposure to dioxin has caused chloracne, deceases in the kidney function, cirrhosis of the liver, stomach ulcerations, nervous system damage, and mental disturbances such as memory and concentration lapses, and depression.

The dioxin in food that derives from 2,4,5-T results from the use of the herbicide on fruit trees, rice crops, and cattle rangeland. Other dioxin in food can arise from the use of pentachlorophenol as a wood perservative in barns, feed bins, fences, and wood shavings used in bedding.

### Contaminated Fish

Partly because of the technical and financial resources required for dioxin analysis, data on dioxin in our food are not voluminous. In 1980, tests by the New York State Department of Health and the Ontario Ministry of the Environment uncovered TDCC in several species of fish from Lake Ontario at levels ranging from three to 40 parts per trillion (ppt). Also, preliminary data from a 1980 U.S. Environmental Protection Agency (EPA) study indicate that 13 percent of American beef fat contains dioxin in the ppt range.

With further reductions of dioxin contamination levels in pesticides, use of dioxin-free pesticides, and stricter controls on the dumping of dioxin-containing wastes, the dioxin levels in food could be reduced.

Finally, the most widespread heavy metal contaminant is lead. The primary source of lead in diet is processed foods packed in tin cans seamed with leaded solder. Here is a prime target for prompt action to rid the food supply of a known toxin, since viable alternatives to leaded cans abound. (*Nutrition Action*, May, 1981).

Pollution of food is *not* the inevitable cost of using chemicals in agricultural and industrial technology, as some people might have us believe. But we must find means of producing food that is free of known health hazards. There are techniques to reduce the escape of toxic chemicals into our environment and their infiltration of the food supply.

It's important that consumers pressure for laws and programs that promote these safer alternatives. ■

---

This article has been adapted from the book, *The Invisible Additives: Environmental Contaminants in Our Food*, published by Doubleday and Co., Garden City, N.Y.

July, 1981 • Nutrition Action

# Hot Dogs

### CSPI Challenges Producers' Claims

It first appeared on the cover of *Eater's Digest* in 1972. And most recently it was prominently featured with a bottle of soda on the "Please, No Junk Food" sign.

More than any other food item, the hot dog has become something of a CSPI symbol for much that is wrong with the American food supply: it contains too much fat and salt, and is preserved with dangerous additives.

Not surprisingly, John L. Huston, the president of the National Live Stock and Meat Board, found our criticism "ill-advised." In a lengthy letter protesting the use of the hot dog on the junk food poster, Huston wrote:

"The truth is, hot dogs contain important nutrients. They have the same essential nutrients found in other red meats, specifically protein, vitamin B-12, thiamin, niacin, and zinc, among others...."

He continued, "Fat may not represent more than 30 percent of a hot dog *by law*, and is present in the hot dog for good reason (it is the main carrier of flavor and also provides a moisture and richness of taste that is enjoyed by most people)...."

"Sodium nitrite is the best product known to man which acts as a preservative to protect against botulism growth and also helps retain color. Which is worse—a trace of sodium nitrite or a chance of being exposed to botulism?"

And finally, Huston concluded, "The salt level in many prepared foods (including hot dogs) is higher than in non-prepared foods, simply because of the way they are processed. However, since salt is not added to the product during or after cooking, the total salt level is no higher than in many of the foods we eat. The main source of salt for most Americans seems to be the salt shaker, not 'hidden sources' like processed foods."

A few facts taken from USDA handbooks tend to undermine Huston's defense of the wiener:

• Fat: While a 45-gram beef and pork frank contains only 13 grams of fat by weight (Huston's 30 percent rule), calories derived from fat amount to fully 82 percent. The American Heart Association and other authorities recommend that people derive less than 30 to 35 percent of their total calories from fat.

• Sodium: One hot dog contributes 504 milligrams (mgs) of sodium—or about 5/8 of a teaspoon of salt—to the diet. The Food and Nutrition Board of the National Academy of Sciences now suggests that salt intake be limited to between 1,100 and 3,300 mgs of sodium a day.

• Protein and Carbohydrates: The typical hot dog contains 1.2 grams of carbohydrate, and 5.0 grams of protein—no more protein than the average hot dog bun. By contrast, the hot dog's popular cousin, the hamburger, offers between 15 and 23 grams of protein per serving, although it too is weighted down with large amounts of fat.

• Sodium Nitrite: Though the Food and Drug Administration (FDA) cleared nitrites on one count recently, the additive can still cause serious problems. It may combine with other chemicals in food or in the stomach to form deadly, cancer-causing nitrosamines.

Fortunately, some manufacturers are beginning to offer consumers healthier alternatives to the hot dog. Several companies are making frankfurters with one-third less fat, by weight. Some companies are producing nitrite-free sausages without resorting to even higher levels of salt. And still others are marketing chicken and turkey franks that contain about 20 percent fat by weight and are 75 percent lower in calories. (That's not to say that these poultry products are nutritional delights. They contain slightly higher levels of salt—about 620 mgs of sodium—and still derive about 70 percent of their calories from fat.)

Instead of promoting a flawed product, why doesn't the National Live Stock and Meat Board influence its friends—newly positioned in high places at the U.S. Department of Agriculture—to push for more healthful processed meats?

—*Greg Moyer and Bonnie Liebman*

---

### Hot Dogs Take A Licking On National TV Show

Our own Michael Jacobson debated the merits of hot dogs and junk foods with a nutritionist from the National Pork Producers Council, Elizabeth Whelan of the American Council on Science and Health, Craig Claiborne of *The New York Times*, and singers Captain and Toni Tennille on a segment of the *Mike Douglas Show* set to air this month. Ralph Nader was Douglas' co-host. The program is scheduled for broadcast July 10 in most large cities, and July 17 or 24 in smaller cities. (By the way, the hot dog lost 4 votes to 2).

# THE POLLUTION STEW

## Section Three

## THE POLLUTION STEW: Industrial Sewers Without Boundaries

Seventy thousand chemicals are in commercial production in the United States. Some of them are in your food. Others in your water. A few toxic chemicals such as polychlorinated biphenyls (PCB), polybrominated byphenyls (PBB), and mercury have been involved in highly publicized food chain poisonings. Freshwater fish are especially exposed to pesticides, the persistent PCB and other pollutants.

The Congressional Office of Technology Assessment (OTA) completed a study in 1979 of environmental contaminants "introduced into food as a result of human activities such as agriculture, mining, and industry." (Excluded from the study were naturally occurring toxins.) An excerpted chapter opens this section. Though the language is cautious, the unanswered questions are clearly ominous.

The "Circle of Poison" tells the story of how Americans can have their coffee, sugar, tea, bananas and other imported food laden with chemical pesticides that are banned in this country. How? These pesticides, which are not prohibited from being sold overseas, are sold abroad and come back into this country through the food we import. The General Accounting Office discovered that during a 15 month period half of all the imported food judged by the FDA to be pesticide-contaminated reached dinner tables without any consumer warning or penalty to those responsible.

Poisons in food also come by way of drinking water sources — lakes, rivers, underground wells that are used as industrial sewers. Two articles on contaminated drinking water and a section of a 1979 Report of the White House Council on Environmental Quality cover this major carrier of contamination. Check your community for this quiet and invisible epidemic. A recent book, *Water Fit to Drink,* by the Rodale Press, Emmaus, Pennsylvania 18047 ($5.95), will give you good, specific directions.

The personal dilemma that may be commonplace in a few years is foreshadowed in the article from *Nutrition Action* on "PCBs and Breast Milk." The colossal cruelty of this corporate pollution may yet provoke a major movement to curb this silent violence against millions of tiny infants and make the polluters compensate them for the risks or damage incurred. It is important to read this article in its entirety to capture the dilemma it assesses.

# Environmental Contamination of Food

> Maintaining an adequate, safe food supply has been a major goal of the Federal Government since 1906, when the first Federal food and drug law was signed into law. Historically, chemicals such as salt, sugar, and wood smoke have been used to preserve foods. Modern food technology relies extensively on the use of chemicals not only for preservation but also to produce appealing colors, flavors, aromas, and textures.
>
> Most developed countries now have food laws designed to permit the use of such chemicals in food under conditions judged to be safe. These chemicals are not considered adulterants or contaminants and are classed as intentional additives. Other chemicals may enter food as a result of their use in food production, handling, or processing. Such substances may be legally permitted if they are unavoidable under good manufacturing practices and if the amounts involved are considered safe. These chemicals are classed as incidental additives. The presence of both these classes of chemicals in food is controlled by regulation.

Environmental contaminants include substances from natural sources or from industry and agriculture. Many of the naturally occurring contaminants in food are of microbiological origin and consist of harmful bacteria, bacterial toxins, and fungal toxins. (Aflatoxin, a contaminant of peanuts and grains, is an example of a fungal toxin or mycotoxin.) The second category of environmental contaminants includes organic chemicals, metals and their complexes, and radionuclides. Only those environmental contaminants introduced into food as a result of human activities such as agriculture, mining, and industry are considered in this assessment.

The environmental contamination of food is a result of our modern, high-technology society. We produce and consume large volumes of a wide variety of substances, some of which are toxic. It is estimated that 70,000 chemicals may currently be in commercial production in the United States and that 50 of these chemicals are manufactured in quantities greater than 1.3 billion lbs per year. Seven percent of this country's gross national product (GNP), $113 billion per year, is generated by the manufacture and distribution of chemicals (1). During the production, use, and disposal of these substances, there are opportunities for losses into the environment. For example, the Environmental Protection Agency (EPA) estimates that there are more than 30,000 chemical and radioactive waste disposal sites. Of these, 1,200 to 2,000 are considered threats to human health (2).

Environmental contamination of food takes two forms: long-term, low-level contamination resulting from gradual diffusion of persistent chemicals through the environment, and relatively shorter term, higher level contamination stemming from industrial accidents and waste disposal.

An example of low-level contamination is polychlorinated biphenyls (PCBs). This group of substances was widely used in transformers and capacitors, as heat-transfer fluids, and as an additive in dyes, carbon paper, pesticides, and plastics (3). Although production was halted in 1977, PCBs remain an ubiquitous, low-level contaminant of many foods, especially freshwater fish.

An example of the second type of contamination is polybrominated biphenyls (PBBs) in

---

Reprinted from <u>Environmental Contaminants in Food</u>, a report by the Office of Technology Assessment, # OTA-F-103, December, 1979.

dairy products and meat. PBBs, a fire retardant, were accidentally mixed into animal feed. Dairy cattle that were fed the contaminated feed produced contaminated milk. The distinctions between the two types of food contamination are not exclusive. For example, PBBs have now become a long-term, low-level contaminant in Michigan because they are very stable and resistant to decay. Animals raised on farms affected by the original feed contamination are now contaminated by the PBB residues remaining in the pastures and farm buildings.

## HOW FOOD BECOMES CONTAMINATED

Chemicals contaminate foods through different routes depending on the chemical and its physical properties, its use, and the source or mechanism of contamination.

Organic substances that have contaminated food have been either industrial or agricultural chemicals. Pesticides are the only agricultural chemicals known to be environmental contaminants in food (see tables 1-3). A pesticide becomes an environmental contaminant when it is present in foods for which the application or use of the substance has not been approved. Livestock, poultry, and fish can be contaminated when application or manufacturing of pesticides occurs in the vicinity or when residues are transported through the environment. Improperly fumigated railroad cars, trucks, ships, or storage buildings used for transport or storage of human food and animal feed are also sources of environmental contamination. The interiors are sprayed or fumigated with pesticides, and if not sufficiently aired, contamination of the food or feed occurs.

The manufacture of organic chemicals produces sludges, gases, and liquid effluents of varying chemical complexities. The usual waste disposal methods (sewage systems, incineration, landfill) are unable to prevent organic residues from entering the environment in spite of Federal laws and corresponding regulations governing disposal. The routes include the atmosphere, soil, and surface or ground water.

Table 1.—Reported Incidents of Food Contamination, 1968-78, by State and Class of Contaminant

| State | Pesticide | Mercury | PCB | PBB | Other | Total |
|---|---|---|---|---|---|---|
| New York | 1 | 1 | 1 | — | — | 3 |
| Idaho | 1 | 2 | — | — | — | 3 |
| South Carolina | — | — | — | — | 1-biphenyl | 1 |
| Minnesota | — | 1 | — | — | — | 1 |
| Louisiana | 8 | 5 | 4 | — | — | 17 |
| Colorado | 1 | 1 | — | — | — | 2 |
| Georgia | 15 | — | — | — | — | 15 |
| Maryland | — | 1 | — | — | 2-petroleum | 3 |
| Texas | — | 1 | — | — | — | 1 |
| New Jersey | 2 | 1 | — | — | 1-β-methoxy napthalene and tetraline | 4 |
| Kansas | — | (1)* | — | — | — | (1)* |
| Missouri | 4 | — | — | — | — | 4 |
| New Mexico | — | 1 | — | — | — | 1 |
| Alaska | (1)* | — | — | — | — | (1)* |
| California | 1 | 3 | — | — | — | 4 |
| Indiana | 3 well-documented 5 other incidents | — | 1 | — | — | 9 |
| Michigan | 13 | 1 | 2 | 1 | — | 17 |
| Virginia | 1 | — | — | — | — | 1 |
| Totals | 56 | 19 | 8 | 1 | 4 | 88 |

*Several conservatively estimated as one.
SOURCE: Office of Technology Assessment.

Table 2.—Reported Incidents of Food Contamination, 1968-78, by State and Food

| State | Dairy | Eggs | Fruit/vegetable | Fish/shellfish | Grain | Game/meat/poultry | Incidents |
|---|---|---|---|---|---|---|---|
| New York | — | — | — | 3 | — | — | 3 |
| Idaho | 1 | — | — | 1 | — | 1 | 3 |
| South Carolina | — | — | — | — | 1 | — | 1 |
| Minnesota | — | — | — | 1 | — | — | 1 |
| Louisiana | 5 | 1 | 1 | 9 | 1 | — | 17 |
| Colorado | 1[a] | — | 1[a] | — | — | — | 2[a] |
| Georgia | — | — | 15 | — | — | — | 15 |
| Maryland | — | — | — | 3 | — | — | 3 |
| Texas | — | — | — | 1 | — | — | 1 |
| New Jersey | — | — | 2 | 2 | — | — | 4 |
| Kansas | — | — | — | — | 1 | — | 1 |
| Missouri | 4 | — | — | — | — | — | 4 |
| New Mexico | — | — | — | — | — | 1 | 1 |
| California | — | — | 1 | 2 | — | 1 | 4 |
| Indiana[b] | 1 | 2 | — | — | — | 1 | 9 |
| Michigan | 7 | — | 2 | 4 | — | 4 | 17 |
| Alaska[c] | — | — | — | — | — | — | 1 |
| Virginia | — | — | — | 1 | — | — | 1 |
| Totals | 19 | 3 | 22 | 27 | 3 | 8 | 88 |

[a]Dieldrin contamination of milk and lettuce, reported as one incident.
[b]Five additional incidents, but not well documented.
[c]Several pesticide incidents, but not well documented, therefore conservatively estimated as one.
SOURCE: Office of Technology Assessment.

Table 3.—Number of Incidents of Environmental Contaminants of Food Reported by Federal Agencies, 1968-78

| Agency | Pesticides | Mercury | PCB's | Other | Food affected |
|---|---|---|---|---|---|
| USDA/FSQS | 39 | — | — | — | Chickens, turkeys, ducks, cattle, swine, lambs |
|  | — | 1 | — | — | Swine |
|  | — | — | 6 | — | Poultry |
|  | — | — | — | 1 (Phenol) | Cattle |
| FDA | 21 | — | — | — | Fish, cheese, pasta |
|  | — | 84 | — | — | Fish |
|  | — | — | 3 | — | Fish, eggs, bakery products |
| Total | 60 | 85 | 9 | 1 | — |

SOURCE: Office of Technology Assessment.

Metals can be released into the environment in several ways. The mining and refining processes produce dust and gases which enter the atmosphere. Metallic salts formed during recovery and refining processes can escape as waste products into surface and ground water. Sewage sludge used as fertilizer on agricultural land also poses a potential food contamination problem. Trace metals present in the sludge can be taken up by crops grown on treated soil. Cadmium is the trace metal in sludge that currently generates the greatest concern.

Radioactivity in food stems from three sources: natural radioactivity, releases from operation of nuclear reactors and processing plants, and fallout from nuclear weapons tests. The primary route by which food becomes contaminated is the deposition of airborne material on vegetation or soil. The subsequent fate of the radionuclide is determined by its chemical and physical nature and whether it is absorbed and metabolized by plants or animals. Natural radioactivity may become a concern when ores containing radioactive substances are mined and processed. The products or wastes may concentrate the radionuclides. Examples of this are uranium tailings, phosphate rock waste, or slags from phosphorus production. Radium may enter the food chain when it dissolves in ground water and is taken up through plant roots.

Nuclear reactors normally release radioactive noble gases that do not contaminate foods. Reactors do contain large inventories of fission products, transuranics, and other activation products. Accidental releases can contaminate vegetation by deposition of particles on leaves and soil, or through water. Gaseous releases would most likely involve the volatile elements such as iodine and tritium, or those with volatile precursors, such as strontium-90 and cesium-137. Aqueous releases would follow failure of the onsite ion exchange cleanup system. Any of the water-soluble elements could be involved. Table 4 summarizes the radionuclide contaminants of significance for foods.

Nuclear waste-processing plants could also have either gaseous or aqueous releases. In this case, the fission products are aged before processing, and iodine and the gaseous precursor radionuclides are not released. Tritium and carbon-14 are the major airborne products, while the waterborne radionuclides are the same as for reactors.

Atmospheric nuclear weapons tests distribute their fission products globally. Local deposition depends on the size of the weapon and the conditions of firing (high altitude, surface, or underground).

### Table 4.—Radionuclide Contaminants of Significance for Foods

| Element or nuclide | Source | Emission | Notes |
|---|---|---|---|
| Uranium | Natural | $\alpha$ | Normally present in small amounts. Significant only when enhanced |
| Thorium | Natural | $\alpha$ | Normally present in small amounts |
| Radium-226 | Natural | $\alpha$ | Member of uranium series. Normally present, metabolized somewhat like calcium |
| Radium-224 | Natural | $\alpha$ | As radium-226, only a member of thorium series |
| Polonium-210 | | $\alpha$ | Members of uranium series |
| Lead-210 | Natural | $\beta$ | Normally present |
| Plutonium | Transuranic activation product | $\alpha(^{238}Pu, ^{239}Pu, ^{240}Pu)$ $\beta(^{241}Pu)$ | Product of nuclear reactions |
| Americium | Transuranic activation product | $\alpha(^{241}Am)$ | Product of $^{241}Pu$ decay |
| Tritium | Natural, also from nuclear reactions | $\beta$ | Low energy, usually in form of water or organic compounds |
| Carbon-14 | Natural, also from nuclear reactions | $\beta$ | Low energy, usually in form of organic compounds |
| Iodine-131 | Fission products | $\beta, \gamma$ | Short half-life, so important only for fresh foods, e.g., milk and leafy vegetables |
| Strontium-89 and -90 | Fission products | $\beta$ | Follows calcium somewhat in metabolism |
| Cesium-134 | Reactor product | $\beta, \gamma$ | Follows potassium somewhat in metabolism |
| Cesium-137 | Fission product | | |
| Mixed fission products | Fission products | $\beta, \gamma$ | Most important are isotopes of zirconium, cerium, barium, rubidium, rhodium. Mostly surface contaminants |
| Iron-59, chromium-51, zinc-65, cobalt-60, cobalt-58, manganese-54, sodium-22, phosphorus-32 | Activation products | $\beta, \gamma$ | Follow stable elements in metabolism |

SOURCE: N. H. Harley, "Analysis of Foods for Radioactivity." OTA Working Paper, 1979.

# MAGNITUDE OF THE PROBLEM

There is little information available on the number of food contamination incidents, the amount and costs of food lost through regulatory actions, or the effects of consumption of contaminated food on health. To obtain information on the extent of the problem, OTA reviewed the literature and sought information from the States and Federal agencies.

## Evidence of Human Illness Resulting From Consumption of Contaminated Food

In evaluating the significance of environmental contaminants in food the key question is whether consumption of contaminated foods poses a health risk. Measurable health effects depend on the toxicity of the substance, the level at which it is present in food, the quantity of food consumed, and the vulnerability of the individual or population. In Japan, foods contaminated with substances such as PCBs, mercury, and cadmium have produced human illness and death. No such mass poisonings have occurred in the United States. However, in cases such as PBBs where a large populace has been exposed, some physiological changes have been noted. But no conclusions can as yet be drawn on the ultimate health effects.

It is known from limited surveys that the U.S. population is exposed to a wide variety of chemical contaminants through food, air, and water. The long-term health effects and the implications of possible interactions among these residues are unknown. A recent literature review of over 600 published studies (4) found that nonoccupationally exposed U.S. residents carry measurable residues of 94 chemical contaminants. Twenty-six of these are organic substances, including twenty pesticides and pesticide metabolites. The remainder are inorganic substances.

Americans also have been exposed to low levels of PCBs, PBBs, mercury, and ionizing radiation through their food. The following sections briefly summarize current knowledge and the extent of uncertainties on the health effects of these environmental contaminants.

## Polychlorinated Biphenyls

PCBs occur in food as the result of environmental contamination leading to accumulation in the food chain, direct contact with food or animal feeds, or contact with food-packaging materials made from recycled paper containing PCBs (5). Several comprehensive literature reviews have been published in the last 5 years detailing the acute and chronic toxic effects of PCBs in animals and humans (5-11).

Human illness has been caused by exposures to PCBs at much higher levels than those that occur in the United States. In the early part of 1968 the accidental contamination of edible rice-bran oil led to a poisoning epidemic among the Japanese families who consumed the oil. The disease became known as Yusho or rice-oil disease. Its chief symptoms were chloracne (a severe form of acne) and eye discharge; other symptoms included skin discoloration, headaches, fatigue, abdominal pain, menstrual changes, and liver disturbances. Babies born to mothers who consumed the rice oil were abnormally small and had temporary skin discoloration. The first symptoms of Yusho disease were registered on June 7, 1968, and 1,291 cases had been reported as of May 1975 (9).

Since the rice oil was also contaminated with polychlorinated dibenzofuran (PCDF), it is difficult to determine from the Yusho data exactly what effect(s) exposure to PCBs alone could have on humans. It has been calculated that the PCDF made the rice oil 2 to 3.5 times more toxic than would have been expected from its PCB content alone (11). Careful records of the 1,291 Yusho patients have been kept to determine possible long-term effects. At least 9 of 29 deaths that occurred as of May 1975 were attributed to cancer (malignant neoplasms), but a causal relationship between PCBs and cancer cannot necessarily be inferred because of the high concentration of PCDF in the oil. The Yusho study, nevertheless, had two important results: first, the information established that PCBs can be transferred from mother to fetus and from mother to child through breast feeding, and second highly chlorinated PCB compounds are excreted more slowly from the body than less chlorinated ones (9).

More recent experiments in animals have demonstrated a variety of toxic effects. Cancers have been produced in mice and rats fed PCBs (6, 12). Monkeys fed levels of PCBs equivalent to the amounts consumed by Yusho patients developed similar reproductive disorders (13-16). Young monkeys nursing on mothers consuming feed containing PCB developed toxic effects and behavioral abnormalities (15-17).

**Polybrominated Biphenyls**

Practically every Michigan resident has been exposed to PBB-contaminated food products. It is estimated that some 2,000 farm families who consumed products from their own PBB-contaminated farms have received the heaviest exposure (18).

Fries (19) studied the kinetics of PBB absorption in dairy cattle and its elimination in milk. If intake of contaminated milk alone is considered, those Michigan residents most severely exposed consumed from 5 to 15 grams of PBB over the initial 230 days of the exposure. Those residents that coincidentally consumed contaminated meat and/or eggs may have received higher total doses of PBB, but the number of such cases is probably small.

Geographically the residents of the lower peninsula, where the original accident occurred, were found to have the greatest levels of exposure. In 1976, the Michigan Department of Public Health conducted a study on PBB concentrations in breast milk. It was found that 96 percent of the 53 women selected from the lower peninsula and 43 percent of the 42 women selected from the upper peninsula excreted PBB in their breast milk (20).

Low concentrations of PBBs also have been detected in animal feed in Indiana and Illinois. Unconfirmed surveys of food throughout the country found extremely low levels below the Food and Drug Administration (FDA) action level in the following States (21):

| State | Food |
|---|---|
| Alabama | Chicken |
| Indiana | Turkey |
| Iowa | Beef |
| Mississippi | Chicken |
| New York | Chicken |
| Texas | Chicken |
| Wisconsin | Duck |

Wolff, et al. (22) reported that serum PBB was higher for males than females. It was suggested that the greater proportional body fat in women may account for this difference, but exposure may also be important. Males may consume more contaminated food or have more direct contact with PBB than females.

The same study found no consistent trends with respect to age. It was observed, however, that young males had greater concentrations of serum PBB than young females. Young females had greater concentrations than older males, and older males had greater concentrations than older females. It was also found that very young children and individuals who had lived on farms less than 1 year had lower serum PBB levels than other groups (22).

Serum PBB concentration is related to the intensity of exposure. Most studies indicate that consumers and residents of nonquarantined farms had significantly lower PBB levels than residents of quarantined farms; however, families on quarantined farms stopped consuming meat and milk from their own animals (20).

In late 1974, the Michigan Department of Public Health conducted a survey to determine if any adverse effects could be correlated with PBB levels in the body. A sample of 165 exposed persons (quarantined farms) and 133 nonexposed (nonquarantined farms) was studied. Medical history interviews and physical examinations were performed on each subject and blood specimens were taken. Blood PBB levels as high as 2.26 parts per million (ppm) were found in the exposed individuals; about half exhibited levels greater than 0.02 ppm. Of the nonexposed individuals, only two showed blood PBB levels greater than 0.02 ppm; 70 percent of the adults and 97 percent of the children exhibited levels of 0.0002 to 0.019 ppm. Comparison of a list of selected conditions and complaints revealed no significant differences in the frequency of illness between the two groups. Physical examinations and clinical laboratory tests disclosed no effects attributable to "chronic" PBB exposure (24).

The effect of PBB exposure on white blood cell (lymphocyte) function of Michigan dairy farmers who consumed contaminated farm products was examined by Bekesi, et al. (25). Forty-five members of Michigan farm families who had eaten PBB-contaminated food for periods of 3 months up to 4 years after the original accident were compared for immunological function to 46 Wisconsin farmers and 79 New York residents. All of the exposed individuals showed reduced lymphocyte function, and 40 percent showed abnormal production of lymphocytes. There were also significant increases in lymphocytes with no detectable surface markers ("null" cells). However, the short- and long-term health implications of these differences are not now known.

Lillis (20) examined Michigan farmers and consumers of dairy products and found that the effect of PBB on humans was mainly neurological in nature. He found marked fatigue, hypersomnia, and decreased capacity for physical or mental work. Other symptoms included headache; dizziness; irritability; and musculoskeletal, arthritis-like complaints—swelling of the joints with deformity, pain, and limitation of movement. Less severe gastrointestinal and dermatological complaints were also encountered.

## Mercury and Methylmercury

Foods are the major source of human exposure to mercury. The mercury concentration in food is dependent on the type of food, the environmental level of mercury in the area where the food is produced, and the use of mercury-containing compounds in the agricultural and industrial production of the food. All living organisms have the ability to concentrate mercury. Therefore, all animal and vegetable tissues contain at least trace amounts (26). Several recent reviews have examined the health effects associated with consumption of mercury (26-28). The results of these reviews indicate that the effects of methylmercury poisoning become detectable in the most sensitive adults at blood levels of mercury of 20 to 50 $\mu$g/100 ml, hair levels from 50 to 120 mg/kg, and body burdens between 0.5 and 0.8 mg/kg body weight (26).

Since the Minamata Bay tragedy in Japan, the effects of chronic exposure to methylmercury have been well-documented. Mercury readily accumulates within the central nervous system (29-31), and clearance of mercury back into the bloodstream is slow (32). Consequently, the central nervous system is considered to be the critical target in chronic mercury exposure. The clinical symptoms of central nervous system involvement include headache, vertigo, vasomotor disturbance, ataxia, and pain and numbness in the extremities (30). The most prominent structural changes of the central nervous system resulting from chronic mercury exposure are diffuse cellular degeneration (30).

In evaluating the teratogenic hazards of mercury exposure to man, the placental transfer of mercury is particularly significant. Levels that are not toxic to pregnant women are sufficient to produce birth defects in their offspring (33-35). Transfer of methylmercury across the human placenta results in slightly higher blood levels in the infant at birth than in the mother (36). Table 5 compares fetal and maternal blood concentra-

**Table 5.—Methylmercury Concentrations in Normally Exposed Populations**

| Location | Concentration (μg Hg/g) | | |
|---|---|---|---|
| | Maternal blood | Placenta | Fetal blood |
| Japan | 0.017 | 0.072 | 0.020 |
| Sweden | 0.006 | — | 0.008 |
| Tennessee | 0.009 | 0.021 | 0.011 |
| Iowa | 0.001 | 0.002 | 0.001 |

SOURCE: Adapted from B. J. Koos and L. D. Longo, "Mercury Toxicity in the Pregnant Woman, Fetus, and Newborn Infant," *American Journal of Obstetrics and Gynecology* 126(3):390, 1976.

tions in normally exposed populations in Japan, Sweden, and the United States.

In humans, the most widely reported fetal risk associated with maternal exposure to mercury is brain damage. The placental transfer of mercury and its effects on the human fetus were first recognized in the 1950's with the well-known outbreak of congenital Minamata disease in the towns of Minamata and Niigata, Japan. By 1959, 23 infants suffering from mental retardation and motor disturbances had been born to mothers exposed to methylmercury during their pregnancies. The clinical symptoms of the infants resembled those of severe cerebral palsy or cerebral dysfunction syndrome. They included disturbance of coordination, speech, and hearing; constriction of visual field; impairment of chewing and swallowing; enhanced tendon reflex; pathological reflexes; involuntary movement; primitive reflexes; superficial sensation; salivation; and forced laughing (30). Only 1 of the 23 mothers exhibited any symptoms of mercury poisoning (32).

### Radioactivity

Ionizing radiation (X-rays, gamma rays, or beta particles with sufficient energy to strip electrons from molecules and produce ions) can produce birth defects, mutations, and cancers (37). These adverse health effects are usually associated with high dose levels delivered at high dose rates.

Such a combination is not ordinarily encountered in food. Previous radioactive contamination of foods has involved relatively small quantities of radioactive elements which have delivered low dose rates (38).

In these situations, the effects of the radiation exposure on health are extremely difficult to evaluate. High dose rates (100 million to 1 billion times background) are estimated to produce 2,600 ionization events per second in cells. Background radiation levels are estimated to produce less than one ionization in the cell nucleus per day (37). Because cells have the capacity to repair damage to their genetic material, repair of ionization damage may occur at low radiation exposure. Higher exposures may overwhelm the cells' repair capacity. Whether any effects are observed in such cases depends on several factors. These include the dose delivered to the tissues, the nature of the emissions, and the metabolism of the cell. The following examples illustrate these points:

- **Strontium-90** in food arouses most concern not only because of its long half-life but also because it behaves in the body in a manner somewhat similar to calcium. The replacement of bone calcium with strontium-90 exposes tissues and cells covering the bone to radiation. In addition, bone marrow is subject to the ionizing radiation from the strontium-90. Thus, cancer of the bone-forming and bone-covering tissue as well as leukemias of the bone marrow blood-forming cells can possibly result.
- **Iodine** is concentrated by the thyroid gland. Radioiodines produced in atmospheric nuclear detonations or released from nuclear power stations are also taken up and concentrated by the thyroid, increasing the risk of thyroid cancer.
- **Tritium**, or radioactive hydrogen, combines chemically with oxygen to form water. Tritium derived from food would be widely distributed throughout the body exposing all tissues to radiation.

The uncertainties surrounding the repair capacities of cells and the irreversible nature of the possible health effects have led to the adoption in the United States of a prudent policy toward low-level ionizing radiation. Since any amount of radiation is potentially harmful, unnecessary exposure should be avoided.

## Number of Food Contamination Incidents

Questionnaires were mailed to the commissioners of health in each of the 50 States and the District of Columbia as well as to Federal agencies. For the 10-year period 1968-78, each was asked to report on the number of incidents of environmental contamination of food that resulted in regulatory action. This survey has limitations. Some States did not answer all questions. The questions were subject to interpretation and misunderstanding. The accuracy and completeness of the answers were dependent on the respondent. The results presented are therefore preliminary and do not necessarily represent complete and comprehensive information on all States responding. Nonetheless, these data are the first to be developed on the extent of environmental contamination of food.

Responses were received from 32 States. Seven of the top ten agricultural States and six of the top ten manufacturing States responded to the questionnaire. The agricultural States in the top 10 were California, Texas, Minnesota, Nebraska, Kansas, Indiana, and Missouri. The manufacturing States in the top 10 were California, New York, Michigan, New Jersey, Texas, and Indiana. Three of these States—California, Texas, and Indiana—are in the top ten for both agricultural and manufacturing production. A fairly representative distribution of States responded from each region of the United States (figure 2).

In the following discussions, an incident is defined as a case in which a Federal or State agency has taken regulatory action against contaminated food. The Michigan PBB episode is reported as one incident because the contamination stemmed from one source and was limited to one State. Mercury contamination is reported as separate incidents because the sources differed (environmental mercury v. industrial waste), the States involved are widely separated, and regulatory actions were taken at different times. Eighteen States reported at least one environmental contaminant incident since 1968 for a total of 88 incidents. All food categories were involved and a variety of substances were implicated (see tables 1-3).

The data provided by States are complemented by the Federal responses. The two Federal agencies responsible for regulating the Nation's food supply reported the number of environmental contamination incidents that they had identified since 1968. FDA had 108 reported incidents, and the Food Safety and Quality Service of the U.S. Department of Agriculture (USDA) had 47 reported incidents (see table 3). The combined Federal and State total number of incidents comes to 243.

Neither State nor Federal responses indicated any significant radionuclide contamination episodes during the 1968-78 period. Extensive Government programs for monitoring radionuclides in food exist. Thus far, radionuclide contamination of food has not been found to exceed the exposure limits recommended in the Federal Radiation Council Protective Action Guides. In most cases, the amount of food contamination in the continental United States has never even approached these limits (39). While atmospheric nuclear testing is less a threat today than before the signing of the 1963 Test Ban Treaty, radionuclide contamination of food is still a concern of both Federal and State governments.

The number of food contamination incidents reported to OTA does not represent the total number that has occurred in the United States, only those in which the Federal Government and 18 State governments have taken regulatory action. Many incidents never come to the attention of State or Federal authorities. This is because local government officials can and do handle environmental contaminant incidents by warning offenders or by condemning contaminated products without informing the appropriate State officials. Also, the farmer whose livestock or poultry has been environmentally contaminated may negotiate directly with the firm responsible for the contamination for financial

Figure 2.—Distribution of States Responding to Questionnaire

The United States would look like this if the sizes of States were proportional to their value of farm production. The shaded States responded to the questionnaire. The States are ranked by 1974 cash receipts.

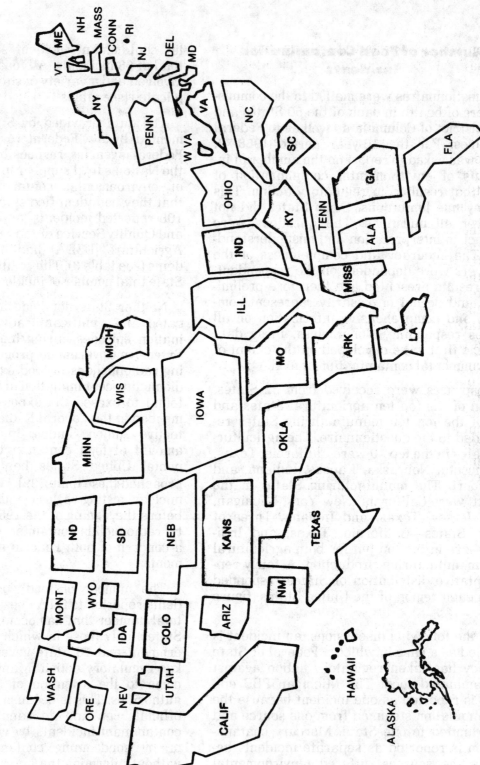

SOURCE: Adapted from the *The American Farmer*, USDA/ERS, 1976, GPO Stock No. 001-019-00325-1.

reimbursement without reporting the contamination to Federal or State officials (40).

### Economic Impact

The economic impact of an incident involving the environmental contamination of food includes the cost of condemned food, health costs, and the corresponding distributional effects and costs. The magnitude of the economic impact is determined by:

- the amount of food contaminated,
- the concentration of the contaminant in food,
- the chemical and toxicological characteristics of the contaminant, and
- the corresponding regulatory action taken on the contaminated food.

The initial regulatory action taken by Federal and State authorities may be the issuance of a warning or the establishment of either an action level or a tolerance. A more detailed discussion of this regulatory action is presented in chapter III. Action levels and tolerances establish a permissible level for the contaminant in food. Any food found to contain concentrations of the substances above this level is condemned and either destroyed or restricted from being marketed.

### Costs of Food Condemned

In addition to the four factors listed above, the cost of condemned food is also affected by its position in the food production and marketing process at the time of condemnation. An action level or tolerance for a contaminant is the most important of the five factors. If no action level or tolerance is set, no food would be condemned and thus there would be no costs incurred. The impact of such a regulation will depend on the exact level of a substance that is allowed to be present in food.

The chemical properties of a contaminant are also important because of the potential for long-term effects on the amount of food affected. Since many contaminants biologically and chemically degrade slowly, their presence in the environment can mean food contamination above the action level or tolerance for many years after the source of the pollution has been stopped. The James River in Virginia, for example, is still closed to commercial fishing several years after kepone discharges into the river have been eliminated. The relative influence for each of these factors on the final cost will vary in each contamination incident.

Estimates of the cost of food condemned through regulatory action are most often expressed in dollars. Consequently, this cost is usually (and incorrectly) cited as representative of the total economic impact. Such costs were collected in OTA's State and Federal surveys. The data, however, only partially reflect the total economic impact for environmental contamination of food in the United States. This is because the cost of condemned food is only one component of the total economic impact of an incident. In addition, few of the incidents reported to OTA included data on the cost of food condemned. OTA estimates from the available data that the total cost of condemned food as a result of environmental contamination in the United States since 1968 is over $282 million (table 6). The only cost estimates used were those clearly stated for an incident by the reporting States or Federal agencies.

**State Estimates.**—Of the 18 States reporting contamination incidents, only 6 provided data on the economic impact in dollar terms. Of those six, Michigan represents 99 percent of the total cost ($255 million) while reporting only 19 percent of the number of incidents in the 18 States. Indeed, Michigan accounts for 90 percent of the total costs reported in the United States while reporting only 7 percent of the incidents that occurred during the 1968-78 period. It must be recognized, however, that 84 percent of Michigan's costs are attributed to the PBB incident. Many incidents reported by State and Federal agencies are considerably smaller than the PBB episode. Thus, the PBB episode is an indication of how severe a contamination incident can be.

Table 6.—Economic Impact of Food Contamination

| | | Reported incidents | Total estimated cost ($) |
|---|---|---|---|
| **State** | | | |
| Idaho | Dieldrin | | $ 100,000 |
| | PCP | | 3,000 |
| Colorado | Dieldrin | | 100 |
| | Mercury | | 3,700 |
| Maryland | Mercury | | 23,000 |
| Texas | Mercury | | 85,000 |
| Indiana | Dieldrin | | 25,027 |
| | Dieldrin | | 250,000 |
| Michigan | Mercury | | 10,000,000 |
| | PCB | | 30,000,000 |
| | PCNB | | 100,000 |
| | PBB | | 215,000,000 |
| | Picloram | | 12,000 |
| | Chlordane | | 2,500 |
| | DDT | | 2,000 |
| | Toxaphene | | 2,000 |
| | Parathion | | 328 |
| | Diazinon | | 13,700 |
| | Pentachlorophenol | | 28,468 |
| | PCB | | 150,000 |
| | Dieldrin | | 12,500 |
| | | | $255,813,323 |
| **Federal** | | | |
| USDA/FSQS | Pesticides | | 18,900,000 |
| | Mercury | | 63,000 |
| | PCB | | 7,450,000 |
| | Phenol | | 350 |
| | | | 26,413,350 |
| Total United States | | | $282,226,673 |

SOURCE: Office of Technology Assessment.

Some States reported the amount of food destroyed without estimating the cost. Kentucky, for example, reported the destruction of 400,000 lbs of milk since 1968 because of pesticide contamination. While such information can be converted into dollars, data on market position and price of product at time of confiscation are not readily available. Many States were unable to provide any estimates on either the cost or the amount of food condemned as a result of reported contamination incidents. New York (with PCBs) and Virginia (with kepone) are two States that could not provide cost estimates for food condemned as a result of environmental contamination. Virginia, however, has initiated a study to determine the economic impact of the kepone incident.

**Federal Estimates.**—Of the two Federal agencies reporting information to OTA on environmental contaminant incidents, USDA's Food Safety and Quality Service (FSQS) reported food condemnation cost estimates. These estimates, however, only cover livestock and poultry—the food products over which FSQS has regulatory authority. FDA, which has regulatory authority over the remaining food commodities, did not estimate costs for reported environmental contamination incidents (70 percent of the Federal total). Thus, a significant proportion of the total costs for environmental contamination incidents requiring Federal action is unknown. Comparison of the two agency responses with the State responses reveals little duplication in the reporting of incidents.

FSQS cost estimates were determined by the number of animals or pounds destroyed multiplied by the market value at the time of confiscation. Since most of these animals were taken at the farm or wholesale level, the market value was the farm or wholesale price. Most of the losses resulting from FDA actions would be based on a wholesale or retail price because the seized products had advanced further in the marketing system. Therefore, their estimated costs would be greater than if they were seized at the production level (generally the case with FSQS seizures).

Summing up, the available data on the cost of condemned food is limited; consequently OTA's $282 million condemned-food estimate is likely to be a gross underestimation of the actual costs. The true cost would be impossible to estimate from this limited sample.

### Health Costs

Health costs are also an important component of economic impact. These costs are incurred by the consumer whose health has or potentially can be affected adversely by a contaminant present in food. These adverse effects can cause illness and death, and the range of effects will vary depending on the toxicity of the contaminant, the concentration of the contaminant in food, and the amount of food consumed.

In this country, the concentration of contaminants has been at levels that have not produced immediate measurable and conclusive effects in exposed populations. Estimates are therefore made for the potential long-term effects on exposed populations from various contaminants in food.

Health costs can be estimated from such projected health effects. Costs would include health care costs for treating illness and burial expenses associated with death. Additional costs would include estimated value of productive days or years lost from work due to the projected illness or death associated with the contaminant in food. All of these health-related costs, however, do not and cannot include the emotional and psychological impacts on those afflicted and their friends and families.

Health costs are not available for previous U.S. food contamination incidents. Approaches and techniques for estimating health costs are discussed in chapter VI.

### Distributional Effects and Costs

Distributional effects and costs involve the various people, groups, and organizations who are economically affected by an environmental contamination incident. Information on the extent and distribution of such effects and costs provides a clearer picture of the total economic impact on society. This information is usually couched in descriptive terms. Those who are economically affected are identified but the extent of the impact is seldom estimated in dollars. The exact distribution of costs from an incident through society is affected by the same five factors that influence the cost of condemned food.

Many of the distributional effects and costs for various types of environmental contaminant incidents are discussed in the following sections. The purpose of this discussion is not to identify all the distributional costs but rather to demonstrate the variety of effects and costs that can result from an incident.

**Producers.**—Food producers are affected economically in different ways by contamination episodes. But all are affected directly when the food they produce is condemned. For example, food found contaminated at the farm level is confiscated and destroyed. This was the case for over 500 Michigan farmers whose dairy herds were partially or entirely destroyed (41). In such cases, farmers either replace their livestock, plant a new crop, or go out of business.

Farmers can be faced with severe economic hardship, since they are not always reimbursed financially for the animals or commodities confiscated. While insurance programs such as the Federal Crop Insurance Corporation are available to cover natural hazards which might destroy crops or livestock, such Federal assistance is not available to farmers for losses from environmental contamination. An injured farmer can obtain a loan at commercial rates or sue the responsible firm for compensation. But the loan and the interest add to a farmer's financial difficulties, and suing for compensation can take time that the farmer may not have.

The commercial fisher is faced with a different situation. If a river, lake, or species of fish is restricted because of environmental contamination, the fisher whose source of income depends on this species or waterway may have few employment alternatives. The alternatives depend to some degree on the extent of the contamination. If the only waterway available in a section of a State or a whole State is closed to commercial fishing because of the contamination, the fisher's source of employment is eliminated until the restriction is ended. Since the restriction can last for years (depending on the chemical stability of the contaminant), the fisher either will have to move to other commercial fishing areas or seek other employment.

Food producers economically affected by the condemnation of contaminated food are likely to incur health costs. This is because many of the producers and their families regularly eat the food that they produce or harvest. Consequently they are exposed to the contaminated food at greater concentrations than the average consumer. This was the case for several farm families in Michigan.

**Firms Held Accountable for Environmental Contamination.**—In most instances blame for a contamination incident can be established. Those accountable are subject to fines and lawsuits. Firms admitting responsibility often try to settle with producers out of court if possible. Most of the compensation is for the economic damages stemming from the destruction of food or loss of employment. Compensation for people whose health has been impaired as a result of eating contaminated food would be sought through civil litigation. Such litigation, however, is rare in this country, since the level of contamination in food is so low that demonstrating the necessary cause and effect is difficult.

Fines or compensation paid by the firms held accountable for the contamination are, in fact, poor indicators of the true costs incurred by the producers. This is because the settlement costs which are frequently negotiated or imposed bear little relationship to the actual costs incurred.

For example, compensation has been provided by Michigan Chemical Corporation and Farm Bureau Services, Inc., to many of the farmers whose livestock and poultry were destroyed following PBB contamination. Michigan Chemical and Farm Bureau Services have together paid more than $40 million in compensation from a jointly established insurance pool (42). In another case involving PCB-contaminated fish meal sold to poultry producers, Ralston Purina Company negotiated compensation for the 400,000 chickens destroyed. The cost of the compensation has not been disclosed (43).

**Governments.**—Federal, State, and local governments also incur costs from an environmental contamination incident. Although the Federal Government and most State governments have agencies with programs to regulate or control food safety problems, these programs usually are not funded to handle the kind of long-term problems created by a PBB or kepone incident. The Michigan Department of Agriculture, for example, estimates it will spend $40 million to $60 million within the next 5 years to monitor and test for PBBs in animals and animal byproducts (44). This is money that could have been saved or spent for other programs if PBB contamination had not occurred. In order to recover its expenses from the PBB incident, the State of Michigan filed a lawsuit against both the Michigan Chemical Corporation and Farm Bureau Services, Inc., claiming more than $100 million in damage (45).

Federal involvement is limited unless the contaminated food is part of interstate commerce. Many of these incidents are not considered by the Federal Government to involve interstate commerce. FDA may provide technical assistance at the request of the State government when a contamination incident is regarded to be a local problem (43). These technical facilities and experts are available to all States through the Federal and regional offices. Additional expenditures by the Federal Government for contamination incidents are limited. Additional State expenditures, however, can be substantial. Federal expenditures are made when Federal regulations are developed and promulgated for particular contaminants in food such as PCB.

**Consumers.**—Consumers can incur costs from an environmental contaminant incident in several ways. The removal of food from commerce could increase prices for that food product or other food products being sold. Thus, the consumer could pay more for food as a result of an environmental contamination incident. In order for this price increase in food to occur, however, a significant amount of a food product or food products would have to be taken off the market. Such prices of food might vary by State or region and affect certain socioeconomic classes differently.

Health costs could increase as a result of the consumption of contaminated food. This would not affect all consumers but rather those who received the most exposure and/or those most susceptible to a contaminant, such as children or senior citizens. While these costs would already be included in estimated total health costs, the distributional effects could indicate those consumers most likely to be affected.

**Indirect Costs.**—Most of the costs mentioned directly stem from an environmental contamination incident. However, indirect or secondary costs can and do occur. For example, a bait and tackle store on a lake that is closed to commercial and sport fishing because of an environmental contamination is likely to suffer economic hardship. Food processors whose normal supply of food has been condemned because of environmental contamination will also suffer economically unless they find new sources of supply. These are just two examples of the many indirect costs which might occur.

## POTENTIAL FOOD CONTAMINATION PROBLEMS

Because a limited number of substances posing health problems already have been identified in food, concern exists that other toxic substances are likely to contaminate food in the future. This concern arises from the number of substances presently being manufactured, used, and disposed of in the United States, and the difficulties in preventing them from entering the environment. New substances developed to meet new needs or to replace known toxic substances may create unexpected environmental problems if not properly controlled. Byproducts of new technologies such as synthetic fuels are also potential environmental contaminants. These are described in appendix A.

There are two methods of objectively assessing possible future contaminants: 1) by sampling the food supply for chemical contaminants and ranking them according to potential hazard and 2) by surveying the universe of industrial chemicals and ranking them according to their potential for entering the food supply in toxic amounts. These methods are discussed in more detail in chapter VII, "Monitoring Strategies."

Of the three categories of environmental contaminants considered in this report, organic chemicals probably pose the greatest potential environmental and food contamination problems. This conclusion is based on the number, volume, and toxicity of the organics manufactured and used in this country (40). Both trace metals and radioactive substances continue to warrant concern, but not as great a concern as organic substances. The extent of food contamination from these substances depends on our success in preventing them from entering the environment.

## CONCLUSIONS AND ISSUES

Data presented here indicate that environmental contamination of food is a nationwide problem of unknown magnitude. Long-term, low-level exposure to toxic substances in food poses health risks that are difficult to evaluate given present techniques. Incidents of high-level contamination of food that cause human illness have not occurred in the United States. However, regulatory actions have been taken to restrict consumption of contaminated food in cases where the potential health risks were considered unacceptable. These episodes have resulted in economic losses when contaminated food was removed from the market.

The following chapters analyze several issues related to the regulation of environmental contaminants in food. These are:

- Is our present regulatory system protecting the public health? (Chapters III and IV)

- Are methods used by the regulatory agencies for estimating health impacts the most appropriate ones? (Chapters III and V)
- Should economic impacts be an explicit part of regulatory decisionmaking? If so, how should economic impacts be evaluated? (Chapters III and VI)
- Should regulatory monitoring be capable of detecting substances as they enter the food chain? (Chapters VII and VIII)

## CHAPTER II REFERENCES

1. Storclz, William J. "C&EN's Top Fifty Chemical Products and Producers," *Chemical and Engineering News* 56(18):33, 1978.
2. Murray, Chris. "Chemical Waste Disposal A Costly Problem," *Chemical and Engineering News*, p. 12, Mar. 12, 1979.
3. National Academy of Sciences, National Research Council, Commission on Natural Resources, Environmental Studies Board, Committee on the Assessment of Polychlorinated Biphenyls in the Environment. *Polychlorinated Biphenyls* (ISBN 0-309-02885-X), 1979.
4. Hammons, A. "Levels of Chemical Contaminants in Nonoccupationally Exposed U.S. Residents," draft manuscript prepared for Health Effects Research Laboratory, Environmental Protection Agency, 1978.
5. U.S. Food and Drug Administration. "Polychlorinated Biphenyls (PCBs)," *Federal Register* 42(63):17487, 1977.
6. Kimbrough, R. D. "The Toxicity of Polychlorinated Polycyclic Compounds and Related Chemicals," *CRC Critical Reviews in Toxicology* 2:445, 1974.
7. Fishbein, L. "Toxicity of Chlorinated Biphenyls," *Ann. Rev. Pharmacol.* 14:139, 1974.
8. Peakall, D. B. "PCBs and Their Environmental Effects," *CRC Critical Reviews on Environmental Contaminants* 5:469, 1975.
9. National Institute of Occupational Safety and Health. *Criteria for a Recommended Standard. Occupational Exposure to Polychlorinated Biphenyls (PCBs)*, 1977 (DHEW Publ. No. 77-225).
10. Kimbrough, R. D., et al. "Animal Toxicity," *Environ. Health Persp.* 24:173, 1978.
11. Cordle, F., et al. "Human Exposure to Polychlorinated Biphenyls and Polybrominated Biphenyls," *Environ. Health Persp.* 24:157, 1978.
12. Kimbrough, R. D., and R. E. Linder. "Induction of Adenofibrosis and Hepatomas of the Liver in BALB/oJ Mice by Polychlorinated Biphenyls (Aroclor 1254)," *Journal of The National Cancer Institute* 53:547, 1974.
13. Allen, J. R., et al. "Residual Effects of Short-Term, Low-Level Exposure of Non-Human Primates to Polychlorinated Biphenyls," *Toxicology and Applied Pharmacology* 30:440, 1974.
14. McNulty, W. P. "Primate Study," National Conference on Polychlorinated Biphenyls, Chicago, 1975.
15. Barsotti, D. A., et al. "Reproductive Dysfunction in Rhesus Monkeys Exposed to Low Levels of Polychlorinated Biphenyls (Aroclor 1248)," *Food and Cosmetics Toxicology* 14:99, 1976.
16. Allen, J. R., and D. A. Barsotti. "The Effects of Transplacental and Mammary Movement of PCBs on Infant Rhesus Monkeys," *Toxicology* 6:331, 1976.
17. Allen, J. R. "Response of the Non-Human Primate to Polychlorinated Biphenyl Exposure," *Federation Proceedings* 34:1675, 1975.
18. Meester, W. D. and D. J. McCoy, Sr. "Human Toxicology of Polybrominated Biphenyls," *Clinical Toxicology* 10(4):474, 1977.
19. Fries, G. F., G. S. Marrow, and R. M. Cook. "Distribution and Kinetics of PBB Residues in Cattle," *Environmental Health Perspectives* 23:43, 1978.
20. Lillis, R., H. A. Anderson, J. A. Valcinkas, S. Freedman, and I. J. Selikoff. "Comparison of Findings Among Residents of Michigan Dairy Farms and Consumers of Produce Purchased From These Farms," *Environmental Health Perspectives* 23:105, 1978.
21. Environmental Protection Agency, Office of Toxic Substances, Special Actions Group. "Assessment of the Hazards of Polybrominated Biphenyls (PBBs)" (draft), 1977.
22. Wolff, M. S., B. Aubrey, F. Camper, and N. Hames. "Relation of DDE and PBB Serum Levels in Farm Residents, Consumers, and Michi-

gan Chemical Corporation Employees," *Environmental Health Perspectives* 23:177, 1978.
23. Cook, H. D., R. Helland, B. H. Vander Weele, and R. J. DeJong. "Histotoxic Effects of Polybrominated Biphenyls in Michigan Dairy Cattle," *Environmental Research* 15:82, 1978.
24. Kay, K. "Polybrominated Biphenyls (PBB) Environmental Contaminants in Michigan, 1973-1976," *Environmental Research* 13:74, 1977.
25. Bekesi, J. G., J. F. Holland, H. A. Anderson, A. S. Fischbein, W. Rom, M. S. Wolff, and I. J. Selikoff. "Lymphocyte Function of Michigan Dairy Farmers Exposed to Polybrominated Biphenyls," *Science* 199:1207, 1978.
26. United Nations Environment Program and the World Health Organization. *Environmental Health Criteria 1: Mercury*, Geneva, p. 23, 1976.
27. National Academy of Sciences, National Research Council. *An Assessment of Mercury in the Environment*, Washington, D.C., 1978.
28. U.S. Environmental Protection Agency, *Ambient Water Quality Criteria: Mercury*, 1979.
29. Lu, F. C. "Mercury as a Food Contaminant," *WHO Chronicle* 28:8, 1974.
30. D'Itri, F. M. "The Environmental Mercury Problem," *CRC Press*, p. 73, 1972.
31. Friberg, L., and J. Vostal. "Mercury in the Environment," *CRC Press*, 1972.
32. Gerstner, H. B., and J. E. Huff. "Clinical Toxicology of Mercury," *Journal of Toxicology and Environmental Health* 2:491, 1977.
33. Olson, F. C., and E. J. Massaro. "Pharmacodynamics and Teratologic Effects of Transplacentally Transported Methyl Mercury in the Mouse," *International Conference on Heavy Metals in the Environment*, pp. 32 and 184, 1972.
34. Weiss, B., and R. H. Doherty. "Methylmercury Poisoning," *Teratology* 12(3):311, 1976.
35. Rizzo, A. M., and A. Furst. "Mercury Teratogenesis in the Rat," *Proceedings of the Western Pharmacology Society* 15:52, 1972.
36. Evans, H. L., R. H. Garman, and B. Weiss. "Methylmercury: Exposure Duration and Regional Distribution as Determinants of Neurotoxicity in Nonhuman Primates," *Toxicology and Applied Pharmacology* 41:15, 1977.
37. National Academy of Sciences, Advisory Committee on the Biological Effects of Ionizing Radiations. *Effects on Populations of Exposure to Low Levels of Ionizing Radiation*, Washington, D.C., 1972.
38. National Academy of Sciences, Food Protection Committee. *Radionuclides in Foods*, Washington, D.C., 1973.
39. Dodge, Chris. "Contamination of Food by Radioactive Nuclides," Congressional Research Service, OTA Working Paper, August 1978.
40. Monitoring Advisory Panel, Sept. 21, 1978.
41. Michigan Department of Agriculture, Information and Education Division. "A Brief Chronology of the PBB Incident" (no date).
42. Anonymous. "Two Firms Fined $4,000 Each for PBB Contamination of Cattle Feed in Michigan," *Toxic Material News*, p. 144, May 24, 1978.
43. Jaroslovsky, Rich. "Contamination of Fishmeal with PCBs Sparks FDA Study of Mishap at a Ralston Purina Plant," *The Wall Street Journal*, p. 46, July 12, 1978.
44. Whitehead, George. Deputy Director for Consumer Affairs, Michigan Department of Agriculture, Jan. 16, 1978.
45. McNally, JoAnn. "Polybrominated Biphenyls: Environmental Contamination of Food," Congressional Research Service, OTA Working Paper, Sept. 21, 1978.

# CHAPTER ONE
# THE CIRCLE OF POISON

*This pesticide barrel in a mountain village in El Salvador is used to collect rainwater that is used for drinking by school children in the adjacent building.*

From the book, Circle of Poison, by David Weir and Mark Schapiro, copyright 1981. Available for $3.95 plus $1.00 shipping from the Institute for Food and Development Policy, 2588 Mission Street, San Francisco, CA 94110.

*Here pesticides are the dish of the day, and one swallows more poison than food. There is not even a living hen or pig, and lately even the children are often sick. Could it be that even the gift that God gives—children—we cannot have?*
                    ALFONSO CASTRO, COLOMBIAN FARMER

*There's no problem with the ban of DBCP within the United States. In fact, it was the best thing that could have happened to us. You can't sell it here anymore but you can still sell it anywhere else. Our big market has always been exports anyway.*
                    EXECUTIVE, AMVAC CORPORATION

*Small shops in Indonesia sell pesticides right alongside the potatoes and rice. The people just collect it in sugar sacks, milk cartons, Coke bottles—whatever is at hand.*
                    LUCAS BRADER, UN FOOD AND AGRICULTURE ORGANIZATION

*Nearly half of the green coffee beans imported into the United States contain various levels—from traces to illegal residues— of pesticides that have been banned in the United States.*
                    FOOD AND DRUG ADMINISTRATION

THIS BOOK documents a scandal of global proportions—the export of banned pesticides from the industrial countries to the third world. Massive advertising campaigns by multinational pesticide corporations—Dow, Shell, Chevron—have turned the third world into not only a booming growth market for pesticides, but also a dumping ground. Dozens of pesticides too dangerous for unrestricted use in the United States are shipped to underdeveloped countries. There, lack of regulation, illiteracy, and repressive working conditions can turn even a "safe" pesticide into a deadly weapon.

According to the World Health Organization, someone in the underdeveloped countries is poisoned by pesticides *every minute*.[1]

But we are victims too. Pesticide exports create a circle of poison, disabling workers in American chemical plants and later returning to us in the food we import. Drinking a morning coffee or enjoying a luncheon salad, the American consumer is eating pesticides banned or restricted in the United States, but legally shipped to the third world. The United States is among the world's top food importers and 10 percent of our imported food is officially rated as contaminated.[2] Although the Food and Drug Administration (FDA) is supposed to protect us from such hazards, during one 15-month period, the General Accounting Office (GAO) discovered that *half* of all the imported food identified by the FDA as pesticide-contaminated was marketed without any warning to consumers or penalty to importers.[3]

*At least 25 percent of U.S. pesticide exports are products that are banned, heavily restricted, or have never been registered for use here.*[4] Many have not been independently evaluated for their impacts on human health or the environment. Other pesticides are familiar poisons, widely known to cause cancer, birth defects and genetic mutations. Yet, the Federal Insecticide, Fungicide, and Rodenticide Act explicitly states that banned or unregistered pesticides are legal for export.[5]

In this book we concentrate on hazardous pesticides which are either banned, heavily restricted in their use, or under regulatory review in the United States. Some, such as DDT, are banned for any use in the United States; others, such as 2,4-D or toxaphene, are still widely used here but only for certain usages. (Table I, on page 79, lists the status of hazardous pesticides which are prohibited, restricted, under review, or unrestricted in the United States.) As we will discuss, even "safe" pesticides which are unrestricted in the United States may have much more damaging effects on people and the environment when used under more brutal conditions in the third world.

In the United States, a mere dozen multinational corporations dominate the $7-billion-a-year pesticide market. Many are conglomerates with major sales in oil, petrochemicals, plastics, drugs and mining.

The list of companies selling hazardous pesticides to the third world reads like a Who's Who of the $350-billion-per-year[6] chemical industry: Dow, Shell, Stauffer, Chevron, Ciba-Geigy, Rohm & Haas, Hoechst, Bayer, Monsanto, ICI, Dupont, Hercules, Hooker, Velsicol, Allied, Union Carbide, and many others. (See Table One.)

Tens of thousands of pounds of DBCP, heptachlor, chlordane, BHC, lindane, 2,4,5-T and DDT are allowed to be exported each year from the United States, even though they are considered too dangerous for unrestricted domestic use.[7]

"You need to point out to the world," Dr. Harold Hubbard of the U.N.'s Pan American Health Organization told us, "that there is absolutely no control over the manufacture, the transportation, the storage, the record-keeping—the entire distribution of this stuff. These very toxic pesticides are being thrown all over the world and there's no control over any of it!"[8]

Not only do the chemical corporations manufacture hazardous pesticides, but their subsidiaries in the third world import and distribute them. (See Table One.)

• Ortho: In Costa Rica, Ortho is the main importer of seven banned or heavily restricted U.S. pesticides—DDT, aldrin, dieldrin, heptachlor, chlordane, endrin, and BHC. Ortho is a division of Chevron Chemical Company, an arm of Standard Oil of California.[9] (See Appendix B.)

• Shell, Velsicol, Bayer, American Cyanamid, Hercules and Monsanto: In Ecuador these corporations are the main importers of pesticides banned or restricted in the United States—aldrin, dieldrin, endrin, heptachlor, kepone, and mirex.[10]

- Bayer and Pfizer: In the Phillipines these multinationals import methyl parathion and malathion respectively;[11] neither is banned but both are extremely hazardous.

The Ministry of Agriculture of Colombia registers 14 multinationals which import practically all the pesticides banned by the United States since 1970.[12] (See Appendix C.) And in the Philippines, the giant food conglomerate Castle & Cooke (Dole brand) imports banned DBCP for banana and pineapple operations there.[13]

### Pesticides: a pound per person

WORLDWIDE pesticide sales are exploding. The amount of pesticides exported from the U.S. has almost doubled over the last 15 years.[14] The industry now produces four billion pounds of pesticides each year—more than one pound for every person on earth.[15] Almost all are produced in the industrial countries, but 20 percent are exported to the third world.[16]

And the percentage exported is likely to increase rapidly: The GAO predicts that during the decade ending in 1984, the use of pesticides in Africa, for example, will more than quintuple.[17] As the U.S. pesticide market is "approaching saturation... U.S. pesticide producers have been directing their attention toward the export potential ... exports have almost doubled since 1965 and currently account for 30 percent of total domestic pesticide production," the trade publication *Chemical Economics Newsletter* noted.[18]

Corporate executives justify the pesticide explosion with what sounds like a reasonable explanation: the hungry world needs our pesticides in its fight against famine. But their words ring hollow: in third world fields most pesticides are applied to luxury, export crops, not to food staples the local people will eat. Instead of helping the poor to eat better, technology is overexposing them to chemicals that cause cancer, birth defects, sterility and nerve damage.

### "Blind" schedules, not "as needed"

BUT THE CRISIS is not just the export of banned pesticides. A key problem in both the industrial countries and the third world is the massive overuse of pesticides resulting from their indiscriminate application. Pesticides are routinely applied according to schedules preset by the corporate sellers, not measured in precise response to actual pest threats in a specific field. By conservative estimate, U.S. farmers could cut insecticide use by 35 to 50 percent with no effect on crop production, simply by treating only when necessary rather than by schedule.[19] In Central America, researchers calculate that pesticide use, especially parathion, is 40 percent higher than necessary to achieve optimal profits.[20]

In the United States the result of pesticide overuse is the

unnecessary poisoning of farmworkers and farmers—about 14,000 a year according to the Environmental Protection Agency (EPA).[21] But if pesticides are not used safely here—where most people can read warning labels, where a huge government agency (the EPA) oversees pesticide regulation, and where farmworker unions are fighting to protect the health of their members—can we expect these poisons to be used safely in the third world?

## An inappropriate technology

IN THIRD WORLD countries one or two officials often carry responsibility equivalent to that of the entire U.S. EPA. Workers are seldom told how the pesticides could hurt them. Most cannot read. And even if they could, labels on banned pesticides often do not carry the warnings required in the United States. Frequently repacked or simply scooped out into old cans, deadly pesticides are often handled like harmless white powder by peasants who have little experience with manmade poisons.

But perhaps even more critical is this question: can pesticides—poisons, by definition—be used safely in societies where workers have no right to organize, no right to strike, no right to refuse to carry the pesticides into the fields? In the Philippines, for example, at least one plantation owner has reportedly sprayed pesticides on workers trying to organize a strike.[22] And, in Central America, says entomologist Lou Falcon, who has worked there for many years, "The people who work in the fields are treated like half-humans, slaves really. When an airplane flies over to spray, they can leave if they want to. But they won't be paid their seven cents a day or whatever. They often live in huts in the middle of the field, so their homes, their children, and their food all get contaminated."[23]

Yet the President's Hazardous Substances Export Policy Task Force predicts that the export of banned pesticides is likely to increase as manufacturers unload these products on countries hooked on the ag-chemical habit. "Continued new discoveries of carcinogenic and other damaging effects of many substances are probable over the next few years," predicts the task force. "In some cases, certain firms may be left with stocks of materials which can no longer be sold in the United States, and the incentive to recover some of their investment by selling the product abroad may be considerable."[24]

## The genetic boomerang

THE PESTICIDE EXPLOSION also has a second built-in boomerang. Besides the widespread contamination of imported food, the overuse of hazardous pesticides has created a global race of insect pests that are resistant to pesticides. The number of pesticide-resistant insect species doubled in just 12 years—from 182 in 1965 to 364 in

1977, according to the U.N. Food and Agriculture Organization. So more and more pesticides—including new, more potent ones—are needed every year just to maintain present yields.

### A circle of victims

BUT ENORMOUS DAMAGE is done even before the pesticides leave American shores. At Occidental's DBCP plant in Lathrop, California, workers discovered too late that they were handling a product which made them sterile. Elsewhere in the United States, worker exposure to the pesticides Kepone and Phosvel resulted in terrible physical and mental damage.

As part of our investigation into the "circle of poison," we looked at these examples of how the manufacture of hazardous pesticides affects American workers. Since companies are allowed to produce pesticides for export without providing health or safety data, there is no way to be sure they are not poisoning their own workers in the process. In fact there is abundant evidence that workers in the industrial countries are indeed suffering from their employers' booming export sales.

We talked with West Coast pesticide workers who complained of inadequate protection—and information—on the job. Even after two hot showers, one group explained, their hands still carried enough toxic residue of an unregistered pesticide that, when they stuck a finger in a fish bowl, the goldfish died.

These workers in pesticide manufacturing plants are the very first victims in the circle of poison. Add to them all the people who load and unload the chemicals into and out of trucks, trains, ships and airplanes; and those who have to clean up the toxic spills which inevitably occur. Then the total number of potential American victims of hazardous pesticide exports becomes very large. In addition, of course, there are the victims whose story is told in this book—third world peasants, workers and consumers—as well as everyone else in the world who eats food contaminated with pesticide residues. *We* complete the circle of victims.

To uncover the story in this book we have had to overcome powerful obstacles. The pesticide industry is a secretive one. The Environmental Protection Agency guards industry production data from the public, press and even other government agencies. The information made available often seems to defy meaningful interpretation.

We filed over 50 requests under the Freedom of Information Act in order to penetrate the industry's "trade secrets" sanctuary inside the EPA and other agencies. In assembling hundreds of tiny pieces of the puzzle, we studied trade publications and overseas magazines and newspapers for evidence of hazardous pesticide sales. In addition, we obtained import figures from a number of third world countries. Finally, we interviewed hundreds of people in industry, government, unions, environmental groups and international organizations. We corresponded with farmers, consumers and environmental groups in the third world.

The story told here is intended not merely to shock and to outrage. Its purpose is to mobilize concerned people everywhere to halt the needless suffering caused by pesticides' circle of poison.

# NOTES

## CHAPTER ONE

1. Proceedings of the U.S. Strategy Conference on Pesticide Management, U.S. State Rept., June 7-8, 1979, p. 33.
2. "Report on Export of Products Banned by U.S. Regulatory Agencies," H. Rept. No. 95-1686, Oct. 4, 1978, p. 28.
3. Ibid.
4. "Better Regulation of Pesticide Exports and Pesticide Residues in Imported Foods Is Essential," U.S. GAO, Rept. No. CED-79-43, June 22, 1979, pp. iii, 39.
5. Frances Moore Lappé and Joseph Collins, *Food First: Beyond the Myth of Scarcity* (New York: Ballantine, 1979), p. 145.
6. "New Pesticides Must Now Be Economic Winners," *Chemical Age*, Feb. 17, 1978; Dr. Jay Young, Chemical Manufacturers Association, telephone interview with authors, Oct. 1979.
7. President's Hazardous Substances Export Policy Working Group, Fourth Draft Report, Jan. 7, 1980, p. 6.
8. Dr. Hal Hubbard, telephone interview with authors, June 1, 1977.
9. Lappé and Collins, op. cit., p. 146.
10. "Listudo de Pesticides Registrados en el Departamento de Sanidad Vegetal," Ministry of Agriculture, Ecuador.
11. "Importacion de Pesticidas," Ministerio de Agricultura y Ganaderia, Costa Rica, 1978.
12. "Plaguicidas de Uso Agricola, Défoliantes y Reguladores Fisiologicos de las Plantas Registrados en Colombia," Ministerio de Agricultura, Colombia, June 30, 1979.
13. (See Lappé and Collins, op. cit., p. 60.)
14. Thomas O'Toole, "Over 40 Percent of World's Food Is Lost to Pests," *Washington Post*, March 6, 1977.
15. Douglas Starr, " 'Pesticide Poisoning Alarming,' says FAO," *Christian Science Monitor*, Feb. 1, 1978.
16. Ibid.; and Lappé and Collins, op. cit., p. 64.
17. "Better Regulation of Pesticide Exports and Pesticide Residues in Imported Food Is Essential," U.S. GAO Rept. No. CED-79-43, June 22, 1979, p. 1.
18. Jeanie Ayres, "Pesticide Industry Overview," *Chemical Economics Newsletter*, Jan.-Feb. 1978, p. 1.
19. Lappé and Collins, op. cit., p. 41.
20. "An Environmental and Economic Study of the Consequences of Pesticide Use in Central American Cotton Production," Final Report, Instituto Centro-Americano de Investigacion y Technologia Industrial (I.C.A.I.T.I.), Jan. 1977, pp. 149, 155, 161.
21. Lappé and Collins, op. cit., p. 67.
22. Osawa Yasuo, "Banana Plantation Workers Strike in the Philippines," *New Asia News*, May 1980, p. 7.
23. Dr. Lou Falcon, telephone interview with authors, May 21, 1979.
24. *Agriculture: Toward 2000*, U.N. Food and Agriculture Organization (FAO), Rome, July 1979, p. 82.
25. 5 U.S.C. §552, Freedom of Information Act.

# Water, Water Everywhere

Jacqueline M. Warren

*Jacqueline Warren is a senior staff attorney with the Natural Resources Defense Council and directs its Toxic Substances Project. She was previously a staff attorney for the Environmental Defense Fund where she worked on drinking water and toxic substances control issues.*

*But is it safe to drink? Faced with the great expense of cleaning up wells contaminated by organic chemicals, communities in California, Connecticut, Massachusetts, Michigan, New Jersey, New York, Tennessee and Wisconsin have been forced to close their wells and seek other sources.*

For half a century it has been said that America has the safest drinking water in the world, but today the facts suggest otherwise. Drinking water supplies throughout the United States have been widely contaminated with industrial chemicals.

Claims about the safety of drinking water stem from the virtual eradication in this country of communicable water-borne disease. However, organic chemicals pose a new and different threat. An objective assessment of the situation reveals a drastic inadequacy in the response of U.S. water supply officials to widespread chemical water pollution resulting from the mushrooming chemical and petrochemical industry of the past three decades.

**SURFACE WATER.** Surface water supplies in the United States are contaminated by more than 700 organic industrial chemicals, heavy metals, pesticides, and other pollutants including vinyl chloride, carbon tetrachloride, dieldrin, trichloroethylene, tetrachloroethylene, chloroform, kepone, PCBs, and benzene. Twenty-two of these chemicals have been identified by the National Academy of Sciences as carcinogens and many others as mutagenic or otherwise toxic. The vast majority have not even been tested for potential toxicity. Moreover, it is estimated that the 700 chemicals represent only a small fraction, perhaps as little as 10 per-

cent by weight, of all the organic chemicals present in drinking water.

Chemical contamination of surface drinking water supplies comes from widespread, multiple sources—industrial discharges, spills and accidents, urban and agricultural runoff, municipal sewage effluents, and chlorination of drinking water and sewage for purposes of disinfection. While untreated surface water is not generally considered pristine water for drinking, the public is not well aware of the fact that drinking water treatment in the United States is limited almost exclusively to disinfection. Usually done by chlorination, drinking water treatment was originally designed to prevent outbreaks of diseases such as cholera and typhoid. Consequently, the overwhelming majority of treatment plants neither remove nor detect the presence of organic chemicals. With very few exceptions, surface water supplies remain completely vulnerable to serious chemical contamination.

**GROUNDWATER.** Groundwater (subsurface water occurring in permeable geologic formations called aquifers) is the primary source of drinking water for 50 percent of the U.S. population. The conventional wis-

## NEW YORK WON'T WAIT

HESITATION AT THE FEDERAL LEVEL has prompted at least one state, New York, to contemplate adopting its own program for alleviating toxic contamination of drinking water. The presence of toxics in aquifers in Long Island, where groundwater is the sole source of supply, and in other parts of the state has been known for several years. Intense local controversies have often followed identification of contaminants such as trichloroethylene and other solvents, PCBs and aldicarb, a potato pesticide. In Glen Cove, Long Island, five of nine wells were closed as a result of tri- and tetrachloroethylene contamination, requiring water purchases from adjacent communities. In Poughkeepsie, which takes its water from the PCB-contaminated Hudson River and has no practical alternate source of supply, conflict has arisen between consumer groups, the City and the State Health Department over the level of health risk to which residents are exposed. Only a few water supplies across the state have been tested for a broad range of possible toxic contaminants, but in most cases some level of unwanted chemicals has been found, usually plastic compounds, solvents or petroleum products.

State Health Commissioner David Axelrod, a longtime advocate of tighter controls on toxics, is considering recommendations to move ahead with a program of identifying and dealing with synthetic organic chemicals. He has already ordered all water supplies in the state to be tested for a broad band of toxics, a process which will take many years at present rates of testing. A report he has received from Langdon Marsh, a former deputy commissioner of the state's Department of Environmental Conservation, recommends that the state not wait for the federal Environmental Protection Agency to establish maximum contaminant levels for the chemicals which are frequently found in New York. Marsh points out that even prior to the change of administration in Washington, EPA had no schedule for adopting standards for chemicals other than six that are found almost nationwide. New York has already found other chemicals in drinking water which are known to be carcinogenic, mutagenic or otherwise harmful to health. In addition to tight standards, the report recommends a broad range of preventive, monitoring and public education actions. Commissioner Axelrod is now considering these recommendations and has been quoted as saying that standards would be issued within a year. "New York traditionally has been in the forefront of assuring public health and the state is continuing to do so," Axelrod said.

The most controversial recommendation in the report is that the level of risk to be used for establishing an appropriate drinking water standard for a particular chemical be one which will result in only

dom is that groundwater, in contrast to surface water, provides a protected source of drinking water, because the ground acts as a natural filter for the removal of impurities. Recent studies have shown, however, that much groundwater is highly contaminated by toxic industrial solvents such as trichloroethylene and tetrachloroethylene and often at much higher levels than those found in drinking water taken from the most polluted surface supplies. For example, the largest concentration of trichloroethylene ever recorded in surface water is 160 *parts per billion* (ppb). By contrast, the President's Council on Environmental Quality recently reported that a level of 27,300 ppb was recently measured in a well in Pennsylvania.

Every day newspapers report another discovery that well water has been rendered unfit for drinking because of improper hazardous waste disposal or industrial activity in an aquifer recharge area. Such widely reported problems have led to the closing of public and private wells in Maine, Massachusetts, Connecticut, New York, New Jersey, Tennessee, Michigan, Wisconsin, and California. There is now serious concern at all levels of government about the long-term viability of our groundwater resources as a drinking water supply.

The sources of organic chemicals in groundwater are numerous. They include disposal of hazardous wastes in landfills and industrial surface impoundments (pits, settling ponds, and lagoons); septic tanks and cesspools and the chemicals used to clean them; municipal waste water; mining activity; petroleum exploration and development; and agricultural, street, and urban runoff. Because groundwater lacks the self-cleansing properties provided surface water by dilution, circulation, and degradation by sunlight and aquatic organisms, groundwater can remain contaminated for centuries.

Although groundwater itself moves slowly, often only 10 or 20 feet a year, the contaminants move in unpredictable *plumes,* the behavior and flow rate of which are difficult and costly to measure. Moreover, once the contamination is detected, few remedies are available, and these are often economically or technically unfeasible.

The best alternative to a contaminated well is a new well. Clean-up involves pumping out, treating vast amounts of the well water, and then returning it to the ground. Other alternatives are bottled water and home water filter devices, but both are expensive and of one-in-one-million chance of dying from cancer in a lifetime of exposure. While of no solace to the individual who happens to be one in a million, the risk level is considered very low by most scientists and government regulators. EPA used a much higher level for setting limits on concentrations of trihalomethanes (THM's) in finished drinking water. The science or art of risk assessment has been much debated and has significant technical and conceptual flaws, but at the moment is considered the best available starting point for establishing acceptable levels of exposure.

The report recognizes that use of a low risk level to set standards may have severe economic consequences to particular water suppliers. According to the report, cost factors should not be used to weaken standards through some kind of mechanistic cost-benefit analysis, which usually results in underestimating difficult-to-quantify health benefits. Tough standards should be applied to all water suppliers with an opportunity for weighing the risks involved in setting the standard against the costs involved in compliance for a particular water supplier. Depending on such factors as the level of contaminants found, the potential adverse health effects, the duration of exposure of affected populations and the degree of certainty of scientific evidence, compliance with the standard could be stretched out over time while new supplies are sought, treatment systems explored and financing arranged. Monitoring of levels of contaminants would be required of any supplier under an extended compliance schedule.

One of the advantages of providing an extension mechanism to deal with severe economic impacts is that it should take the focus of litigation away from the scientific evidence underlying the standards themselves. In a case decided by the the U.S. Supreme Court in 1980, an OSHA standard on airborne benzene was overturned because the health data underlying it was found insufficient by a plurality of the justices. That meant that the standard was unenforceable everywhere, even where compliance would not have been particularly burdensome. Changing the focus of legal challenge to the denial of a time extension should concentrate the courts' attention on the regulators' action in a concrete case. Even if the health department's judgment were overturned in that case, the standard would remain applicable to everyone else.

New York's approach to the bedeviling problem of dealing with the risks of synthetic organic contamination in drinking water bears watching as a possible precedent for action in other states. It ultimately may be followed at the federal level when interest in regulating organics reawakens at EPA.

questionable efficacy. Bottled water is subject only to sporadic surveillance by the Food and Drug Administration, and by law need only meet existing EPA drinking water standards which do not consider the vast majority of organic chemicals that have been found in drinking water. Consequently, bottled water may in fact exceed the chemical quality of tap water. Home filter devices, with a few exceptions, have been shown in recent studies by the Environmental Protection Agency to be inefficient generally in removing organic chemicals. Moreover, unless an individual homeowner can have the water routinely sampled and analyzed to ensure that the device is maintaining its removal efficiency, the investment, which can amount to several hundred dollars, is not worthwhile. Nevertheless, consumers are spending $400 to $500 million a year for bottled water or home water filters.

Faced with the great expense of cleaning up contaminated wells, many communities have been forced to close their wells and seek other sources of supply. For example, 39 public wells were closed in the San Gabriel Valley, California, in January 1980. Although the wells supplied water to 400,000 people in 12 cities, California public health officials ordered them closed, because of high levels of trichloroethylene, an industrial solvent found to be carcinogenic in experimental animals. Similar contamination and well closings have occurred in New York (especially on Long Island), New Jersey, Pennsylvania, Connecticut, Massachusetts, and

*Careful siting of residential development, road construction, industrial activity and waste disposal facilities is critical to the prevention of groundwater contamination.*

Maine. In every case, the contaminants were organic chemicals in high levels, many known to cause, or suspected of causing, cancer or genetic mutations. Among the chemicals most often found in high concentrations in wells are trichloroethylene, 1,1,1-trichloroethane, tetrachloroethylene, carbon tetrachloride 1,2-dichloroethane and vinyl chloride. The similarities in patterns of contamination suggest that the problem is widespread.

To illustrate the potential magnitude of only one aspect of the problem, a recent survey of 176,647 industrial waste impoundments conducted by the federal Environmental Protection Agency showed that 95 percent of the sites were not being monitored for groundwater contamination and that 70 percent of the impoundments were unlined. Thirty percent of the unlined industrial impoundments were found to "overlie usable aquifers and are underlain by unsaturated zones which freely allow downward movement of any liquid wastes escaping from the impoundment." Thus, both the extent and the number of groundwater contamination incidents can be expected to increase in the foreseeable future.

**THE SAFE DRINKING WATER ACT.** Drinking water has traditionally been a matter of state and local jurisdiction. In the late 1960s and early 1970s, however, various agencies of the federal government began to be concerned about the extensive presence of organic chemicals in drinking water supplies throughout

the country, and the apparent inability or unwillingness of state and local authorities to address the problem. Congress responded in 1974 by enacting the Safe Drinking Water Act, authorizing the Environmental Protection Agency to establish uniform national standards for drinking water.

Further impetus for a federal drinking water law had been provided by studies of public water supply systems showing great variations among states in the quality of drinking water and of water treatment practices. As set forth in the House Report on the Safe Drinking Water Act, 56 percent of the 969 systems studied showed physical deficiencies in the protection of groundwater sources and in disinfection capacity; 77 percent of the plant operators were inadequately trained, and most systems were unprotected against contamination from sewage cross-connections; 79 percent of the systems were not inspected annually by state or county authorities; and 69 percent did not even analyze for half of the contaminants listed in 1962 by the U.S. Public Health Service in drinking water standards for interstate carriers. In addition, surveillance of drinking water quality was so poor that 90 percent of the systems covered by EPA's Community Water Supply Study "had no idea what the chemical quality of their drinking water was." Other widespread problems noted in the House report were the lack of adequate, inexpensive monitoring or measurement methods; the proliferation of small water systems which do not support well trained full-time operators and necessary equipment; inadequate health effects research; and the increasing demand for drinking water at a time of increasing pressure to dispose of contaminants in ways that may endanger drinking water quality.

The Safe Drinking Water Act established a regulatory framework within which the Environmental Protection Agency sets drinking water standards called "maximum contaminant levels" or prescribes water treatment techniques which are uniformly applicable to community water systems across the county. The EPA is also authorized to establish minimum national requirements to ensure that underground injection of wastes does not endanger groundwater supplies used for drinking. And, lastly, the agency may designate certain groundwater sources as "sole source aquifers," i.e., providing the sole source of drinking water for a given area, and may then withhold federal financial support from contaminating activities in the area. As with other environmental statutes such as the Clean Air Act and the Clean Water Act, primary enforcement responsibility is delegated to the states, once their state programs are approved by EPA. But unlike the Clean Air Act and the Clean Water Act, the Safe Drinking Water Act gives EPA neither the necessary grant money nor sanctions to provide incentives for the development and enforcement of strong programs by the states.

The absence of significant federal financial assistance or compliance initiatives has left its mark on the federal drinking water program. Despite clear signals from Congress that organic chemicals were high on the list of contaminants to be addressed under the Safe Drinking Water Act, EPA has made only slow and piecemeal efforts to deal with this problem. The first set of drinking water standards issued by EPA in 1975 did not include organics, except for a handful of pesticides and herbicides contained in the old 1962 Public Health Service standards. The many carcinogenic chemicals and other toxic substances known to be widespread in surface water supplies were not included.

Following a lawsuit by the Environmental Defense Fund in 1975 and a change in administrations, EPA in 1978 proposed a comprehensive regulation to reduce the level of organic chemicals in drinking water. First, the agency proposed a maximum contaminant level of 100 *ppb* for trihalomethanes—chemicals produced in water treatment plants by the chlorination process. Trihalomethanes include the carcinogens chloroform and bromoform, as well as bromodichloromethane and dibromochloromethane. Of these, chloroform has been found in every sampled system that chlorinates for disinfection, and detection of the other trihalomethanes has been almost as common.

The second aspect of EPA's organics proposal was a requirement that water supplies drawn from a source contaminated to a certain degree by synthetic organic

From *Water Quality in a Stressed Environment*. Burgess Publishing Company, 1972.

*How leachate from a landfill can contaminate groundwater aquifers with toxic substances.*

chemcials install removal technology such as granular activated carbon filters. Activated carbon has been used successfully in the water treatment, waste treatment, air treatment, food processing, beverage manufacturing, and sugar refining industries. Currently, more than 60 water treatment plants in the United States use activated carbon for taste and odor control, as do more than 20 municipalities in Europe for the control of trace organic chemicals of health significance.

The organized water supply industry in the United States, led by the American Water Works Association and an *ad hoc* group of water utilities calling itself the

> *..."cradle-to-grave" hazardous waste management programs that most states are developing ... should help to prevent future drinking water contamination problems by controlling the disposal of toxic wastes.*

Coalition for Safe Drinking Water, objected strenuously to EPA's proposal to control organics in drinking water. Viewing the proposal as unnecessary, an insult to their professional pride, and unduly expensive, they challenged the preventive health rationale for the proposed requirements, as well as the safety and efficacy of available removal technologies. By persistent and strident opposition the water suppliers succeeded in persuading EPA to withdraw the treatment technique proposal. The agency promulgated the 100 *ppb* trihalomethane standard, however, and is currently defending its action against a challenge brought by the American Water Works Association in the U.S. Court of Appeals for the D.C. Circuit. The water utilities also took their objections to Congress, where a serious effort to weaken the Safe Drinking Water Act was narrowly defeated in the last session of Congress.

At the same time that EPA, the water utilities and environmentalists were engaged in skirmishes over the proposal to address the problem of organic contaminants in drinking water, evidence was accumulating that the problem extended far beyond surface water sources. Neither of the approaches to organics control contained in the EPA proposal (numerical standards or treatment techniques) is appropriate for groundwater. Under the Safe Drinking Water Act, monitoring is only required after standards are promulgated, and remedial action is taken only after those standards have been exceeded. Similarly, treatment techniques only need be applied after a problem has been detected. Once concentrations of toxic chemicals in groundwater exceed standards, however, the damage is done and remedial action is unlikely to be available or affordable.

Groundwater protection requires an entirely different approach. Groundwater contamination must be prevented by controlling polluting activities in sensitive aquifer recharge areas. Careful siting of residential development, road construction, industrial activity and waste disposal facilities is critical to the prevention of groundwater contamination. To this end, EPA is in the process of devising and circulating for comment a comprehensive groundwater protection strategy to be recommended for adoption by the states.

Implementation of such a strategy through zoning, restrictions on underground injection of wastes, and regulation of hazardous waste disposal will provide some protection for vulnerable groundwater supplies. However, the damage from existing landfills, impoundments, abandoned hazardous waste sites and similar contamination sources cannot be easily remedied. Recently enacted legislation to create a federal "superfund" to clean up such sites should be helpful in minimizing the damage from future Love Canals, but only a long-term commitment by states and local communities to maintain the integrity of our remaining unsullied groundwater resources will be able to prevent future groundwater contamination.

Even if such a commitment were made, our knowledge of the existing state of groundwater quality is deficient. We must evaluate the quality of groundwater supplies, learn where the critical aquifer recharge areas are located, and determine how groundwater resources are to be utilized in the future. It is a massive undertaking, and the ongoing activities and exigencies of modern industrial society are likely to conflict directly and often with requirements for groundwater protection.

Drinking water has been a neglected resource for many years. Now that the severity of the problem is becoming increasingly apparent, efforts are underway to clean up polluted sources and to prevent future degradation of both ground and surface waters. Despite the enormity of the task and the many competing demands on our resources and attention, there is some basis for cautious optimism. First, the comprehensive EPA program to control industrial discharges of toxic pollutants into water, obtained by the Natural Resources Defense Council, in settlement of a series of lawsuits in 1976, is due to be completed by 1984 for the 21 major industrial categories affected. This program will significantly reduce the quantity of toxic pollutants currently discharged into the nation's waterways and therefore lessen the concentrations of organic chemicals and other toxic substances in surface drinking water supplies.

Second, "cradle-to-grave" hazardous waste management programs that most states are developing under the Resource Conservation and Recovery Act should help to prevent future drinking water contamination problems by controlling the disposal of toxic wastes. Such programs are expensive to administer, however, and the projected reduction of federal financial assistance to the states may drastically curtail the implementation of effective hazardous waste management programs. The potential impact of the cut in federal aid could be dramatic. New York State, which has been developing a state hazardous waste management program widely acclaimed as "bold" and "forward-looking," presently depends upon the federal government to meet 80 percent of the costs of its present hazardous waste management efforts. The new program is estimated to cost four times as much while the percentage of federal aid is expected to decrease markedly.

The strongest basis for optimism is the public's growing awareness of the problem of the risks posed by exposure to hazardous substances in drinking water, and of the reasons why their water supplies have become tainted or even unusable. The demand for action to remedy existing problems has prompted Congress to enact a major new federal spending program, "superfund," at a time when most other federal programs were being cut back. The change in administration is unlikely to lessen the outcry unless some constructive actions are taken to assure the public that their water supplies are again safe to drink. □

# BE ALERT TO YOUR DRINKING WATER

by

Ralph Nader

Water, a resource usually taken for granted, has become a precious commodity, especially in parts of the country struck by severe droughts. Since drought reduces freshwater flows, the water available for drinking supplies contains an unusually large component of waste water, including industrial and municipal effluents. For example, at some points, two years ago, effluents composed over half of the flow of the Passaic River in New Jersey, which provides drinking water for over two million people.

To handle these impurities, the Passaic Valley Water Commission increased chlorination during that time by 300 percent. Water systems have long been able to control waterborne diseases such as typhoid and cholera through chlorination of the water. But chlorine creates another problem. Chlorine combines with other organic chemicals in the water to produce chloroform, as well as other carcinogenic (cancer-causing) or toxic chemical compounds. An Environmental Protection Agency (EPA) survey of 80 cities found chloroform in the drinking water of all of them.

Our drinking water, which is used in preparing food from the processing plants to your kitchens, contains many other hazardous chemicals as well. According to government research, only about ten percent by weight of the chemical contaminants in drinking water have even been identified. More will be found as analytic techniques improve and further research is done. Those contaminants already identified include nitrates, arsenic, barium, cadmium, lead, chromium, and mercury. In addition, a myriad of organic chemicals, including 23 known or suspected carcinogens, enter our water supply, according to the National Cancer Institute. These substances include vinyl chloride, benzene, carbon tetrachloride, PCBs, the pesticides dieldrin, kepone, and heptachlor, and many other toxic chemicals.

These and other industrial waste products are regularly dumped, accidentally spilled or are part of run-offs into water that ultimately finds its way to your tap. In this country today, there are tens of thousands of hazardous waste disposal sites. Who can forget the tragedy of Love Canal, New York? Over 900 families there were directly threatened by the industrial wastes on which their homes were built. Not far from that location, dioxin, perhaps the most deadly chemical ever synthesized, was found close to a drinking water source for the Niagara Falls community. Other glaring examples exist of deadly waste threatening our water supplies:

* In one 130-mile stretch of the Mississippi River, 50 of the nation's largest chemical and petro-chemical plants discharge their waste;

* In western Pennsylvania, massive amounts of sulfuric acid in waste water from abandoned coal mines flowed from the underground water table into the Monongahela River;

* In Waterloo, Iowa, a laboratory company dumped over 250,000 gallons of waste containing arsenic. In the coming years, this poison will show up in the drinking and irrigation water supplies of nearby towns;

* At Hanford, Washington, over 500,000 gallons of radioactive wastes have leaked into the ground and may find their way to the Columbia River;

* Recently, EPA officials warned that drinking water supplies of hundreds of communities in 15 western states may be dangerous for long-term use. The threat, they say, comes from the uranium waste that runs into streams from uranium mines.

Much of the nation — including virtually all Americans living in rural areas — depends on ground water, as opposed to surface water, for their drinking supplies. Groundwater is a fragile resource since it can remain soiled for hundreds of years if polluted. But throughout the country, hazardous wastes from municipal or industrial dumps have seeped into groundwater supplies, poisoning them with cancer-causing chemicals present in concentrations even higher than those found in surface waters. Groundwater wells in at least 25 states have been capped because of dangerous pollutants. All of the major groundwater sources on Long Island — which supply some 3 million people — have been contaminated, as have approximately one-third of the water systems in Massachusetts. As one Congressional committee put it, "Toxic chemical contamination of groundwater supplies in several areas of the country has reached alarming proportions."

Evidence about the danger of polluted drinking water is substantial. Right now, enough is known to conclude that cities such as Miami, New Orleans, Louisville, and Cincinnati have significant health problems tied to the contamination of their drinking water. A recent study by President Carter's Council on Environmental Quality reports that recent scientific research in five states has "strengthened the evidence for an association between rectal, colon and bladder cancer" and chlorinated drinking water supplies.

In 1974, Congress, with the backing of consumer groups, passed the Safe Drinking Water Act, giving the EPA the responsibility of establishing national drinking water standards. Congress took this action in response to broad public approval; but, surprisingly, the victory was not an easy one. The water supply companies, with few exceptions, pressed for weaker legislation.

The battle continued. In 1977, the EPA set the first national standards for safe water. They were very mild and did not cover certain chemicals. What's more, variances from these regulations were readily available. In 1978, the EPA proposed to modestly strengthen the standards. The result was a hot controversy.

The waterworks lobby (misnamed the "Coalition for Safe Drinking Water") renewed its fight against measures to improve your drinking water. The industry maintains that cleaning up our water would cost too much. They also say that current technology isn't sufficient for them to meet the standards. And they claim that the health risks of polluted water are vastly overrated. Two years ago, the American Waterworks Association went to court to weaken EPA regulations to control chloroform and other pollutants created by chlorination. The EPA just recently reached a settlement with the AWWA which would allow communities to receive a variance from these regulations, without having to employ some very effective technologies, such as the granular activated carbon filters originally suggested by EPA.

Yet, according to Dr. Arthur Upton, former director of the National Cancer Institute, chemicals known to cause cancer can do so even when present in extremely low concentrations — quantities measured in parts-per-billion or even parts-per-trillion. And Dr. Gerard Rohlich of the National Academy of Sciences says, "Low-level ingestion over a lifetime may seem to be a subtle problem. But that does not make the problem any less real."

The current situation is grim but it can be overcome. With an investment over the next ten years of about $3 billion, less than the cost of four days to run the Defense Department, the quality of our drinking water can be vastly improved. For we have available the best technology in the world to defend our drinking water. Carbon filtration systems can remove chlorine-created pollutants from drinking water; the State of New Jersey recently allocated funds to install such a system for drinking water drawn from the Passaic River. The systems are already widely used in Europe.

Progress will only happen if you make it happen. You can start a citizen's committee for clean drinking water in your community by enlisting the aid of a few concerned neighbors or friends. Then write to the Office of Drinking Water Supply, U.S. Environmental Protection Agency, Washington, D.C. 20460, for complete materials on your rights given you under the Safe Drinking Water Act. Once you understand the basic regulatory framework, your group can assemble the facts about your local drinking water supply, and begin the process of public education and working for engineering improvements.

Groups such as yours will be of critical importance in counteracting powerful industries which are continually trying to persuade Congress to weaken the law. They are at it again in 1982. These interests will be sure to make their voices heard shortly when the Safe Drinking Water Law comes up for renewal. Some are already pressing the Reagan Administration to roll back the regulations governing the disposal of hazardous wastes that can threaten groundwater.

You might also want to contact the Clean Water Action Project, 1341 G Street N.W., Washington, D.C. 20005, or the Natural Resources Defense Council, 122 East 42nd Street, New York, N.Y. 10168. They'll provide practical information and advice. The CWAP keeps abreast of developments in Congress and will provide you with tips about local organizing. The NRDC can also give you information about the healthfulness of water in America. Both of these groups can help you in your fight for clean water. Above all, write and ask your Senators and Representatives to state in some detail where they stand on the need to pursue cleaner drinking water programs. Hearing from the folks back home gives your legislators a backbone in Washington and lets them know the voters are watching them.

From <u>Environmental Quality-1979</u>, the Tenth Annual Report of the Council on Environmental Quality:

## DRINKING WATER

The Safe Drinking Water Act (SDWA), passed in 1974 and amended in 1977, requires EPA to establish federal standards for drinking water, protect underground sources of drinking water, and establish a joint federal-state system for assuring compliance with the resultant regulations.

SDWA envisions that the states will exercise primary enforcement responsibility (primacy) for drinking water programs, with EPA assuming this task only when a state is unable or unwilling to meet minimum requirements contained in the regulations. In order to attain primacy under the act, a state must establish drinking water standards and procedures for variances and exemptions at least as stringent as the national regulations; adopt and implement an adequate enforcement program; maintain records and submit reports as required by the Administrator; and establish an emergency response plan and a program for plan review.

Most U.S. citizens drink water from large publicly owned systems drawing on surface water supplies. However, the greatest number of public water supply systems in our country are ground water systems, even though a minority of citizens use them. Significantly, while large communities are able to operate and maintain water supply facilities properly, small communities often cannot. Therefore, ground water, filtered and covered by several layers of earth, has traditionally been considered the best source of water for small communities. However, as discussed earlier, contamination of ground water has become a significant drinking water problem in some areas.

In the next few years, the nation will have to address three basic and difficult problems concerning safety of drinking water:

* Carcinogens in source and finished water-- There is continuing debate on how best to eliminate carcinogenic contaminants in water, air, and food. Several such compounds have been identified in sources of drinking water, and even more have been identified in finished waters leaving drinking water treatment plants.
* Aging distribution systems-- The pipes in the water supply distribution systems of many large East Coast cities are old. It has been suggested that there may be large scale interruptions of water supply due to breakdowns in the future. The estimated costs for preventive action are large, perhaps as high as $50 to $100 billion nationwide. Other preliminary reviews indicate that the cause of distribution system failure may not be so much the age of the pipes as it is the inadequacy of maintenance of other urban infrastructures such as roads. Potholes alone cause extreme pressure on underground pipes as traffic travels the roadways.
* Drinking water supplies in water starved areas-- Much of the West experiences water shortages. Traditionally, all consumer-related water has flowed through the same pipes, thereby requiring all water used in the home to be of drinking water quality. However, much of this water need not be potable. Dual water systems allowing the reuse of treated water for certain activities may become not only desirable but necessary. Of course, duplication of water systems could also be tremendously expensive.

November, 1980 • Nutrition Action

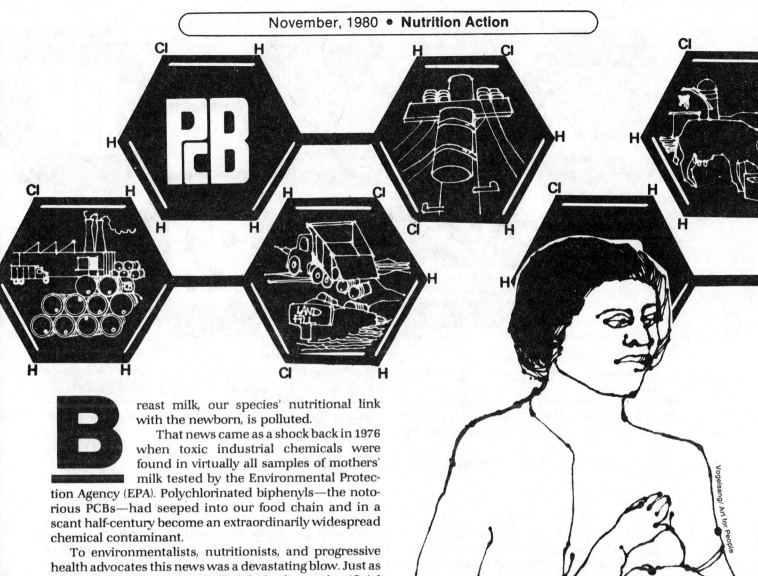

# PCBs and Breast Milk

## Mothers Reconsider Choice of Nursing Over Bottled Formulas Given PCB Threat

**By Marcella R. Mosher and Greg Moyer**

Breast milk, our species' nutritional link with the newborn, is polluted.

That news came as a shock back in 1976 when toxic industrial chemicals were found in virtually all samples of mothers' milk tested by the Environmental Protection Agency (EPA). Polychlorinated biphenyls—the notorious PCBs—had seeped into our food chain and in a scant half-century become an extraordinarily widespread chemical contaminant.

To environmentalists, nutritionists, and progressive health advocates this news was a devastating blow. Just as America was weening itself off bottle-feeding and artificial supplements, the purity of the "ideal food" seemed seriously compromised.

The public learned that PCBs—shown to cause cancer in laboratory rats and mice—were widespread in the natural environment. Air, water, soil, plants, and animals were all touched. Coincidentally, a study from the University of Wisconsin Primate Research Center reported that offspring of monkeys nursed with PCB-laden milk exhibited skin disorders, showed signs of hyperactivity, and appeared to have learning deficiencies. Within a year, three of the six infant monkeys had died.

The public outcry then, as now, was one of anger mixed with frustration. No woman can quickly or totally cleanse her system of this toxic industrial product. Nor can she totally avoid ingesting some amount of the substance from the food she eats.

People slowly realized that future generations were destined to be chemically "branded" with one of the most toxic and persistent synthetic chemicals ever invented. The PCB tragedy became the classic example of what happens when toxic chemicals are let loose in the environment without strict governmental controls.

But the possible consequences of PCBs' effects are compounded for infants who breast-feed. Research has shown that breast-feeding tends to pass on higher concentrations of PCBs and other toxins than had existed in the mother. As a woman lactates, she draws on nutrients

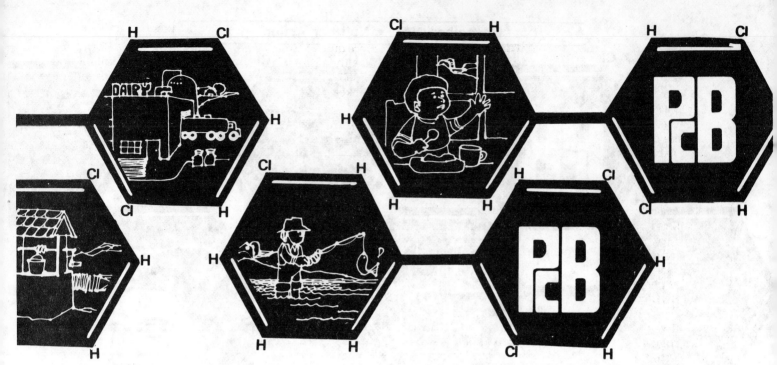

stored in her fatty tissue. And it is in these cells where the pollutant accumulates.

Once the PCBs get into the blood, they quickly find their way into the mother's milk. Ironically, one of the best methods for excreting the poisons is through breast milk.

The Wisconsin primate study showed that it took only three months for the breast-fed monkeys to reach the contamination level of their mothers. By the end of the lactation period, the levels in the babies had far exceeded the levels of the mothers.

Aware of this new scientific information, parents are put in the position of gambling with their infant's well-being. A mother can elect to breast-feed given the ample nutritional, sensory, and psychological benefits, but risk exposing her baby to higher levels of PCBs. Or she can opt for bottle-feeding and settle for the lower nutritional content of the supplements in order to guard the infant from the toxins.

## How Did It Happen?

PCBs belong to a group of chemicals known as chlorinated hydrocarbons. This highly stable organic compound arrived on the scene in 1927 when Theodore Swann, of the Federal Phosphorus Company, first used biphenyl in the refining of lubricating oils. Two years later, Swann invented polychlorinated biphenyls and sold the process to the Monsanto Company.

Since 1930, 1.25 billion pounds of the stuff have been manufactured for use in this country. Over half of all PCBs ever made are still in use, though the 1976 Toxic Substances Control Act allowed EPA to order all further production halted. About 90 percent of the rest lies in dumps, landfills, soil, water, and living organisms. The EPA estimates that about 10 million pounds contaminate the biosystem each year through seepage, spills, and evaporation. For years PCBs were routinely discharged into rivers, streams, and sewers.

In its 50-year history, industry has found no shortage of uses for this versatile material. It burns only at very high temperatures, conducts heat well, refuses to dissolve in water, and resists corrosive chemicals. Only in fats will the PCB molecule enter solution. The compound has been used in everything from transformers and power capacitors, adhesives, washable wall coverings, upholstering materials, and flameproofers to early versions of carbonless copy paper.

PCBs are actually many times more toxic and much slower to break down into less harmful forms than the better known insecticide pollutants: DDT, chlordane, dieldrin, and mirex. The toxic effect of the biphenyls was first seen in 1943 among workers handling electrical equipment who had developed skin lesions and liver damage. In 1966, PCBs were found in birds in Sweden.

PCBs created more headlines in the early 1970s when striped bass caught in the Hudson River near New York City registered dangerously high concentrations. By 1974, the U.S. Department of Interior's Fish and Wildlife Service identified PCBs in 90.2 percent of freshwater fish sampled across the entire United States.

Each step up the food chain tends to concentrate the toxins. This process is known as "bio-accumulation." For example, fish have been discovered with PCB levels many million times that of their watery environment. Also, cow's milk contains a much higher PCB level than the corn on which the animal feeds. Breast-feeding, in effect, sends bio-accumulation into overdrive.

## Who Are the Victims?

It falls to the tiniest and weakest members of the human race to withstand the greatest concentrations of PCBs. The studies of Dr. James Allen, formerly of the Wisconsin Primate Center, and new research detailing the transfer of PCBs through breast milk underscore the dangers for the newborn in a post-PCB world.

But another category of victims include those people exposed by accident or through occupation to unusually high dosages of PCBs. The real life experience of a Japa-

> ## Regulators Mum on Risks
>
> In its report to the nation, the Surgeon General's *Healthy People* barely acknowledges the PCB threat in mother's milk while endorsing breast-feeding. Chapters on healthy infants, pregnancy, and infant care make no mention of the trade-offs and risks of breast-feeding.
>
> Only in a chapter on toxic substances is there a brief description of the PCB problem. It concludes... "PCBs have been found in tissues of humans and in the milk of nursing mothers. When fed experimentally to nursing monkeys, PCBs have led to serious injury of offspring." Period. No recommendations. No advisories.
>
> When contacted, a spokesperson in the Surgeon General's Office said that stronger warnings were not appropriate until other agencies in the government take a firm position on the matter.
>
> We suggest, instead, that the risks be aired and serious studies begin immediately.

nese community serves as a frightening example of the awesome power packed in this toxin.

In 1968, residents on the Japanese island of Kyushu unwittingly ate large amounts of PCBs when a cooling pipe leaked into a vat of rice oil at a food processing plant. The level of contamination (2000 to 3000 parts per million) reached 200 times what is considered safe by the U.S. Food and Drug Administration. Officially, 1,291 people contracted the "rice oil disease" characterized by acne-like rashes, headaches, nausea, diarrhea, loss of hair, loss of libido, fatigue, numbness, and menstrual disturbances.

Thirteen children were born to PCB-exposed women. One fetus was stillborn, four were small for gestational age, ten had dark skin pigmentation, four had pigmented gums, nine had conjunctivitis (inflammation of the membrane around the eye), and eight had neonatal jaundice. Nine years after the accident, a follow-up study of some of the children breast-fed by exposed mothers showed "slight but clinically important neurological and developmental impairment," according to a report of the research prepared by Dr. Walter J. Rogan of the National Institute of Environmental Health Sciences, and others.

Allen's rhesus monkeys fared little better. His research involved feeding controlled amounts of PCBs to female monkeys. Their offspring exhibited the same skin disorders as the victims of the "rice oil disease." The baby monkeys failed to make the developmental progress shown by the control group and several died at an early age.

A study led by Dr. Hirokadzu Kodama of Nagoya, Japan, finds comparable effects in humans. He discovered that mothers with average exposure to PCBs are still likely to pass on levels higher than their own to nursing offspring thanks to bio-accumulation.

Still, some researchers caution against making direct comparisons between monkeys and humans, because the monkeys have less subcutaneous fat in which to store the PCBs. In the animals, the toxins assault the vital organs more quickly. But with this caveat in mind, one can compare the levels of PCBs ingested by a nursing infant and those which were shown to affect the health of the monkeys.

### PCB Contamination in Breast Milk

| Subject | Toxins in milk | Contamination by weight* |
|---|---|---|
| Monkey | 16 ppm PCBs on a fat basis. (Toxicological effect observed at Wisconsin Primate Center) | 64 micrograms of PCB per kilogram of body weight. |
| Infant (human) | 1.8 ppm PCBs on a fat basis. (EPA national average) | 10 micrograms of PCB per kilogram of body weight. |
| Infant (human) | 10.6 ppm PCBs on a fat basis. (Worst case uncovered by EPA study) | 60 micrograms of PCBs per kilogram of body weight. |

*These figures are based on the assumption that baby monkeys weighed 500 grams and drank 100 cubic centimeters (cc) milk per day, while human babies weighed six kilograms and drank 1,000 cc milk per day.
Data drawn from Harris, "Health Risks and Benefits of Nursing or Bottle-feeding: The Limits of Choice", a chapter in the forthcoming book, *Strategies for Public Health*.

The average exposure infants receive when adjusted for body weight is one-sixth that shown to adversely affect the baby monkey's health. In the worst case found by EPA's nationwide sampling, a human infant was receiving virtually the same dose as the monkey. Normally, regulatory agencies use a safety factor of 100 to set a "safe" standard for humans based on the results of the animal studies. Here, the concentrations are on the same order of magnitude, if not actually equivalent when translated to the common denominator of micrograms per kilogram of body weight.

The EPA figures are shocking in another sense: the average levels of PCBs in human breast milk is seven times higher than the amount permitted in cow's milk by the Food and Drug Administration. And it is about ten times what the FDA recommends as an acceptable daily allowance for infants.

In a telephone interview, Allen underscored the need for parents and the pediatrician to know about the physiological changes that may be occurring in infants exposed to high levels of PCBs. He said autopsies of the monkeys indicated abnormalities of the lymph system, bone marrow, thymus, and cortex. In short, he saw irregularities in some of the organs most important to normal growth and development.

Despite these ominous signs, Allen does not suggest women return to bottle-feeding their babies: "The values of breast-feeding outweigh the deleterious effects," he said. "I would advise my wife to breast-feed unless we lived in an area of high exposure to PCBs."

## The Benefits of Breast Milk

During the 1970s, scientists undertook new studies to better understand the attributes of breast milk. They discovered that mother's milk contains:

- a thyroid hormone which babies cannot produce for themselves;
- adequate amounts of iron and zinc, long thought to be deficient in breast milk;
- antibodies and white cells to fend off disease.

For the mother, nursing immediately after birth acts as an anti-hemorrhaging agent while the uterus contracts. Nor is any drug, such as DES (a carcinogen), necessary to relieve the discomfort of breasts temporarily engorged with milk.

Of all the benefits claimed for breast-feeding, the ones least susceptible to measurement are the psychological and emotional. Through skin contact and sensory stimulation, a woman who enjoys nursing may be passing on a real sense of security and attachment to her baby. No one can say for sure how valuable this is for the mother-child relationship.

## What Kind of Balance Sheet?

One person who has consistently focused on both the risks and benefits of breast-feeding is Stephanie Harris, a chemist by training, who worked on pesticide issues for Ralph Nader's Health Research Group. Aware of the generally high levels of pesticides in animal products, Stephanie became a vegetarian three years before the birth of her first child. She grows most of her own food organically, filters her water, and takes careful steps to avoid toxic chemicals. Due to a family history of allergies, she chose to nurse both her children, though she knows her milk contains some PCBs.

Stephanie Harris decries the inadequate information available to pregnant mothers about the poisons in their bodies. It was her persistence and concern that helped force the issue before the regulators in 1976. Then, as now, Harris says, there is no inter-agency governmental group responsible for both the monitoring and analysis of the PCB threat in breast milk.

"It is an issue that slips through the cracks," Harris commented sadly.

Meanwhile, parents like Stephanie Harris are faced with difficult choices. She offers women who may become pregnant these precautions for limiting an infant's exposure to PCBs:

- Eliminate from the diet freshwater fish, like salmon and carp, and bottom-feeding estuarine fish, like catfish, flounder, and sole.
- Eat more fresh vegetables, especially organic vegetables, and grain products since more PCBs are ingested through animal and dairy products.
    Note: don't expect diet to make a difference overnight. Studies show that PCBs remain in the body for years after the initial exposure. All that you can do is stop increasing the PCBs to your system.

---

### CSPI Calls for Action

Breast-feeding infants with milk contaminated by PCBs poses an acute dilemma for new parents and a generalized risk for society-at-large. The Center for Science in the Public Interest urges:

- The National Institutes of Health to hold conferences and workshops aimed at identifying the research needs in this little-understood field. We would expect Congress and NIH to follow through with funds to commission such research.

- A national conference sponsored by a consortium of governmental agencies to educate the public about the long-term implications of PCB contamination.

- The Institute of Medicine of the National Academy of Sciences to undertake a study weighing risks and benefits of both breast-feeding and bottle-feeding.

---

- Lose weight *before* pregnancy to reduce the fatty tissue that keeps PCBs in the body. Once pregnant, maintain your weight. Weight reduction during pregnancy only releases more contaminants from the fat cells into the blood stream and eventually through the placenta.
- Avoid breast-feeding if you know you are exposed to unusually high levels of PCBs where you work or live.
- Consider breast-feeding one or two weeks to reap the known benefits of colostrum, the secretion preceding milk, then shift to bottle-feeding.
- Mix breast and bottle-feeding throughout the lactation period depending on the amount of PCBs you suspect yourself of carrying.

(Unfortunately, no inexpensive and reliable tests exist for detecting PCBs in mother's milk.)

Stephanie Harris sums up the dilemma best:

"The choice [a mother makes] is based upon speculation and hypothesis rather than established scientific fact. It is rare that a clear 'right' decision exists. . . . Yet the continuing health and well-being of the infant hangs in the balance."

---

### Suggested Readings

Allen, J.R. et al., **Fd. Cosmet. Toxicol. 15**, 401-410. 1977.

Goldman, A.S. et al., **J. Pediatrics 82**, 1082-1090. 1973.

Harris, S.G. and Highland, J.H., **Birthright Denied: The Risks and Benefits of Breast-feeding**. Environmental Defense Fund. 1977.

Harris, S.G., **Strategies for Public Health**, ed. Lorenz, N.S. and Devra Lee Davis. Van Nostrand Reinhold, N.Y. 1980.

Kodama, H. et al., **Arch. Env. Health 35**, 95-100. 1980.

Rogan, W.J. et al., **The New England Journal of Medicine 302**, 1450-1453. 1980.

# DEFENDING AND STRENGTHENING THE FOOD SAFETY LAWS

**Section Four**

## DEFENDING AND STRENGTHENING THE FOOD SAFETY LAWS

From a health and safety standpoint, American consumers face a three-pronged aggression. First is the mounting pollution and chemicalization of the food supply. Second is the Reagan Administration's anti-regulatory bias against applying law and order to negligent or criminal corporate actions. And third is the giant food processing and soft drink lobby's well-financed drive to seriously weaken the food safety laws. Senator Orrin Hatch (R-Utah) has introduced legislation heavily shaped by food and drink industry lawyers, to make it difficult to prevent the use of dangerous additives and easier to approve them. Consumer advocate William Schultz of Public Citizen's Health Research Group explains and exposes Hatch's bill (S. 1442) in step-by-step detail. The Center for Science in the Public Interest has a comparative chart of how already banned additives would fare under the Hatch proposal, if enacted.

Sen. Hatch defends his bill in a *Washington Post* article and Attorney Schultz responds. Also included in these readings is an article by the *Washington Post* about the revolving door phenomenon whereby ex-FDA officials, now representing the industry, are attacking the present, modest food safety laws. A *Post* editorial offers its view of the situation.

Documentation of reduced FDA law enforcement is contained in a Health Research Group report by Dr. Sidney Wolfe and Allen Greenberg. Notice also the stomach-grabbing examples near the end of the report of the kind of enforcement action which decreased in 1981.

Last year the Health Research Group and the Center for Science in the Public Interest, after years of urging the FDA to stop the illegal marketing of 23 color additives used in food, drugs or cosmetics which have never been proven safe as required by law, filed suit against the Food and Drug Administration. That complaint and some background information are included in this section. These materials also illustrate the critical importance of consumer protection groups in Washington who work in this way and who ask you to help them from time to time.

# THE FOOD SAFETY AMENDMENTS OF 1981:
## A CASE OF LEGISLATIVE MISBRANDING

Presented at the Food and Drug Law Institute Annual Educational Conference, December 15, 1981, by William B. Schultz, Public Citizen Litigation Group

### OVERVIEW

In June of this year, Senator Orrin Hatch introduced S. 1442, which is a comprehensive bill that would thoroughly change the regulation of additives in the United States. Although called the "Food Safety Amendments of 1981," S. 1442 would actually undermine the safety of our food supply, impairing the public health. The theme of the bill, which was written by and for the food industry which relies on chemical additives, is "easy on - slow off." If adopted, it would make it easier to market new chemical additives by requiring the Food and Drug Administration ("FDA") to act on additive petitions within an unrealistically short period of time, and in many cases without the evidence necessary to determine whether the additive is safe.[1]

The bill also would make it harder, indeed practically impossible, to take unsafe food additives off the market, since the FDA could not act without first jumping over several procedural hurdles. During the time agency action was delayed (which could last many years), consumers would be exposed to unsafe chemical additives.

Thus, the bill is mislabeled because its title, "The Food Safety Amendments of 1981," does not accurately describe the effect of the bill. Rather than improving the safety of our food supply, S. 1442 would:

* redefine the statutory term "safe" and could allow the introduction into the food supply of new chemicals that can cause serious injuries;

* repeal the Delaney clauses which prohibit the intentional addition of carcinogens to foods;

* allow consideration of the costs of savings to industry in deciding whether a food additive is safe;

* require the use of scientific advisory committees to delay removal from the market of unsafe chemical additives;

* require the use of a host of new procedures which would make it almost impossible to withdraw FDA approval of an unsafe additive;

* require the FDA to allow new uses of additives before the safety of those uses has been evaluated.

The most disturbing aspect of the amendments is that they are, without exception, one-sided. Every new provision will benefit industry, but there is no provision which offers any additional protection to consumers. Because S. 1442 would jeopardize the safety of the food supply, I believe that it should not be adopted.

---

[1] Throughout this paper, the term "additives" is used to refer to food additives, color additives, and animal drugs which leave residues in human food. These substances are subject to the requirements of the Federal Food, Drug and Cosmetic Act, administered by the FDA, and three statutes administered by the U.S. Department of Agriculture ("USDA").

## SECTION-BY-SECTION ANALYSIS

### Introduction

The Food, Drug and Cosmetic Act divides food substances into two broad categories. The first is "foods" which include all food substances ingested by humans. 21 U.S.C. § 321(f). The second category is intentional additives such as food additives, color additives, and animal drugs.[2] Under the Act, these additives are subject to more stringent regulation than food substances which are not additives.

All foods and food additives must meet the requirements of sections 402 and 406 of the statute. 21 U.S.C. §§ 342, 346. In general, a seller of a food substance which is not a food additive (such as corn) may simply sell the product without obtaining approval from the FDA. Section 402 prohibits the sale of adulterated foods, which includes foods that contain a "poisonous or deleterious substance which may render it injurious to health" (emphasis supplied). 21 U.S.C. § 342(a)(1) However, an unavoidable substance will not render a product adulterated unless it "ordinarily renders the product injurious to health" (emphasis supplied). Id. Section 406 requires the FDA to set tolerances for such unavoidable substances in order to protect the public health. 21 U.S.C. § 346.

Intentional additives (food additives, color additives and animal drugs) are also subject to separate, more stringent provisions. In general, these additives may not be used unless first proven safe. Moreover, whereas the FDA may not remove a food from the market under section 402 unless it meets its burden of showing that the food is adulterated, it may decline to approve an additive if the manufacturer has not met its burden of proving that the additive is safe.

The different treatment given to additives is appropriate for several reasons. First, most additives used in foods (for example, preservatives) have substitutes, and thus the public typically has no stake in whether a particular additive is approved. Second, many of these substances are chemicals with no substantial history of use, and they should be tested before being introduced into the food supply. Third, in contrast to a food, an additive is often not identifiable in the food. Since the consumer cannot identify the animal drug residues in meat, or often the specific color additive in candy, it is appropriate to require a higher standard of safety than the one applicable to foods which can be readily identified and avoided. By requiring that in general additives be proven safe prior to use, and prohibiting the use of new additives when the evidence on safety is inconclusive, the statute errs, if at all, on the side of safety. The justification for striking this balance is that, in general, additives are neither beneficial nor essential.

Despite the justifications for stringent regulation of additives, the FDC Act exempts two categories of additives from the premarket approval requirement. First, it exempts food additives which are generally recognized as safe ("GRAS") and animal drugs which are generally recognized as safe and effective ("GRAS/E"). 21 U.S.C. §§ 321(s), 321(w), 348, 360b. These provisions, which apply to additives which have been used for a substantial period of time, allow the exempted substances to be sold without prior approval. However, a substance will lose GRAS protection if a new study raises questions about its safety, in which case it may not be sold until proven safe under the provisions of the Act applicable to additives. See Weinberger v. Hynson, Westcott & Dunning, 412 U.S. 609, 632 (1973)(interpreting an analogous provision applicable to drugs as requiring an "expert consensus" that the product is safe). Second, the provision which covers food additives contains a "grandfather" clause which exempts from the statute's safety requirements substances marketed prior to 1958 (the effective date of the Food Additives Amendment) pursuant to a government sanction or approval. However, prior-sanctioned substances are subject to the adulteration provisions generally applicable to foods. 21 U.S.C. §§ 342, 346.

\* \* \* \*

---

2  The Act defines "food additive" to exclude "color additives," and regulates the two substances separately. 21 U.S.C. § 321(s)(3).

A review of the statutory scheme shows that there are three ways to weaken the safety protections of the FDC Act. First, one could dilute the definition of "safe." Second, one could exempt additional substances from the safety standard. The final approach would be to relax the requirements for obtaining approval of an additive, or, conversely, to make it more difficult for the FDA to withdraw its approval of an additive later found to be unsafe. S. 1442 adopts all three approaches.

## I. THE SAFETY STANDARD

### A. The Definition Of Safe

The FDC Act does not define the term "safe," although it requires all substances which it defines as "food additives," "color additives," and "animal drugs" to be "safe" prior to sale to the public. 21 U.S.C. §§ 348, 360b, 376.[3] However, based on the legislative history of the statute and more than 20 years of experience with the concept, the FDA has defined safety as "a reasonable certainty in the minds of competent scientists that [a chemical additive] will not be harmful to the public under its proposed conditions of use." Summary of Commissioner's Decision on Aspartame, Food and Drug Administration. Docket No. 75 F-0355 (July 15, 1981), p.1.

Section 103 of S. 1442 would redefine "safe" to mean "the absence of significant risk under the intended conditions of use." The problem with this definition is that it would raise doubts about what kind of safety would be required for chemical additives because the term "absence of significant risk" is so ambiguous. Thus, the phrase could be interpreted in legislative history or regulations to allow additives which pose serious health risks. For example, under the new definition, a 1/100,000 risk might be deemed as "not significant," since any individual's risk is so small. However, the risk is very significant to the consumer who develops cancer. In a nation of 200 million people, 2,000 people would develop cancer from consuming a substance with 1/100,000 risk.[4]

---

[3] The term "safe" is also used in at least three other places in the Act. Certain substances which would otherwise be regulated as food additives are exempted from the premarket approval requirements of the statute if they are generally recognized as "safe," whereas substances which would otherwise

Moreover, it is unclear whether the FDA could take cumulative exposures into account under the proposed definition of "safe." For example, the risk to an individual exposed to 100 chemicals carrying a 1/100,000 risk would be 1/1,000, and the cumulative exposure would cause 200,000 cancers. This example highlights the uncertainty inherent in the concept of significant risk.[5]

In addition, the bill covers only "<u>intended</u> conditions of use," and implicitly excludes unintended conditions of use. It is unclear whether moderate over-consumption of a product would qualify as an unintended condition of use, although the FDA currently requires manufacturers to show that chemical additives are safe for consumers who consume large amounts of a substance.[6]

### B. The Anti-Cancer Clauses

The FDC Act in general gives the FDA broad discretion in deciding what factors to take into account in determining whether an additive is safe. The one exception is if the additive is a human or animal carcinogen, in which case the Delaney clauses prohibits the FDA from approving it as a food additive, animal drug, or color additive. 21 U.S.C. §§ 348(c)(3), 360b(d)(1)(H), 376(b)(5)(B). Since scientists in 1960 were not yet able to identify additives which were animal carcinogens but which did cause cancer in humans, Congress adopted the Delaney clauses to prohibit all additives which have been found to cause cancer in animals. The provision gives the FDA broad discretion to determine the appropriateness and validity of the scientific tests that are used to identify animal carcinogens which are unsafe for human consumption.

Since additives are generally interchangeable or unnecessary, the Delaney clauses lower the incidence of cancer by eliminating many food additives which cause cancer. The clauses are particularly important in prohibiting the introduction of new additives, but they may also be used as a basis for withdrawing a license for an approved additive which has been shown to cause cancer.

[4] Of course, the science of quantitative risk assessment is not nearly as precise as these numbers suggest, and the actual cancers caused could be greater or less than those predicted, depending on the accuracy of the assumptions which are used in the analysis.

---

be animal drugs are exempted if they are generally recognized as "safe and effective." 21 U.S.C. §§ 321(s), 321(w). In addition, the Act uses the term "safe" in connection with the regulation of pesticides. 21 U.S.C. § 346a. Presumably a change in the statutory definition of safe would affect those provisions as well.

Sections 112(f), 124(c), 131(c) of S. 1442 purport to amend the Delaney clauses, but in fact S. 1442 would repeal the anti-cancer clauses. Under the bill, the Delaney clauses would not apply if the FDA found that the new additive "does not present a significant risk to human health." This tracks the new definition of safe, and means that the Delaney clauses, as amended, would add nothing to the new general safety provision. The result would be no different if the Delaney clauses were repealed.

Any justification for repealing the Delaney clauses depends on the assumption that it is possible to identify animal carcinogens which are safe for humans, even if they cause cancer in animals. In fact, the weight of scientific opinion is that this is impossible today just as it was in 1960, and therefore there is no justification for repealing the Delaney clauses.

In addition to repealing the Delaney clauses, S. 1442 would require the FDA to approve human carcinogens if it found that the substance's economic or other benefits outweigh its risks. This is also an impossible task. Thus, the primary impact of S. 1442 would be to embroil the FDA in disputes about whether an animal carcinogen causes cancer in humans, and about how it should perform the impossible task of weighing those risks against possible benefits. At best, the result would be to waste agency resources; at worst, it would mean exposure of consumers to additional carcinogens which are wholly avoidable and possibly very dangerous, further diminishing the safety protections of the FDC Act.

C. Phase Out Authority

S. 1442 contains a third provision which would weaken the safety protections of the FDC Act. Section 107(b) of the bill would give the FDA authority to phase out the use of an additive which it had determined to prohibit on the ground that the additive posed a danger to the public health. The theory of the provision is that, unless a phase-out of the additive's use is permitted, manufacturers will have to bear the cost of

---

[5] Section 409(c)(5) of the FDC Act directs the FDA to consider "the cumulative effect of [the food] additive" being evaluated, but it is unclear under what, if any, circumstances the FDA may consider cumulative exposures of different additives.

[6] Even the current statute does not require the manufacturer to show that a product is safe at every conceivable level of use, and the FDA apparently does not require manufacturers to show that the additives would be safe if consumed at excessive levels.

changing to new additives, as well as the loss of the stock already on hand. The problem with the provision is that it sacrifices the health and safety of consumers to advance the economic interests of manufacturers. Moreover, under the current statute, the FDA exercises limited phase-out authority by allowing substances to be sold during the rule making proceedings, and prior to the effective date of regulations banning substances, which is typically 60 days after the issuance of a formal rule. The courts have upheld this practice. Public Citizen v. Schmidt, [1976 Transfer Binder] Food Drug Cos. L. Rep. (CCH) ¶ 38,075 (D.D.C. 1976), affirmed sub nom. Public Citizen v. Kennedy, 564 F.2d 600 (D.C. Cir. 1977 table); see Merrill and Hutt, Food and Drug Law 389 (1979).

The only example cited to support the inclusion of this provision in the bill is nitrites. In 1978, the Carter Administration requested Congressional authority to phase out the use of nitrites on the ground that, even if nitrites were a carcinogen, they should remain on the market until a substitute preservative was found because of their alleged health benefits in protecting against botulism. Consumer groups have argued that nitrites are not necessary to prevent botulism, but regardless of whether there was a justification for phasing out the use of nitrites, that single example is not a sufficient basis for broad phase-out authority for all chemical additives, since most additives do not have even an arguable health benefit (e.g., color additives), or are interchangeable with other, safe additives (e.g., preservatives).

Any authority to phase out the use of an additive should be carefully scrutinized because such authority would allow the FDA to override the health and safety provisions of the FDC Act. The provision in S. 1442 has no limitation because it is not limited to a specified period of time, or to specified health-related purposes. The only limitation in S. 1442 is the requirement that the phase out of a chemical additive "not present an imminent danger to health" is wholly inadequate since it creates a standard which is not defined and which has no precedent in the FDC Act. Under this concept, the FDA might be able to conclude that a widespread risk of cancer 20 years hence would not present an "imminent danger to health," and thus this new standard would be have no limitations at all.

II. THE ADULTERATION STANDARD

As discussed earlier, sections 402 and 406 of the FDC Act prohibit the sale of adulterated food, and the FDA may stop the shipment of such food by proving that it contains a substance which may be, or ordinarily is, injurious to human health. 21 U.S.C. §§ 342, 346. See pp. 5-6, supra. Section 108 of S. 1442 would add a proviso to section 402(a)(1), 21 U.S.C. § 342(a)(1), which would require the FDA to perform a risk assessment in order to prove adulteration under the statute.

The risk assessment would include, "among other relevant factors, the nature and extent of the risk" (emphasis added). It is unclear whether factors other than risk, such as economic cost, may be considered. In addition, section 108 would change the current categorization of herbs and spices, treating them as foods rather than food additives. Thus, a spice added to food which "may" render it injurious to health would not render the food adulterated. Instead, the FDA could prohibit use of the spice only if it "ordinarily" renders the food injurious to health.[7]

With respect to the risk assessment part of the amendment, it should first be noted that section 402 currently requires the FDA to assess the risks of foods and food additives in determining whether they may be or ordinarily are injurious to health. The risks of a substance are typically proven through expert testimony if the case is litigated in court. Thus, to the extent that the purpose of the amendment is to require the FDA to take health risks into account in determining whether a food is adulterated, it is wholly unnecessary.

Moreover, the fundamental problem with the amendment is that it apparently would require the FDA to produce a risk assessment document prior to proceeding in court to remove adulterated foods from the market. The preparation of such a document could delay agency action, and is wholly unnecessary since the evidence of adulteration would have to be presented to the court in any event. Second, it is unclear whether the decision would be based only on data pertaining to the risks of the substance. The phrase "among other relevant factors" suggests that the economic interests of the industry might be relevant, which would be a major departure from the current statute and could result in adulterated food remaining on the market solely to protect commercial interests.

Section 108, which would treat herbs and spices (hereinafter referred to as "spices") as unavoidable substances in foods rather than intentional food additives, departs from the safety standards of the current statute. Whereas the FDC Act deems a food "adulterated" if it contains a spice which "may" render it injurious to health, under S. 1442 a spice could not render a food adulterated unless the spice is "ordinarily injurious," a standard which would make it more difficult for FDA to remove some potentially dangerous additives from the market. There is

---

[7] S. 1442 accomplishes this change in the regulation of spices by declaring that "basic and traditional foods" shall not be deemed to be added substances (section 108), and defining "basic and traditional foods" as spices and certain raw agricultural commodities (section 106).

no justification for prohibiting the FDA from treating spices as added substances subject to the "may render injurious" standard.[8]

Section 110 would amend section 406 of the FDC Act, which requires the FDA to promulgate tolerances for substances which are added to foods, but which nevertheless are unavoidable.

21 U.S.C. § 346. Under section 406, for example, the FDA should set tolerances for aflatoxin (a carcinogen caused by mold and found in corn and peanuts) or other food contaminents. The amendment would weaken the FDC Act in a number of ways. First, it would delay the promulgation of a tolerance by requiring the FDA to produce a risk assessment, which would be further delayed by the requirement that a food safety committee be consulted. See pp. 11, 12, infra. Second, the FDA would be required to take into account the effect of setting a tolerance on "the cost and availability of food," although there is no indication of how the agency is supposed to balance health risks against cost.

Third, the amendment would give a manufacturer who is unhappy with a tolerance the right to obtain judicial review of the agency's decision and to have the judge decide anew all the scientific issues already decided by the agency. This provision is particularly troublesome because it would eliminate the deference which courts traditionally give agencies on scientific matters. It would put a manufacturer unhappy with the FDA's decision in the best possible position, since a company unhappy with a tolerance could often litigate the issue before "home-town judge" who could make factual findings which would be difficult to reverse on appeal. Under principles of collateral estoppel, the FDA might be bound by those findings in other courts.[9]

Fourth, the Amendment creates a different standard for prior-sanctioned substances, which are substances that are exempted from the safety requirement of the Food Additives Amendment pursuant to a grandfather clause. This provision applies to substances which were used prior to 1958 pursuant to a sanction or approval granted by the FDA or the USDA.

---

[8] The change in the treatment of spices is more significant than it might first appear. Under the current Act, most spices must either be "generally recognized as safe" or must be approved as "safe" food additives, 21 U.S.C. §§ 321(s), 348. However, S. 1442 would exempt spices from the definition of food additive (section 102(b)), and thus they would be subject only to the adulteration requirements of sections 402 and 406, 21 U.S.C. §§ 342, 346, which are further weakened by this bill.

21 U.S.C. § 321(s)(4); see Public Citizen v. Foreman, 631 F.2d 969 (D.C. Cir. 1980). The amendment would allow the FDA to set tolerances for substances for which "there is no reasonably practicable substitute." This provision might be used to allow the addition of a substance to the food supply if it imparted a unique taste or other trivial characteristic to food even if the substance was a potent carcinogen and was intentionally added. This provision would also weaken health and safety protections of the Food, Drug and Cosmetic Act.

### III. FOOD ADDITIVES.

Sections 111-117 and 120 would substantially amend Section 409 of the FDC Act which establishes the standards currently applicable to food additives. 21 U.S.C. § 348. Under section 409, a substance defined as a "food additive" may not be sold unless it has first been proven to be "safe." A company desirous of using the additive must prove safety, but once the additive is approved, that company has a license which entitles it to market the additive. If new evidence shows the additive to be unsafe, the Secretary may revoke the licence, 21 U.S.C. § 348(h), but the holder of the license would have a right to a hearing, during which time he ordinarily could continue to market the additive. Even under the current statute, the procedures necessary to revoke a license can take two to four years to complete.

As has been previously discussed, certain substances are exempt from the definition of food additive, including substances which are generally recognized as safe ("GRAS") and substances which are prior-sanctioned. While the prior-sanction exemption is permanent, GRAS substances are exempted only as long as they continue to be generally recognized as safe. Once they lose their GRAS status, they may not be sold unless they have been proven safe pursuant to the licensing provisions applicable to other food additives. However, even after questions have been raised about the safety of a GRAS

---

[9] For example, when the meat industry challenged the USDA's regulation allowing nitrite-free bacon to be called "bacon," it sued the USDA in Iowa where it found a favorable judge who concluded that this labeling regulation created a health hazard. National Pork Producers Council v. Bergland, 484 F. Supp. 540, 547 (S.D. Iowa 1980). Upon appeal, the Eighth Circuit reversed, according the USDA the traditional deference which courts have given federal agencies since the adoption of the Administrative Procedure Act, 35 years ago. National Pork Producers Council v. Bergland, 631 F.2d 1353 (1980). However, under the amendment, the District Court's findings about health hazards would have to be affirmed unless clearly erroneous, since S. 1442 would allow the court to accord the agency no deference whatsoever.

substance, the FDA often takes several years to consider the evidence, and during this time it places the substance on an interim food additive list. See 21 C.F.R. § 180. Thus, for example, the FDA did not act to eliminate the use of saccharin as a food additive for many years after questions were first raised about its safety, and today the FDA has not taken any action with respect to caffeine even though serious questions have been raised about its safety. Neither saccharin nor caffeine was ever approved as a food additive.

S. 1442 would weaken the regulation of food additives in a variety of ways. Although the devices used vary, in each case the bill will make it easier to get new additives on the market (typically by weakening the current safety requirements), or they will make it more difficult to remove additives which have been found to be dangerous.

### A. Easy On

Section 112 would weaken the safety standards of the statute by requiring the FDA to sanction new uses of food additives which had not been proven safe on an "interim basis" if the FDA found that it "appears, from initial review of the petition . . ., that such use is safe." Thus, an additive such as aspartame, which is currently approved for a limited use as a dietary sweetner in foods, could be approved for widespread use (in diet sodas, for example) simply on the basis of an "appearance" of safety. Moreover, there is no limit placed on the interim listing, and it could be continued indefinitely while the petitioner is undertaking testing necessary to show that the additive is safe. This provision alone could seriously undermine the public health and safety.

For new additives, S. 1442 retains the requirement that the FDA rule on applications within 180 days. 21 U.S.C. § 348(c)(2). However, under the current statute, that time limit is treated more as a goal, which may or may not be met depending on the FDA's other commitments. Section 116 of S. 1442 would make the 180 day time limit mandatory and enforceable in court. This provision could have the unfortunate effect of diverting FDA resources now committed to protecting public safety (e.g., removing unsafe drugs and food additives from the market) to the less important task of evaluating new food additives. In addition, section 116 would set the court's priorities, requiring the FDA to answer the complaint within 10 days rather than the usual 60 days, and requiring the court to expedite the case.

While giving the FDA less time, S. 1442 would require the agency to do more work in evaluating an application to market a new food additive. Section 112(e) requires an "assessment" of the risks of the additive. Whereas the current Act requires the applicant to submit the data relevant to the safety of the additive, including, presumably, an assessment of the risks, S. 1442 may require the FDA to undertake the time-consuming task of preparing a formal risk assessment document.[10]

S. 1442 also creates a device for giving the applicant a second chance if the FDA denies a food additive petition. Under section 113, the applicant or any interested user of the additive may require the FDA to refer the matter to a food safety committee after the agency has made its decision. If a referral is appropriate, it should be made simultaneously with the FDA's evaluation of the food additive petition. The food safety committee, which is also used to delay removal of unsafe additives, is composed of scientists as well as non-voting representatives of consumer and industry interests. (Section 120). Industry representatives should not be members of the committee since their participation could destroy the committee's confidentiality.

The bill would also affect the introduction of new animal feeds. Under the current statute, animal feeds with cancer-causing ingredients are prohibited unless the FDA finds that no residue of the carcinogen in any edible portion of the food will be found "by methods of examination prescribed or approved by the Secretary by regulations." 21 U.S.C. § 348(c)(3). S. 1442 deletes the language directing the Secretary to prescribe methods for detecting animal residues (Section 112(e)). This could result in the use of less sensitive methods to detect carcinogenic substances in meat and thus in greater exposure to consumers.

B. Slow Off

The bill contains several provisions which would incorporate lengthy delays into the process of removing additives found to be unsafe from the market. Richard Cooper, former Chief Counsel at FDA, has "conservatively" estimated that under S. 1442 it would take the FDA nine years to ban an additive under one of the Delaney clauses. The bill would provide that, any time the FDA proposes to remove an additive, the companies which manufacture or use the product may require the agency to refer the matter to a food safety committee, and, if the basis for removing the additive is that it causes cancer, the agency must refer the issue to the committee. (Sections 113, 112(f)). While the matter is pending before the committee, the FDA may not act on the petition.[11] If the committee is able to

---

[10] Section 112(e) states that the FDA shall "base [its] decision on all relevant factors including" the assessment of risks, whereas the current Act states that the FDA must "consider among other relevant factors" the safety of the substance, 21 U.S.C. §348(c)(5). The reason for this change is unclear.

within the 120 days allowed by the bill, which seems unrealistic, and if the FDA could act on the committee's decision within 60 days, which seems even more unlikely, then the referral to the committee would delay agency action for seven months. In fact, referral to the committee can be expected to delay agency action by at least one year.

In addition to prohibiting the agency from acting while the food safety committee is considering the matter, the bill would build in another delay for food additives which have "a substantial history of use and no reasonably practicable substitute." Section 112(e) requires the FDA to consider a variety of non-health factors such as "the nature and extent of consequences resulting from use," "cost," and "availability and acceptability in food" in deciding whether to withdraw approval for such an additive. In addition to requiring the FDA to allow the use of unsafe additives, this provision would delay action on any additive which could not meet the new standard because consideration of these cost factors could be made only _after_ the FDA had completed the first rulemaking proceeding, and would necessitate a second and separate rulemaking proceeding. Each rulemaking could be expected to take approximately two years. When the time required for hearings and appeals both at the FDA and in federal court is added, the time required to ban an additive would be many, many years.

Finally, section 117 would give manufactureres the right to interim status for GRAS substances. The FDA devised the interim food additive list to allow the marketing of GRAS substances where questions had been raised about the additive's safety. 21 C.F.R. § 180. The interim list has been used only a few times, and today only five substances are on the list. However, section 117 of S. 1442 would make the interim list available for all additives, and would make interim listing a matter of right for GRAS additives, or other additives which have a "substantial history of use in food" unless the FDA finds that use of the substance "constitutes an imminent hazard to health." This provision would seriously weaken the safety protections of the statute since it requires the FDA to allow an additive to remain on the market pending administrative proceedings, even if the additive is likely to cause cancer or otherwise impair human health.

[11] The FDA may remove the additive from the market if it finds "there is an imminent hazard to health" (section 113), but it is unlikely that this finding could ever be made because the health risks from food additives are typically long term in nature. Even in the area of drugs, where the hazards are often far more "imminent," a similar provision has been used only once since it was adopted 35 years ago. _Order of HEW Secretary Joseph Califano Suspending Approval For Phenformin_ (July 25, 1977); _see_ 21 U.S.C. § 355(e).

## IV. EXEMPTIONS TO THE DEFINITION OF FOOD ADDITIVE

### A. Basic And Traditional Food Additives

Another way of weakening the statute is to exempt classes of additives from the definition of "food additive" and hence from the requirement that the additive be proven "safe"(or be GRAS) prior to sale, as well as from the Delaney clauses. Section 102(b) exempts one category from the requirements applicable to food additives, namely "basic and traditional foods." Instead of showing that "basic and traditional foods" used as food additives are "safe" prior to sale, manufacturers could use these substances as additives unless the FDA proved they rendered the food adulterated under section 402, 21 U.S.C. § 342, or were poisonous and deleterious, requiring a tolerance pursuant to section 406, 21 U.S.C. § 346. As has been discussed previously, other sections of the statute substantially weaken the adulteration provisions.

Section 106 would define the "basic and traditional foods" exemption as including common foods such as potatoes, which could be theoretically regarded as a food additive if mixed with carrots in a stew. This aspect of the amendment is not objectionable, although it is also unnecessary in light of the fact that the FDA has never interpreted the Act to cover such uses of foods.

Second, the amendments would cover processed raw agricultural commodities, such as the processing of apples into applesauce, as long as the processing is "by a method that is generally recognized . . . as not significantly changing the properties of such raw agricultural commodities." This provision also solves a problem which does not exist since the FDA does not currently regulate apples used in applesauce as a food additive. Finally, as drafted, the provision may not even exclude processing which created new, known risks to health.

Finally, the amendments would exclude spices from the definition of food additive in order to authorize the sale of spices which have been used for a long period of time even if not GRAS and not proven safe. Essentially, the bill would treat all spices like prior-sanctioned food additives. As previously discussed (p. 8, supra), the amendments to sections 402 and 406 would further weaken the statute since spices would be allowed unless they are "ordinarily" injurious. Even a potent carcinogen might not be considered "ordinarily" injurious, and therefore the current section 402 standard, which deems spices adulterated if they "may" be injurious, should be retained.

## B. Food Contact Substances

Sections 102(a) and 102(b) of S. 1442 would also exempt "food contact substances" from the definition of "food additive," further weakening the Act. "Food contact substances" are substances such as packaging which come in contact with food and migrate to the food. The justification for the amendment is that the definition of food additive in the FDC Act could be construed to cover minute amounts of substances which migrate from food packaging to foods. As scientists are able to measure smaller and smaller amounts of substances (parts per billion and possibly parts per trillion), food packaging materials which migrate in small amounts could fall within the definition of food additives. The argument is that regulating packaging materials as food additives will burden the industry, which would have to obtain approval prior to using old packaging materials, and the FDA, which will have to expend resources reviewing the food additive petitions, while not offering any additional protections in terms of the public health.

However, the provisions of the FDC Act currently applicable to food additives resolve these problems in a reasonable fashion. Under the statute, food packaging is not a food additive unless it migrates to the food. 21 U.S.C. § 321(s). Even if it migrates, the packaging is not a food additive if it is generally recognized as safe at the level of migration. 21 U.S.C. § 321(s); Monsanto Co. v. Kennedy, 613 F.2d 947 (D.C. Cir. 1979). Moreover, under Monsanto v. Kennedy, the FDA may exempt packaging material which migrates to foods in small amounts on the ground that, in light of the quantity which migrates, use of the material is safe. 613 F.2d at 955-56. This amendment, however, would place such packaging outside the regulatory requirements applicable to food additives. On the other hand, under Monsanto, a non-GRAS packaging material which migrates in significant amounts to food, or in quantities which are unsafe, is a food additive which must meet the requirements of the statute. Because the line drawn by Monsanto is workable, there is no justification for amending the statute.

Instead of the current scheme which requires that food packaging, that migrates in significant quantities to food and is not GRAS, be shown to be safe prior to use, section 119 of S. 1442 would merely require a food packaging manufacturer to notify the FDA of its intent to market a particular food contact substance. Ninety days later, it could sell the product unless the FDA had first notified it that the registration will not be effective based on a finding that use of the substance presents a significant risk of harm to the public health.

First, the contents of the pre-market notification are substantially less than those required in a food additive petition. There is no requirement that the applicant submit data showing that use of the packaging material is safe. Instead, manufacturers need only submit a summary of such data, and thus the FDA has no opportunity to look at the slides or other underlying evidence. See subsection (b)(2) of the new provision added by section 119. Second, there is no requirement that the applicant submit information about the substance's chemistry or about the production processes. This information is also important so that the FDA can assess the safety of the substance. See sections 409(b)(2)(A), 409(b)(3) of the FDC Act, 21 U.S.C. §§ 348(b)(2)(A), 348(b)(3).

The new section would also require the FDA to act within 90 days; otherwise the food contact substance could automatically be used. This time limit is unrealistic for the same reasons as the time limits under the food additive provisions are unrealistic. However, section 119 would limit the FDA's ability to protect the public health in a much more serious way. Whereas under the current law, the manufacturer has the burden of showing a food additive, including a food packaging material, is safe, under the new amendments the Secretary would have the burden of showing that use of the product entails a risk to public health. See subsection (c) of the new provision added by section 119. This might mean that if the manufacturer presented no data on safety, and the Secretary[12] knew of none, the substance would go to market within 90 days because the FDA would be unable to carry its burden of establishing that the use of the substance presented a significant risk of harm to the public. There is simply no justification for requiring users of new food packaging materials, which migrate in significant amounts to foods, to meet a lower standard than that required for new food additives.

The problems with the new standard which would be applicable to food packaging are highlighted by subsection (c)(2) of the new provision which would prohibit the Secretary from barring the sale of a new food packaging unless he finds that "the chemical nature of the substance, and the general knowledge, or lack thereof" raise "concerns of a significant risk of harm that can only be resolved by further data." The FDA may not even know the chemical nature of the substance, or the extent of exposure, and it might be very difficult to make this finding. Under the scheme of the statute, then, the additive could be used in food.

Section 119 would also give the FDA broad authority to exempt substances from even the minimal pre-market notification requirements if it finds that notification is "not necessary to protect the public health." This provision has no standards whatsoever and it could be used to exempt all packaging materials from the statute. In addition, section 119 continues the exemption for substances which are generally recognized as safe, as well as those which are prior-sanctioned. There is no basis for exempting such substances from pre-market notification.

Once again, the provision contains judicial review provisions with new advantages for the industry. Review of the denial of the petition would be in the district court instead of the court of appeals, and the court could take additional evidence, apparently even if that evidence had not been presented to the Secretary. In addition, the court could order that the evidence be taken before the Secretary, but the Secretary's decision would be treated only as a "recommendation" and apparently would be given no presumption of validity. See subsection (d)(4) of the new provision added by section 119. These provisions have advantages for the industry but could potentially burden the courts with scientific decisions which are ordinarily left to the FDA.

Although section 119 would make it very easy to market a new food packaging material, once the packaging is on the market, it would be as difficult to remove as a food additive. Removal could be done only through a regulation which would be accompanied by all the procedures, including referral to the food safety committee, required for revoking a food additive regulation. The imbalance of the statute, in terms of the safety of the public and economic desires of the industry, is best shown if one considers a food contact substance which is marketed pursuant to a pre-market notification which the FDA did not act on within 90 days. Even if shortly thereafter the FDA learned that the substance was very dangerous, it would have to go through years of procedures in order to remove the substance from the market, unless it could show that there was an imminent hazard.

V. ANIMAL DRUGS

The Food, Drug and Cosmetic Act regulates animal drugs both to insure that the drugs are safe and effective for use in animals, and because residues of those drugs can appear in human food. Because use of animal drugs is intentional,

---

[12] Both the FDC Act and S. 1442 delegate decision-making authority to Secretary of Health and Human Services rather than the Commissioner of the FDA. However, the Secretary has traditionally delegated his authority to the Commissioner, although recently Secretary Schweiker withdrew part of that delegation, reserving for himself the final decision on important matters. See 21 C.F.R. § 5.1, et seq.

and those drugs become part of human food, the standards applicable to animal drugs under the FDC Act are similar to the requirements which apply to food and color additives. Section 201(w) defines a "new animal drug" as "any drug intended for use for animals" which is not generally recognized as safe and effective ("GRAS/E"), and section 512 requires premarket approval for all such new animal drugs. 21 U.S.C. §§ 321(w), 360b. Section 512 also contains a Delaney Clause prohibiting the use of carcinogenic animal drugs which leave residues in human food that are detectable by methods prescribed by the FDA. 21 U.S.C. § 360b(d)(1)(H). A separate subsection of section 512 regulates animal drugs to be used in animal feeds. 21 U.S.C. § 360b(m). The Act also requires that each batch of antibiotic animal drugs be certified by the FDA prior to use. 21 U.S.C. § 360b(n).

Sections 104 and 121-130 of S. 1442 would amend the statutory provisions applicable to animal drugs in a way analogous to the amendments to the food additive provisions. The amendments would: make the 180-day period for acting on new animal drug applications enforceable in a court proceeding which could be expedited (section 128); require that a risk assessment, which might have to be performed by the FDA, be part of the determination of whether to approve an animal drug (section 124); require the FDA to consider evidence on economic cost in deciding the safety issues relevant to approving or withdrawing an animal drug (section 124); allow the applicant to have an application which the FDA is about to deny referred to a food safety committee, or to delay a decision to withdraw approval of an animal drug by the same referral procedure (sections 123, 124); effectively repeal the Delaney Clause (section 124); authorize the FDA to grant interim status to animal drugs which had been marketed, without approval, on the grounds that they had been GRAS/E, even after questions about the drug's safety had been raised; give GRAS/E animal drugs the same procedural protections given to animal drugs which have been approved (section 125); and, finally, require the FDA to leave approved new animal drugs for which questions had been raised on the market if the agency determines that there is a "reasonable certainty that . . . no harm to the public health will result" (id.).

There are three other amendments which apply specifically to animal drugs. First, section 122(c) of S. 1442 would amend section 512(b)(7) of the FDC Act, 21 U.S.C. § 360b(b)(7), which currently requires that a new animal drug application include "a description of practicable methods for determining the quantity, if any, of such drug in or on food . . . ." This requirement, that the applicant identify the regulatory methods used, is important to the question of how much of the animal drug will be left in human food, and ultimately to the question of whether the animal drug is safe. Nevertheless, section 122(c) of S. 1442 would relieve the applicant from this provision unless a description of regulatory methods is specifically required by the FDA.

Second, section 130 of S. 1442 would repeal section 512(m) of the Act, which currently requires premarket approval for the use of animal feed containing a new animal drug. 21 U.S.C. § 360b(m). In its place, section 130 would substitute a provision which would merely require the medicated animal feed user to register its establishment with the FDA. Because the registration apparently would be effective upon filing, this amendment would deprive the FDA of the opportunity to review the proposed uses of medicated animal feed prior to the time the feed is fed to animals, and therefore it would delete the health protections of the FDC Act.

Finally, section 104 and 130 of S. 1442 would repeal section 201(w)(3) and 512b(m) of the FDC Act, 21 U.S.C. §§ 201(w)(3), 512b(m), which require the FDA to separately certify each batch of antibiotic animal drugs. The importance of batch certification of antibiotic animal drugs is unclear, but at a minimum the FDA should retain discretion to require batch certification of antibiotic animal drugs.

## VI. COLOR ADDITIVES

The FDC Act also adopts standards for color additives which are very similar to the standards applicable to food additives. Like section 409, 21 U.S.C. § 348, which requires that food additives be proven to be safe prior to sale, section 706, 21 U.S.C. § 376, requires that color additives be proven "safe" before sale. Section 706 also contains a Delaney Clause which deems any carcinogenic color additive to be unsafe. 21 U.S.C. § 376(b)(5)(B).

The one difference between the regulation of food additives and color additives is the treatment of additives which were on the market when the two provisions were adopted. In 1958, when it adopted the Food Additives Amendment, Congress "grandfathered" food additives which were being marketed at that time pursuant to a sanction or approval by the FDA or the USDA, excluding those additives from the premarket approval requirements of the Act. However, dissatisfied with the grandfather exemption, in 1960 Congress adopted a different approach for color additives. Instead of exempting additives which were on the market in 1960, Congress placed those additives on a provisional list, and temporarily exempted them from the new requirements while testing on their safety was conducted. It limited the provisional list to 2-1/2 years, but authorized the FDA to extend that period where "consistent with the objective of carrying to completion in good faith, as soon as reasonably practicable, the scientific investigations necessary for making a determination" as to whether the color additive is safe. Pub. L. 86-618, 77 Stat. 399, section 203(a)(2).

During the past 20 years, the FDA has granted numerous extensions to the provisional list, so that today 23 of the most commonly used color additives (out of approximately 200 in 1960) remain in interim status. However, the FDA has recently adopted a schedule which would require that all testing of additives on the provisional list be completed within two years. 21 C.F.R. § 81.27. After testing is completed, the FDA will then decide, within one year, whether each color additive is safe. If it is, the color additive will be permanently listed; otherwise, the color additive will be removed from the market.

The additives on the provisional list are significant to the safety of the food supply, since they include six of the nine color additives used in foods.[13] For many years, consumer groups have complained that the FDA's delays in resolving the safety of these additives violate the FDC Act because Congress intended that the additives be provisionally listed for only 2-1/2 years, or slightly longer if additional time was needed to complete testing. See Health Research Group v. Califano, Civil No. 77-0293 (D.D.C. 1977); McIlwain v. Hayes, Civil No. 81-0555 (D.D.C.)(pending).

S. 1442 would amend the FDC Act provisions applicable to color additives in much the same way as the bill would amend the food additive provisions. The following provisions are analogous to the amendments to the provisions applicable to food additives, and would impair the safety of the food supply for the reasons previously discussed: section 132, which would make the 180-day limit on acting on a color additive petition enforceable in a court proceeding which would be expedited; section 131(a), which might require the FDA to perform a risk assessment in deciding whether to approve a color additive; section 131(c), which would require the FDA to consider evidence on economic cost in deciding the safety issues relevant to withdrawing its approval to market a color additive; section 131(d), which would allow the manufacturer to have its application to market a color additive referred to a food safety committee if the FDA is about to deny the application, or to delay a decision to withdraw approval of a color additive by the same referral procedure (section 376 of the FDC Act, instead, authorizes the FDA to refer issues to an advisory committee); section 131(c), repealing the Delaney Clause's prohibition on carcinogens; and the provision in section 132 which would require the FDA to list color additives on an interim basis after safety questions had been raised.

There are two other provisions which require separate comment, and which would seriously undermine the safety protections of the current Act. First, subsection (d)(8) of the new provision added by section 132 would delay final

---

[13] Three of those additives are on the permanent list for some uses and on the provisional list for others.

action on color additives on the provisional list for an even longer period of time by requiring the FDA to follow all the Act's procedures applicable to permanently listed additives, including the new delaying procedures in S. 1442. In light of the fact that the 23 color additives have remained on the provisional list for 20 years, while their safety has remained unproven, amending the Act to build in additional delays would be unconscionable.

Second, subsection (d)(4)(B) of the new provision added by section 132 of S. 1442, which would require the FDA to permit use of a color additive on an interim basis if the additive is already approved for a different use and if it "appears" to the FDA that the new use will be safe, is an even more serious impairment of food safety than the analogous amendment to the Food Additives Amendment to the FDC Act. Whereas food additives are only used in foods, color additives may also be used in drugs and cosmetics. Thus, in contrast to food additives, the fact that a color additive has been approved for another use is no indication that it is safe for a different use. For example, a color additive may be "safe" for use in externally applied cosmetics, or drugs where the amount ingested is minute, but not for foods. The fact that the FDA has approved only six chemical color additives for use in foods indicates that most color additives are not safe for use in human food, and thus the permissive addition of new uses could seriously impair the public health.

## VIII AMENDMENTS TO ACTS ADMINISTERED BY THE USDA

In addition to the requirements of the FDC Act, poultry meat and egg products are subject to the requirements of separate statutes administered by the USDA -- the Poultry Products Inspection Act, the Meat Inspection Act, and the Egg Products Inspection Act. Both the poultry and meat acts give the USDA authority to find a food adulterated if it contains a substance (other than a pesticide chemical, food additive or color additive) which the Secretary finds would "make such article unfit for human food." 21 U.S.C. §§ 453(g)(2)(A), 458(a)(2), 601(m)(2)(A), 610(c). In addition to contaminants, these provisions cover prior-sanctioned and GRAS substances which are not classified as food additives under the statute, as well as animal drugs. These statutes do not undercut the FDC Act, but may add new safety requirements.

Section 201-203 of S. 1442 would amend these three statutes to conform with the amendments to the FDC Act. The proposed provisions for risk assessments and phase-out authority are objectionable for the same reasons that those provisions in the FDC Act are objectionable. The amendments would also delete the USDA's broad authority to deem substances (other than pesticides, food additives and color additives) "unfit for use in human food," thus further weakening the prohibitions against the sale of unhealthy food.

## CONCLUSION

The FDC Act strikes a careful balance between the public's need for assurances that food is safe, and the industry's desire to introduce more and more chemicals into the food supply. In general, the purpose of those chemicals is to make food look better (e.g., color additives), last longer (e.g., preservatives), cost less to produce (e.g., animal drugs), or taste different. Because the additives have no important health benefits (e.g., benefits comparable to prescription drugs), the FDC Act generally prohibits additives whose safety is unproven as well as additives which are carcinogenic or otherwise harmful. S. 1442 would undermine the fundamental tenent of the FDC Act that chemicals should be proven safe prior to introduction into the food. Thus, its title, "The Food Safety Amendments of 1981," represents an example of legislative misbranding.

The United States has the world's strongest food safety laws, and, as a result, its safest food supply. For this reason alone, we should be very reluctant to tamper with the FDC act. Yet S. 1442 would significantly weaken almost every safety standard applicable to food. Because adoption of the bill would increase health risks to consumers, solely to advance the financial interests of some food companies, it should not be adopted.

## Food Safety Legislation
## Fact Sheet

**Prepared by**
**Center for Science in the Public Interest**

For more than 75 years, the food safety provisions of the Food, Drug and Cosmetic Act have protected Americans from dangerous and poorly tested food additives, as well as from harmful contaminants found in packaging materials, pesticide residues and other substances. Throughout this century, as new dangers have been discovered, the level of protection provided by the FD&C Act has been extended. In 1958, for example, Congress amended the Act to protect Americans from additives that are shown to cause cancer. The FD&C Act now bans most cancer-causing additives used in food. The law's great strength is its clear standard of safety that discourages compromising by Federal agencies.

Yet, some segments of the food industry are now proposing to end this trend towards providing greater protection from unsafe food additives and shift the focus of the FD&C Act to ensuring the economic well-being of the industry.

**INDUSTRY PROPOSALS TO AMEND THE FOOD SAFETY LAWS WILL MAKE OUR FOOD SUPPLY LESS SAFE**

\* S. 1442 and H.R. 4014 would change the definition of "safety" under the FD&C Act so as to allow additives *which have been proven to cause cancer, birth defects and other problems* to be used in food.

* According to these bills, if an *additive currently in use* were found to cause cancer or other disease the Food and Drug Administration (FDA) would have to consider other factors directly relating to industry profitability before proposing that the substance be removed from the market.

* To ban an additive, the FDA would have to meet a series of burdensome procedural requirements which may delay the removal of an additive from the market for more than nine years. Such delays make a mockery out of the law and waste agency resources.

* These bills could result in the approval of a flood of *new additives,* each of which by itself might only cause an "insignificant" amount of cancer. The effect would be felt in the future when an entire generation of Americans is forced to bear the burdens of a shortsighted decision made to benefit an economically healthy food industry.

* Ultimate decisions regarding the safety of an additive would be based upon unproven risk assessment techniques. A National Academy of Sciences report which utilized various risk assessment models to determine the risk of cancer posed by saccharin found that results varied five millionfold — from 0.0007 cases to 3,640 cases per 50 million people exposed to saccharin — depending on the risk assessment model which was used.

## STRONG REGULATION OF FOOD ADDITIVES IS ESSENTIAL TO THE PUBLIC HEALTH

* There is no safe dose of a carcinogen. Virtually all cancer experts agree that if large amounts of a chemical additive cause cancer, small amounts of that same additive would also cause cancer, but less frequently. A particular individual's sensitivity to a carcinogen cannot be determined.

* Most additives are not harmful. However, identification and elimination of those that are must be done as swiftly as possible. New dangerous chemicals must be kept out of the food supply.

* There is no single standard of "acceptable risk" that can be applied to determine whether the use of some dangerous additives should be allowed. Estimating health risks is fraught with uncertainties. Risk/benefit analyses often produce widely varying and arbitrary results. Both risks and benefits are impossible to quantify accurately in most cases.

## THE FOOD SAFETY LAWS MUST BE STRENGTHENED, NOT WEAKENED

* Additives in use prior to 1958 are virtually exempt from food safety laws. This giant loophole should be closed. Such additives should be retested in accordance with contemporary standards.

* Testing of possibly unsafe *color additives,* which have remained on a "provisional list" for more than twenty years, must be expedited. Congress, in 1960, passed legislation which contemplated that these color additives would remain in use for only 2½ years before they were either banned or proven safe by the manufacturer. Color additives provide a minimum benefit to society. Those additives which cannot be tested and proven safe in a reasonable period of time should be removed from the marketplace.

* Food additives should be *licensed* for ten years, not approved permanently. Improvements in testing procedures are constantly being made and could uncover dangers that were overlooked by previous tests.

* Additives should be tested for their effects on *behavior.* Some food additives and other chemicals have significant impacts on the central nervous system. Routine testing for behavioral effects will identify culprits.

* The testing regimen for food additives should be broadened to include tests for *interactions* between commonly used additives, food ingredients, alcohol, and drugs.

# Center for Science in the Public Interest
1755 S Street, N.W. • Washington, D.C. 20009
(202) 332-9110

## DEPC (BANNED: 1972)

### PRESENT LAW

DEPC had been used as a preservative in beverages.

However, FDA became aware of data which showed that the additive was "Theoretically shown to be capable of combining with other ingredients in beverages to form urethan," a potential carcinogen.

Industry requested time to present additional evidence that the additive did not pose a danger. The FDA however decided that the health risk was great enough to warrant prompt removal of the additive from the market. The additive was banned 6 months after the FDA proposed its findings (37 Fed. Reg. 15426).

### POSSIBLE ACTION UNDER S. 1442

Because the tests on DEPC indicated a theoretical risk, the FDA might not have been able to demonstrate that the additive posed a "significant risk" to humans as required by S. 1442. If such a determination was made, S. 1442 would require that the matter be referred to the Food Safety committee because carcinogencity was involved.

In accordance with suggestions in the section by section analysis of S. 1442, The Food Safety Committee might have rejected the FDA findings of significant risk because DEPC posed a cancer risk shown only in combination with other ingredients, not by itself.

If the Food Safety Committee approved the FDA findings, the additive would still be subject to other provisions of the bill such as phase out authority, which could delay its removal from the market.

### RESULT OF CHANGE

This dangerous additive might still be approved as "safe" by the FDA. Manufacturers would have no incentive to develop or use existing substitutes and consumers would be subjected to an avoidable risk of cancer.

# Center for Science in the Public Interest
1755 S Street, N.W. • Washington, D.C. 20009
(202) 332-9110

## CYCLAMATES (Banned: 1969)

### Present Law

Cyclamate was on the GRAS list and had been used as an artificial sweetener in soda water. However, on the basis of animal studies, the additive was shown to cause cancer.

FDA chose to have the results of the studies reviewed by the National Cancer Institute, the National Academy of Sciences and other consultants. These groups confirmed the FDA findings and FDA ordered that the additive be taken off the GRAS list and be prohibited from use in food.

Cyclamate used in food was removed from the market over a five month period (31 Fed. Reg. 17063).

### Possible Action Under S. 1442

The FDA would have to determine if the animal studies demonstrated that cyclamate was a danger to human health.

Assuming this was technically possible, S. 1442 would require the FDA to refer this finding to a "Food Safety Committee" that would reexamine the issue. If FDA's finding was confirmed by the Committee, S. 1442 might then require that FDA place the additive on the "interim list" and conduct more testing. No substance has ever been banned once it has been placed on the interim list by the FDA.

If the additive was ever taken off the interim list, S. 1442 might require that the FDA perform a risk-benefit analysis, that would involve weighing the health risks of cyclamate against its potential usefullness as a dietary aid.

### Result of Change

Cyclamate might still be in use. If the FDA ultimately attempted to take cyclamate off the interim list and ban it from the marketplace, it would first have to determine whether the health benefits of cyclamate outweighed its health risks. Because these factors could not be accurately quantified, FDA would have to base its final decision on a subjective determination of the risks and benefits of cyclamate.

# Center for Science in the Public Interest
1755 S Street, N.W. • Washington, D.C. 20009
(202) 332-9110

## SAFROLE (Banned: 1960)

### Present Law

Safrole is the primary component of oil of sassafras, a flavoring agent that had been used in rootbeer beverages for a significant length of time. Long term feeding studies conducted by the FDA demonstrated this additive to be a weak hepatic (liver) carcinogen. FDA voluntarily had the data reviewed by the National Academy of Sciences which confirmed the agency's findings.

Representitives of the beverage industry were kept apprised of the progress of the laboratory experiments. As the data were confirmed, industry began to seek substitutes for the flavoring agent and eliminated the use of safrole. The additive was ultimately banned in 1960 (25 Fed. Reg. 12412).

### Possible Action Under S. 1442

Based upon the long term feeding studies, the FDA would have to decide if safrole posed a "significant risk" of cancer to humans. Because the studies showed safrole to be a weak carcinogen, S. 1442 might have required FDA to keep this additive on the market.

If the FDA determined that the substance was a "significant risk", S. 1442 might require the agency to perform a risk-benefit analysis because the additive had been in use for a substantial period of time and manufacturers could claim that no "reasonable practical" substitute flavoring agent was available.

In performing this analysis, FDA would be required to weigh the risk of cancer from safrole against the effect of banning it on the "acceptability" of rootbeer.

### Result of Change

It is possible that safrole would still be used in foods. Even if the FDA found that the additive posed a "significant risk", it might have decided on the basis of the criteria set out in S. 1442 that safrole should not be banned because doing so would adversely affect the "acceptability" of rootbeer. Consumers would continue to be subjected to a known carcinogen even though manufacturers were able to utilize a substitute flavoring.

# Center for Science in the Public Interest
1755 S Street, N.W. • Washington, D.C. 20009
(202) 332-9110

## VIOLET # 1 (BANNED: 1973)

### PRESENT LAW

Violet #1 is a dye that was used in many common foods and in the ink used to stamp the USDA's familure inspection symbol on meat. In accordance with the color's "provisional listing" i.e., not permanently approved as safe, a number of multiple acute and chronic studies on laboratory animals were conducted. The studies provided conflicting conclusions about the safety of the substance.

FDA referred the matter to an advisory panel which recommended that the color remain of the market pending further testing. The FDA first concurred with the panel's findings but later, on the basis of new studies which showed that Violet #1 caused cancer in laboratory animals, concluded that the additive should be banned.

Violet #1 was finally prohibited from use in food but manufacturers were allowed to use up existing stocks (38 Fed. Reg. 9077).

### POSSIBLE ACTION UNDER S. 1442

Even though the safety of Violet #1 had never been proven, the FDA would have to follow all the procedural steps necessary to ban a permanently listed color additive, i.e., one whose safety was established.

This means that FDA would have to conduct a risk assessment and find that the additive posed a "significant risk" of cancer to humans. In addition, the FDA might have to conduct a risk-benefit analysis which would require that factors relating to industry profitability be weighed against the risk of cancer to the public. In addition, the removal of Violet #1 from the market might be further delayed by the imposition of a lengthy phase out program.

### RESULT OF CHANGE

Manufacturers of color additives that had never been proven safe would receive all the "procedural protections" that are granted to manufacturers of color additives that have been proven safe. These "procedural protections" would include all the same delaying steps that the FDA would be required to follow before banning a food additive, even though color additives generally provide relatively less tangible benefits to society. The final result would be that Americans would be exposed for a longer period of time, than at present, to a cancer-causing chemical.

# Center for Science in the Public Interest
1755 S Street, N.W. • Washington, D.C. 20009
(202) 332-9110

**SO, THE FDA HAS EVIDENCE TO WARRANT BANNING A FOOD ADDITIVE, THEN WHAT? UNDER S. 1442, THE CASE WILL DRAG ON FOR AT LEAST EIGHT YEARS. HERE'S WHY:**

**Year One**
**Year Two**
**Year Three**

Risk Assessment

- Requires FDA to speculate how much disease will result from use of an additive and to permit its use if the threat to the public health is "insignificant."

**Year Four**

Advisory Committees:

- Would allow food manufacturers to force FDA to refer findings that an additive is dangerous to panel of non-government scientists.

**Year Five**
(Additives placed on the Interim List have remained there for more than ten years)

Interim Status (Further Testing):

- Requires that GRAS substances stay on the market pending additional research *after* FDA and advisory committee determine that the safety of the additive is questionable.

**Year Six**

Risk/Benefit Analysis:

- Must be performed before restricting the use of additives which have a substantial history of use and which have no readily available substitutes.

- Would require FDA to consider factors relating to industry profitability before banning an additive.

**Year Seven**

Phase-outs:

- Even if all other steps are met, additives would not be immediately removed from the market but phased out gradually.

**Year Eight**

Judicial Review of Science:

- Final agency actions could be appealed to Federal Court, further delaying effective dates and enforcement of restrictions.

# Saccharin, Bacon and the Law
*by Orrin G. Hatch*
*The Washington Post,* November 10, 1981

There is quiet on the food-safety front. No imminent crises. No public alarms. No important food substances about to be removed from our dinner tables. On July 25, Congress eagerly agreed for the third time in four years to keep the Food and Drug Administration from banning saccharin during the next two years. In addition, the scientific evidence on nitrites has been re-examined and initial fears about its potential toxicity have been assuaged, allowing bacon to remain on our breakfast tables, at least for the moment.

Yet it is an uneasy quiet. The momentous task of re-examining the broader questions of food safety policy remains. Four years ago, amid the tumultuous public reaction against the proposed saccharin ban, Congress defused the situation by postponing revision of the food safety laws pending a study by the National Academy of Sciences. This study has since been conducted, and the NAS has recommended an overhaul of current food safety laws. In the view of NAS, "the law has become complicated, inflexible and inconsistent in implementation. . . [and] is inadequate to meet changing and increasing problems of food safety." Under the Carter administration, the FDA's Task Force on Food Safety advocated significant revision of the statute. The General Accounting Office recently completed a survey of former FDA commissioners that revealed a strong consensus for major reform.

But what reform? I recently introduced S. 1442, the Food Safety Amendments of 1981, as one way to revise and update these laws.

Unfortunately, past debate on food safety has been less than dispassionate. Too many consumer groups and food processors have been raining invective on each other.

At the center of this storm rides the Delaney Clause, a 1958 amendment requiring a ban on any substance shown to induce cancer in man or animals, regardless of any benefits associated with its use. No exception is made for substances found in such infinitesimal amounts as to present the most insignificant risk to humans.

Combatants in earlier battles of the food safety war have been fighting either for the Delaney Clause or against it. But what is needed, in fact, is some fine tuning.

The basic purpose of the Delaney Clause represents sound public policy — no substance should be added directly to food if it is a carcinogen. But how should substances like saccharin be regulated? Applying a rigid, inflexible rule to a substance that has been in use for decades and for which there are no available substitutes is shortsighted at best. For diabetics and others who cannot eat sugar, cutting off their only viable substitute sweeetener is unfair.

Reasoned decision-making is even more crucial in matters affecting the supply and marketing of our basic foods. Natural nutrients found in familiar foods such as potatoes, onions and black pepper have been implicated as potential carcinogens. Essential nutrients such as selenium, Vitamin D and calcium, substances required in our human diet, act in high doses as carcinogens. The logical extension of implementing the current law would lead to FDA's banning these most basic foods.

Fortunately, the FDA has chosen the more prudent course of action and ignored the statute. Yet the law should not be merely an option or suggestion that a federal agency can accept or reject with impunity. Laws are normative, which means they not only tell us what we can and cannot do, but set standards reflecting what a consensus of us in the community considers reasonable policy.

Right now, the Delaney Clause as a law that provides for summarily banning foods without any recourse or appeal is not reasonable.

Our revision of law should keep pace with scientific advancements. Scientists can now detect traces of potentially harmful substances at parts per trillion — a million-fold increase in analytical sensitivity compared with the methods available more than 20 years ago, when the Delaney Clause was conceived, and the law last examined. Traces of such substances are being found throughout our food supply in very minute quantities. Attempting to eliminate all these food substances threatens the quantity of our national food inventory while promising only the most nominal impact in our common cause to improve the public health.

Keeping our food suply safe is of the highest priority. Yet some degree of risk is inherent. Zero risk has proved to be impossible to achieve. The FDA must focus its resources on eliminating those risks that have a potential for actually harming humans. Trivial or speculative risks should not be regulated.

Revision of the food safety law is only the first step in assuring the American people a safe and plentiful food supply. Increasing our knowledge by supporting research in the fields of nutrition and disease prevention is of equal importance. Only if sound and reliable information is available can consumers, industry and the FDA make informed and rational decisions.

*The writer, a Republican senator from Utah, is chairman of the Committee on Labor and Human Resources.*

# A Lot of Baloney About Delaney
*by William B. Schultz*
*The Washington Post,* November 21, 1981

Sen. Orrin Hatch's recent column ("Saccharin, Bacon and the Law," op-ed, Nov. 10) restates two myths that continue to cloud the debate about the Delaney Clause, the provision of the Food, Drug and Cosmetic Act that bans the use of food additives found to cause cancer. Sen. Hatch then uses the myths in an attempt to generate support for his bill (S.1442) that would effectively repeal the Delaney Clause, as well as undermine other food safety laws.

The first myth is that Delaney applies to "any substance shown to cause cancer in man or animal." Sen. Hatch claims it applies even to onions and potatoes, which he implies are carcinogens.

The Delaney Clause does not cover *every* substance; it covers only those chemicals and substances that are *added* to foods, and certain animal drugs. Because these additives are either generally interchangeable (for example, most preservatives) or are relatively unimportant (for example, color additives), banning those that cause cancer makes sense.

The second myth perpetuated by Hatch is that the Delaney Clause covers infinitesimal amounts of substances that can migrate from food packaging to foods. More than two years ago, the U.S. Court of Appeals for the District of Columbia Circuit ruled that the Food and Drug Administration may approve a food packaging material even if it contains trace amounts of a carcinogen, as long as the amount that migrates to food is "so negligible as to present no public health or safety concerns" (*Monsanto v. Kennedy*). The same principle applies to trace amounts of other substances in food.

Anyone who has studied the senator's proposal — and anyone who does not represent a food processor — would dispute his statement that the bill simply gives some "fine tuning" to the Delaney Clause. As Richard Cooper, former chief counsel of the FDA, has concluded, Hatch's bill would actually "gut" the anti-cancer clauses of our present law.

Hatch's proposal would also water down other food safety standards and create procedural hurdles that would tie the FDA in knots. Cooper has estimated that, under Hatch's bill, the FDA would require at least nine years to ban a food additive.

Congress has found it necessary to make an exception to the Delaney Clause only once in 20 years, which proves the provision has served us well. That occurred four years ago when Congress directed the FDA to leave saccharin on the market despite the FDA's findings that saccharin is a carcinogen.

During the next several months, Congress will be faced with a question of whether the single example of saccharin is sufficient justification for completely overhauling the laws that until now have given us one of the safest food supplies in the world.

*The writer is an attorney with the Public Citizen Litigation Group, which was founded by Ralph Nader.*

# FDA Alumni Lead Drive on Food Safety
*By Paul Taylor*
*The Washington Post,* October 18, 1981

When Stuart Pape was an up-and-coming young lawyer and a special assistant to the commissioner of the Food and Drug Administration, he wrote a paper proposing that the nation's food safety standards be relaxed.

The memo, commissioned after the FDA had suffered credibility damage in its unsuccessful effort to ban saccharin, wound up on a shelf somewhere in the bureaucracy, and Pape wound up leaving government service.

But he hasn't stopped drafting new food safety law. And from his new perch as a lawyer with the lobbying firm of Patton, Boggs & Blow, he is finding that you can sometimes move public policy further and faster by attacking from the outside.

As counsel to the National Soft Drink Association, Pape is the lead lawyer for a high-powered ad hoc industry group, studded with alumni of the FDA, that set out late last year to capitalize on a friendly political climate and write the most sweeping revisions in the food safety laws in 75 years.

Besides Pape, the group includes Peter Barton Hutt, formerly chief counsel at FDA; Richard Silverman, a former FDA associate chief counsel for enforcement, and, for good measure, former congressman Paul Rogers (D-Fla.), who developed a reputation as a strong proponent of health regulation in his years as chairman of the House Commerce subcommittee on health and the environment. All are now employed by food industry asssociations or law firms representing industry clients.

Because of the depth of their expertise in the arcane world of food law, the legislative committee they have been working with is overmatched, in the view of several neutral observers who say that the industry spokesmen have dominated the bill-drafting process from start to finish.

Their lawmaking has focused on the same issue Pape was dealing with at the FDA in the late 1970s. They want to relax what they view as the inflexibility of the Delaney Clause, which bans from the market any foodstuff found to contain a carcinogen, no matter how small the quantity.

Early on, the industry decided to have the National Soft Drink Association, which felt most aggrieved by the saccharin episode, and the American Meat Institute, still smarting from the nitrite scare of 1979, lead the drafting effort. "Either we went that route," says Pape, "or we got a huge haul of lawyers into one room and had them all run their meters while they shot their mouths off."

Still, the process was painstaking and, to accomodate the vast variety of interest within the industry, the scope of the bill kept growing. By the time the group was ready to take it to the Hill in March, its draft had been transformed into a virtual rewrite of the basic food safety codes. In addition to watering down Delaney, it called for adding a series of procedures, that would make it more difficult for the FDA to remove a suspect food substance from the market and less difficult for the industry to introduce one.

On the Hill, the group chose to knock on the door of Sen. Orrin Hatch (R-Utah), the new chairman of the Senate Labor and Human Resources Committee, whose record on the subject was to industry's liking.

Hatch was interested. He had his committee staff work on the draft for four months, relying heavily on the expertise provided by the industry. "At one point we had something like 15 lawyers in one room, all screaming at each other," said chief of staff Steven Grossman. "Our staffer finally had to walk out and let them settle it among themselves."

In June, Hatch formally introduced the food safety amendments of 1981 (S.1442), a complex 70-page bill that one former FDA chief counsel, William Goodrich, says "defies reading and understanding." Public hearings are due next month.

The most controversial provision of the bill is the relaxing of the Delaney clause. Industry claims that the clause, enacted 23 years ago, has been rendered an anachronism by scientific advances that permit the detection of harmful substances down to a part per trillion. It is routinely ignored now by the FDA, they say, because applying it would lead to the banning of such basic foods as potatoes, black pepper, nutmeg and spinach — all of which have been found to contain potential carcinogens in trace amounts.

To consumer groups, however, Delaney has become a rallying point. Gutting it "takes the government down the road of allowing a little bit of cancer in the food supply — the only question is how much," according to Michael Jacobson, executive director of the Center for Science in the Public Interest.

Consumer groups are crying foul not only at the substance of the bill but also at the drafting process. "Industry was basically allowed to write its whole dream list into the bill," said William Schultz of Public Citizen. "We weren't even allowed to see the draft before it came out."

No one disputes the pro-industry thrust of the bill, but food lobbyists and congressional staff members alike take umbrage at the charge that there was something lopsided about the way the legislation was put together.

"If the consumer groups didn't have much input," said Grossman, "it's because they weren't willing to work with us. We gave them plenty of chances. They just haven't come to accept the fact that there is a broad consensus for change."

Stronger words come from Howard Roberts, vice president for science and technology at the National Soft Drink Association and a former acting director of the FDA's Bureau of Foods:

"I resent the notion that the only honest person in this town is an activist. They're the ones who are being morally irresponsible in their approach. They're in the business of selling panic. That's how they make their living, that's how they sell their newsletters. Their natural tendency on every issue, especially if the word cancer is involved, is to cry wolf.

"This isn't a matter of some devious industry group sitting in a smoke-filled room and ginning up some bill that they're going to run through the Hill in a week."

Roberts predicted the legislative process would take years. Hatch has taken pains to describe the draft as merely a starting point for discussion.

Still, the consumer groups fret that industry, by having been so heavily involved in the draft of such a major piece of legislation, "has managed to swing the whole terms of the debate into their corner," in the words of Ellen Haas of the Community Nutrition Institute.

That, of course, was the whole point of the industry drafting effort. But the industry lawyers insist that their ability to shape the process derived from nothing more sinister than their intimate knowledge of the laws in question.

"Food law is at least as arcane as the tax code, maybe more so," says Pape. "It's a blend of law, social policy, science, medicine and psychology."

Adds Peter Barton Hutt, "I would say there aren't more than five people in the country who really understand it." By consensus of their colleagues, the five would include Pape and Hutt, a former general cousel of the FDA who has co-authored a book of case law on the subject. Hutt is now a partner at Covington & Burling, where he represents, among other food industry clients, the Grocery Manufacturers Association.

Pape and Hutt have had major differences this past year over how to proceed with the code revisions. The former favored the legislative approach, the latter argued for administrative reform ("Peter Hutt," said one admiring antagonist, "is one of those lawyers who thinks you can get the sun to rise in the west by administrative change.").

Whatever their disagreements, Pape, Hutt and their fellow FDA graduates were, in the view of several outside observers, able to run circles around the committee staffers they were working with. At the Hatch committee, the staff aide in charge of drafting the bill was a part-time volunteer consultant who was simultaneously pursuing a pediatric internship at Johns Hopkins.

"It's no wonder they got hornswoggled by the industry on this one," said Thomas Grumbly, a House staffer who used to be an executive assistant to the FDA commissioner. "They were just overwhelmed."

As Grumbly's comments illustrate, not all FDA alumni endorse the bill. Richard Cooper, a former FDA chief counsel who now works for the firm of Williams & Connolly, says it "has some useful ideas, but on the whole it goes too far in removing public protections." Goodrich, who now is head of the Institute of Shortening and Edible Oils, says he, too, likes parts of the bill, but adds, "the repeal of Delaney won't fly."

The current leadership of the FDA will not take a position on the bill until a Cabinet-level review of food safety law is completed, perhaps next month. Meantime, consumer groups have been meeting regularly with FDA staffers, and are eager for them to join the fray. "The career people tend to be a stabilizing influence," said Schultz, who is optimistic that what he views as the more onerous provisions of the bill will eventually be removed.

"My views now are consistent with the ones I had there. That's true for Peter and everyone else working on this. While we may have trade association clients, we don't and they don't want to do anything that damages the FDA's ability to do its job."

# Changing the Food Laws
*Washington Post Editorial*
*October 22, 1981*

Led by the National Soft Drink Association and the American Meat Institute, the food industry is pushing for a sweeping revision of the food safety laws. Consumer groups are opposing what the industry is proposing, and congressional staff members are receiving an endless stream of lawyers, lobbyists and regulators. Yet, oddly, as the sides prepare for a battle, there seems to be general agreement among those who know it best that the law, as applied by the Food and Drug Administration and interpreted by the courts, has on the whole worked pretty well.

The food supply, it turns out, is not as relatively uncomplicated a thing as it might seem. Food additives, for example, are not just substances like preservatives, coloring agents, emulsifiers and so forth that are deliberately added to food. There are substances "added" to the food supply while it is growing: animal drugs and hormones in meat, for example. There are also toxic substances that may migrate into food from its packaging: lead from cans, for instance. And there are natural or man-made contaminants, such as pesticides, PCBs and mercury, that enter the food supply from the general environment. To top it all, naturally occurring foods and all their individual components are themselves considered to be additives when they, or their components, become part of processed foods — peanuts in peanut butter for instance, or nutmeg in eggnog.

This broad legal definition of a food additive accounts for much of the opposition to the controversial Delaney clause that bans the use of additives found to cause cancer in human beings and animals. As it becomes harder and harder to define what "causes" cancer — scientists now distinguish, for example, between direct carcinogens, cocarcinogens, promoters and more — and as rapidly advancing technology provides the capacity to detect substances present in infinitesimally small amounts, the 20-year-old Delaney amendment appears to be out of step with current knowledge.

No one, however, has come up with a good substitute. The food industry's suggestion that only substances shown to present "significant risks to health" be banned is premature: there are not yet agreed methods for calculating such risks or for determining which among them is significant. It is an invitation to endless litigation. It would, for example, allow the use of inessential additives such as coloring agents — even after they had been shown to be carcinogenic in animals.

Moreover in only two cases — saccharin and nitrites — has the Delaney clause actually created much controversy. Both of these could have been accommodated by relatively minor changes allowing the FDA some flexibility — such as a phase-out while alternatives are sought rather than an immediate ban — on cases where an additive provides a clear benefit to consumers.

In the coming debate, the burden should be on those who propose changes to demonstrate just where the current law does not work. Are there, for example, additives or substances that should be on the market that the current law does not allow to be sold? Americans now enjoy the world's most varied, and probably its safest, food supply. Congress shouldn't act unless it is sure it is making things better.

# DECREASED FDA LAW ENFORCEMENT DURING 1981

A Report By:

Public Citizen Health Research Group

Sidney M. Wolfe, M.D.
Director

Allen Greenberg
Staff Associate

## SUMMARY

An analysis of two hundred and sixty FDA Enforcement Reports for the last five years shows that during the first year of the Reagan Administration, there has been a marked reduction in the number of FDA enforcement actions. Compared with either 1980 or the average for 1977-1980, there was a sharp drop in the number of seizures combined with other enforcement actions such as recalls, injunctions and prosecutions. These actions were taken to remove adulterated, contaminated and mislabelled foods, unapproved or unsafe drugs, and other products under FDA jurisdiction. In 1981, there were only 577 total enforcement actions compared with an average of 1041 enforcement actions per year for the years 1977-1980, a decrease of 45%. Seizures of all products fell from an average of 369 per year to 129 in 1981, a 65% decrease. Enforcement activities involving food decreased from an average of 383 per year to 203 in 1981, a 47% fall and enforcement activities involving drugs decreased from an average of 393 per year to 181, a 54% fall.

Although polls show the public wants more regulation, not less, especially for FDA-regulated products, the Reagan Administration appears to be doing just the opposite. It is only a matter of time before the adverse health impacts of decreased enforcement of the food and drug laws become evident and force the government to get back on the backs of the industries they are supposed to regulate.

January 28, 1982

# INTRODUCTION

Ronald Reagan stated--while campaigning for president in the Spring of 1980--that "there are areas that properly belong to the federal government. For example, I have no quarrel with there being an FDA....That's fine." But he also promised to get government off the backs of industry and, after the election, confirmed he would keep that promise.

Around the time of the election, a variety of people--including FDA officials, drug industry executives and drug stock analysts--predicted that the Reagan FDA would be much better for business and, implicitly, worse for consumers. One drug company executive said:

"I think the past few commissioners may have been overly influenced by consumer activists and that the FDA may not have been fair to the industry....The FDA, under Mr. Reagan, would be more pro-business."[1]

An FDA official said FDA, under Reagan, would be "more business oriented," "less activist" and that "activists within the agency are not too keen on the prospect of a Reagan presidency; these people want the agency to have more of a 'cops and robbers philosophy'."

When it comes to policing the safety of the food we eat, the medicines we use and other FDA-regulated products, the FDA "activists" desire for a "cops and robbers" philosophy is essential. The passage of the Pure Food and Drug Act of 1906 was a strong protest against the deaths and injuries caused by the unsafe food and drugs marketed by industries who had been allowed to "police" themselves.

If 1981 is any indication, the Reagan Administration is well on its way to bringing back adulterated foods and unsafe or unproven drugs and medical devices as a result of its promise to get government off the backs of industry. The very significant decrease in the total number of FDA law enforcement activities-- seizures, complaints for injunctions, prosecutions and recalls-- during 1981 (from an average of 1041 per year in 1977-80 to 577 in 1981) bodes poorly for the health of this country even though it must please the disemburdened industries. Although Mr. Reagan does not object to "there being an FDA," he appears to object to its vigorous policing of the food industry, the drug and medical device industry and other regulated industries.

---

[1] Chemical Business: October 20, 1980.

## METHODOLOGY

Each week, FDA publishes an "Enforcement Report" which lists regulatory actions including prosecution, seizures, injunctions and recalls. The following is FDA's explanation of these actions:

PROSECUTION: A criminal action filed by FDA against a company or individual charging violation of the law. Prosecutions listed below have been filed with a court but not yet tried or concluded.

SEIZURE: An action taken to remove a product from commerce because it is in violation of the law. FDA initiates a seizure by filing a complaint with the U.S. District Court where the goods are located. A U.S. marshal is then directed by the court to take possession of the goods until the matter is resolved. The date listed is the date a seizure request is filed, not the date of seizure.

INJUNCTION: A civil action filed by FDA against an individual or company seeking, in most cases, to stop a company from continuing to manufacture or distribute products that are in violation of the law. Injunctions listed have been filed with the court but not concluded.

RECALLS: Voluntary removal by a firm of a defective product from the market. Some recalls begin when the firm finds a problem, others are conducted at FDA's request. Recalls may involve the physical removal of products from the market or correction of the problem where the product is located.

To determine the total number of each type of enforcement action for a given year, we added up the number of actions in each of the 52 weeks for that year.[1] By adding up all the weekly reports <u>dated</u> 1980, for example, a reasonably accurate count of 1980 enforcement activities can be calculated. However, actions occurring late in 1980 may not get into the <u>Reports</u> until early 1981. Conversely, some activities, especially recalls, which occur late in 1979 may not be reported in a weekly FDA report until early 1980. To correct for this, <u>we have placed the enforcement actions in the year in which they occurred rather than the year they were reported</u>. For all years except 1981, the record is complete. In order to compare 1981 more precisely with earlier years, we have extrapolated to a larger number than actually was reported as of January 1, 1982. This extrapolation

---

[1] The only weekly report (of the 260 reports) not available for the five years was from May 10, 1978.

is based on the degree of incompleteness of reporting in the same categories for 1980. For example, in 1980, 86% of the total number of enforcement actions ultimately reported in the FDA Enforcement Report were listed by January 1, 1981. The incomplete number, actions reported by January 1, 1981, is 892. Since the final number of actions for 1980 is 1037, we calculated that 86% of all 1980 enforcement actions were reported by January 1, 1981. 892 is 86% of 1037. In arriving at the figures for 1981 we assumed that the 499 1981 enforcement actions reported by January 1, 1982 also represented 86% of the actions that will ultimately be reported. Thus, the number listed in Chart 1 on the following page, 577, was determined by calculating that 499 is 86% of 577.

For all other calculations of 1981 (Est.) data, we have similarly used the 1980 degree of incompleteness for that category, assumed it would be the same in 1981 and made the same upward increase in the number used for 1981 from the number reported by January 1, 1982.

# DECREASE IN FDA LAW ENFORCEMENT DURING 1981

1. TOTAL ENFORCEMENT ACTIVITIES: SEIZURES, RECALLS, COMPLAINTS FOR INJUNCTION AND PROSECUTION ACTIONS FOR ALL FDA-REGULATED PRODUCTS.

IN 1981, THERE WAS A 45% DECREASE FROM THE AVERAGE FOR 1977-1980.

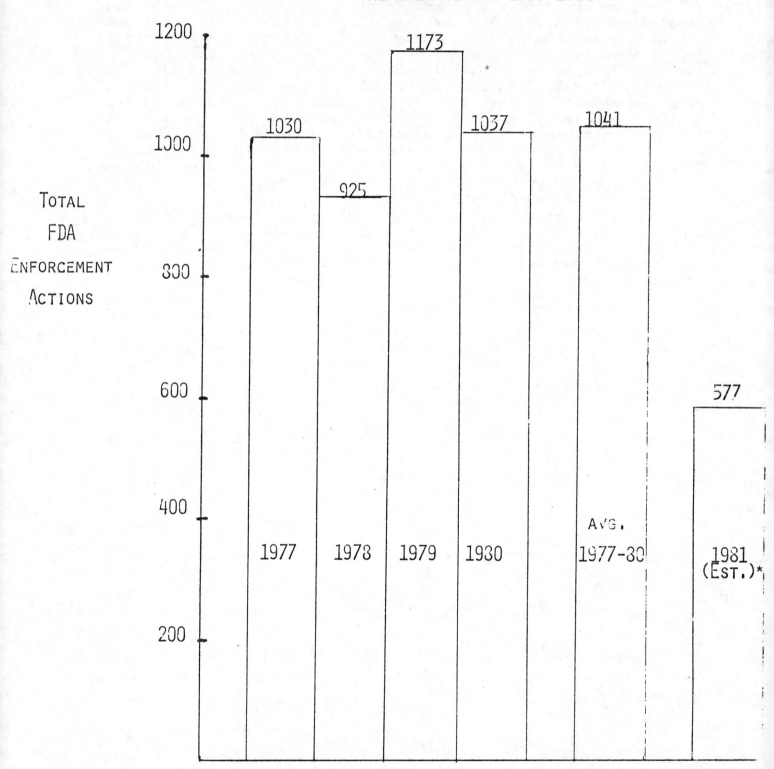

*(BY JAN. 1, 1981, 86% OF TOTAL ENFORCEMENT ACTIONS FOR 1980 HAD BEEN REPORTED. ASSUME SAME 86% OF 1981 ACTIONS REPORTED BY JAN. 1, 1982. THE ACTUAL NUMBER REPORTED BY JAN. 1, 1982 WAS 496. 577 IS THE ESTIMATE WHEN REPORTING IS COMPLETE.)

SOURCE: FDA ENFORCEMENT REPORTS: 1977-1981.

RESULTS

1. Total Enforcement Activities For All FDA-Regulated Products

As seen on the previous page, there was a significant decrease in total enforcement activities in 1981 in comparison with the average for the previous 4 years. From an average of 1041 actions per year in 1977-80, there was a 45% decrease to 577 in 1981. (From 1980, with 1037 actions, there was a 44% decrease in 1981.) The main categories of products which are the subject of these actions, in order of decreasing numbers for 1980 are drugs, foods, medical devices and cosmetics.

2. Seizures For All FDA-Regulated Products

In 1981, as seen in Chart 2 (page 7), there was a 65% decrease in the number of seizure actions filed with U.S. Federal District Courts. The average number per year, for 1977-1980 was 369 and in 1981, the number decreased to 129. (From 1980, with 360 seizures, there was a 64% decrease in 1981.)

3. Recalls of All FDA-Regulated Products

In 1981, as seen in Chart 3 (page 8), there was a 31% decrease in the total number of recalls from the average of 643 per year for the previous 4 years to 441 for 1981. (From 1980, with 658 recalls, there was a 33% decrease in 1981.) This is particularly interesting in light of the striking decrease in the number of seizures. Although a large number of "voluntary" recalls are accomplished as an alternative to the threat of seizure, some are actually voluntary. The Reagan Administration's pro-business notion of voluntary activities as a preferable alternative to a seizure-type of "cops and robbers" activities has certainly not been reflected in an increase in the number of such recalls to compensate for the 65% decrease in the number of seizure actions. Rather, there has also been a decrease in "voluntary" enforcement activities--recalls--as well.

4.   Total Enforcement Actions For Drugs

As seen in Chart 4 (page 9), there was a 51% decrease in the total number of enforcement actions for drugs in 1981 from the average of 393 per year in 1977-1980 to 181 in 1981. (From 1980, with 416 drug enforcement actions there was a 56% decrease in 1981.)

5.   Total Enforcement Actions For Foods

In 1981, there was a 47% decrease in the total number of enforcement actions for foods from the average of 383 for the previous 4 years to 203 in 1981. (From 1980, with 337 actions, there was a 40% decrease in 1981.) In the case of foods, there appears to be a slight downward trend during the 1977-1980 interval. However, even if it continued at its average decrease of 34 fewer per year in 1981, there still would have been over 300 actions last year as opposed to 203.

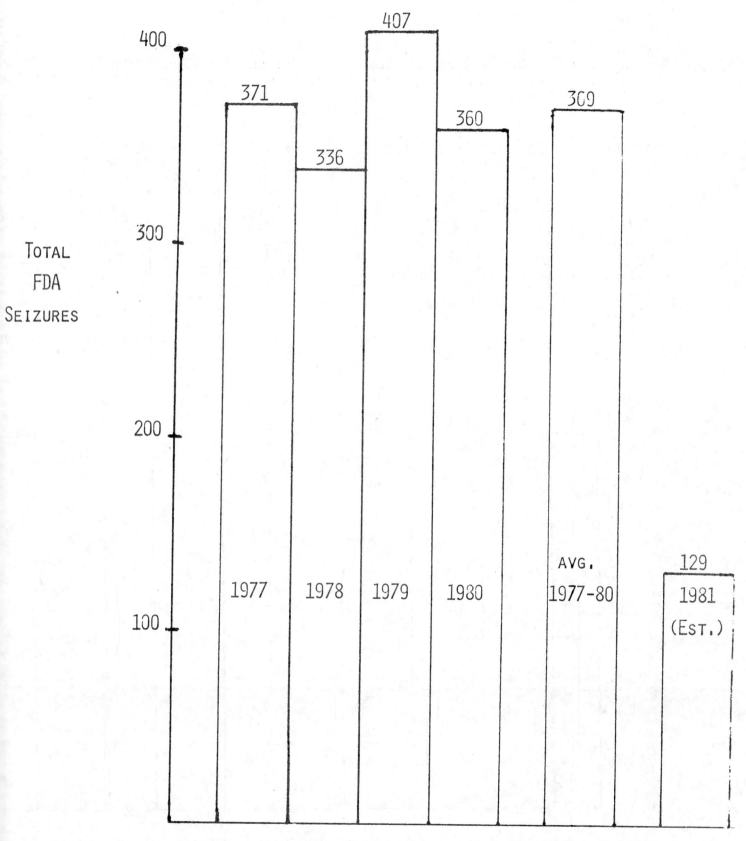

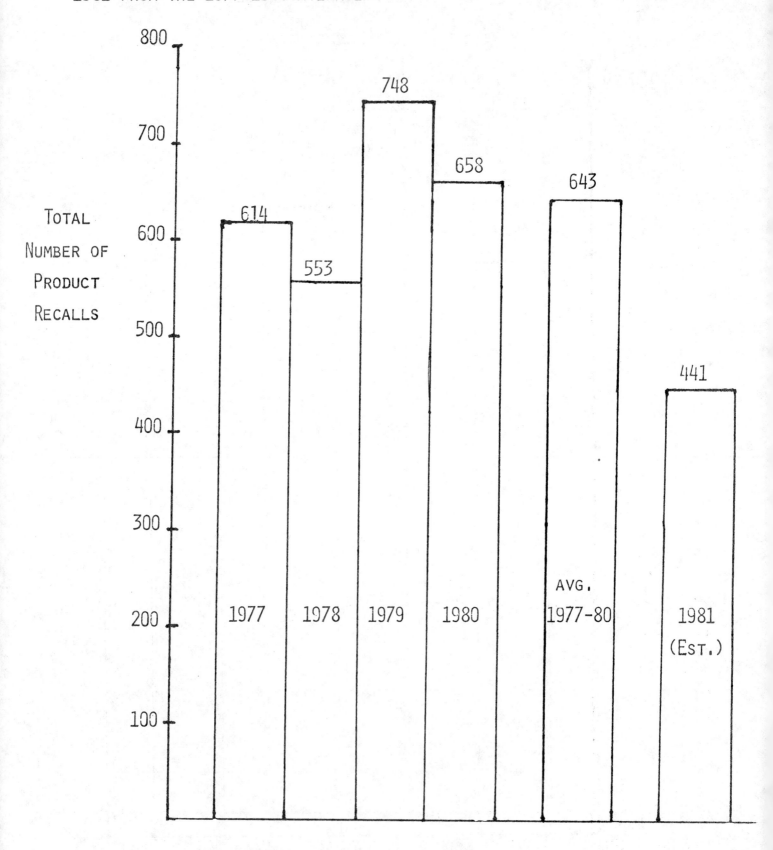

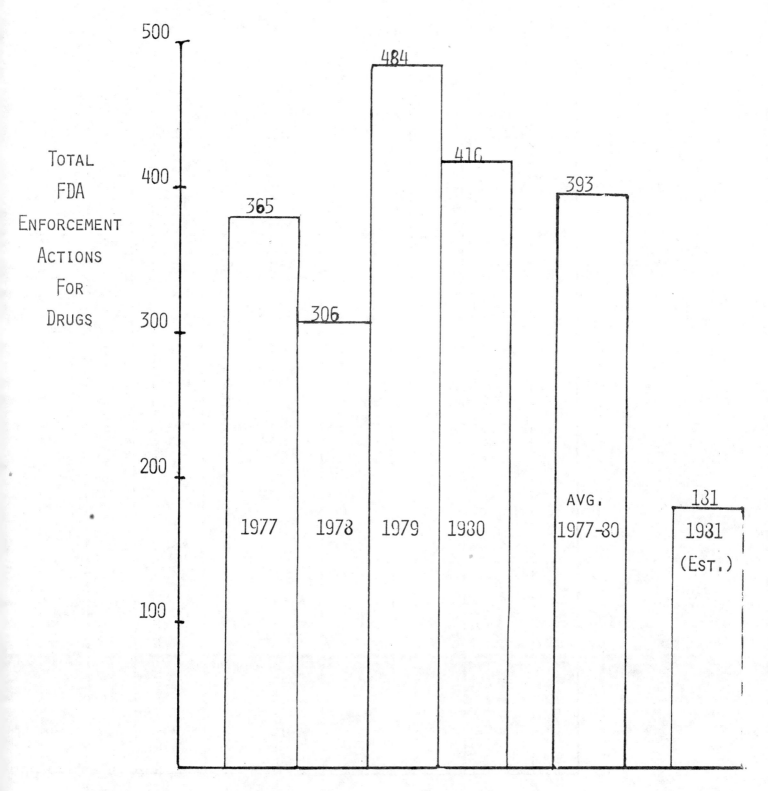

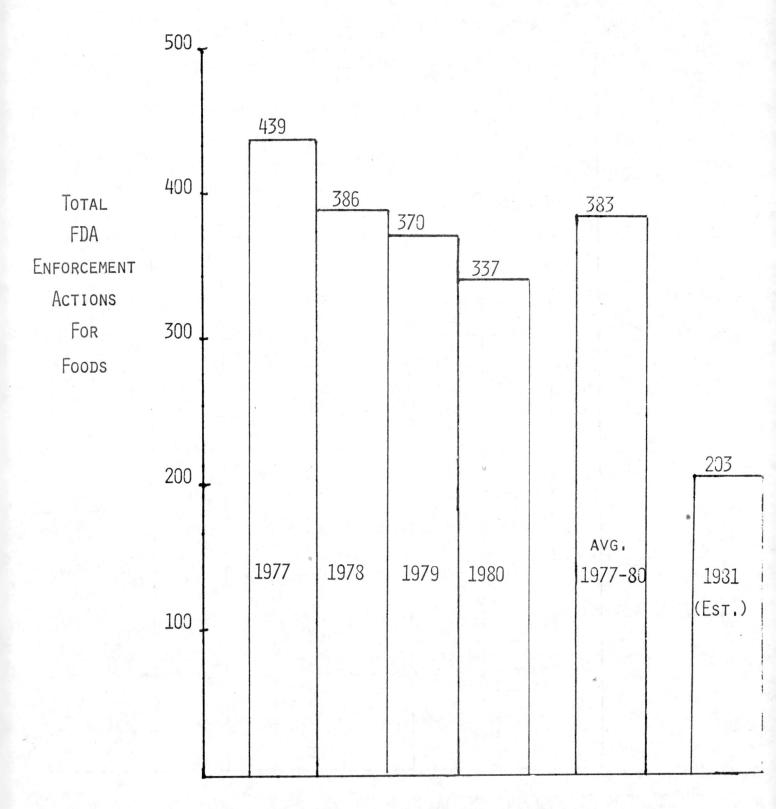

1979-1980 EXAMPLES OF THE KIND OF ENFORCEMENT ACTIONS
WHICH DECREASED 45% IN 1981

FOODS:

    Cottonseed Meal: adulterated, "contaminated with (the carcinogen) aflatoxin."

        Action: seizure filed December 18, 1979.

    Dried Bananas: "may contain pieces of wire mesh screen."

        Action: recall initiated November 13, 1979.

    Beer: "glass contamination."

        Action: recall initiated November 27, 1979.

    Cream Pies: contaminated with bacteria (staph).

        Action: recall November 26, 1979.

    Tomatoes: "adulterated, unfit for food because of deteriorated can linings."

        Action: seizure filed January 25, 1980.

    Rice & Raisins: "adulterated, products contaminated with insects, insect larvae, insect pupae..."

        Action: seizure filed November 26, 1979.

    Brazil nuts: adulterated, "product contaminated with rodent urine, rodent excretory pellets and were rodent gnawed."

        Action: seizure filed February 22, 1980.

    Olive oil: bottles labelled as olive oil contained camphor spirits which could cause convulsions or death if ingested.

        Action: recall February 21, 1980.

Olives: "contaminated with clostridium tetani bacteria and may possibly be contaminated with clostridium botulinum.

    Action: Class I recall March 10, 1980.

Mushrooms: "FDA has confirmed pre-formed Type B Botulinum Toxin in Swollen Cans."

    Action: Class I recall May 31, 1980.

Fish Oil: "adulterated-product contains PCB's (polychlorinated biphenyl in excess of the temporary tolerance of 2.0 parts per million."

    Action: seizure filed June 26, 1980.


DRUGS:

Nitrous oxide (anesthetic): adulterated, not as pure as labelled.

    Action: seizure filed November 16, 1979.

Dextroamphetamine sulfate: quality and purity below standard.

    Action: seizure filed March 10, 1980.

Dexamethasone (a steroid): injectable drug found non-sterile.

    Action: Class I recall August 13, 1980.

Most of the drug examples are seized or recalled either because the drug was manufactured without following good manufacturing practice regulations or because the drug was marketed without FDA approval.

## CONCLUSION

There was a significant decrease in 1981 in the number of FDA enforcement actions reported in FDA's own <u>FDA Enforcement Reports</u>. The inquiry into this was generated by comments from various people about the low morale in FDA legal offices and amongst FDA compliance personnel due to the prevailing "philosophy" of the Reagan Administration to be more "pro-business." The decreased amount of enforcement actions has one of two explanations: one unlikely, one more likely. It is unlikely that the decreased number of seizures and recalls of foods, drugs and other FDA-regulated products is due to a sudden and drastic improvement in the safety and quality of products since Mr. Reagan was sworn in. A much more likely explanation is that many adulterated foods and unapproved and/or dangerous drugs of a type which were seized or recalled during 1977-1980 are now being consumed or used because of a decreased amount of FDA regulatory activity.

It is ironic that as the Reagan Administration starts paving the way back to a more "pro-business" attitude with less "cops and robbers" activity and increased health risks to the American public, the public clearly wants more regulation, not less. A November 2, 1981 memo from FDA Associate Commissioner for Planning and Evaluation Gerald Barkdoll to the FDA Policy Board reported the results of a 1981 Roper Public Opinion Survey about FDA. Since 1979, when a companion survey had been done, a larger percent of those polled thought there was not enough regulation by FDA.

"The sentiment in 1981 for more regulating rather than less (29 percent versus 14 percent) is a significant net change from the nearly balanced opinion in 1979 (24 percent versus 20 percent," the memo stated.

This was seen even more strikingly amongst employed females, 34% of whom thought there was not enough regulation versus just 13% saying there was too much.

The memo ended by saying that "the results of the 1981 survey and the rising expectations of all groups may have substantial implications for the agency during the next few years."

As the public expresses a desire for more regulation in 1981 than in 1979, the Reagan Administration is giving them less regulation in 1981 than in 1979. The rising expectations of the public for more food and drug safety regulation is a much clearer mandate than the mythical mandate of President Reagan and his business cronies for "getting the government off the backs of business." It will not take many examples of negative health impacts from the increasing amount of unsafe and unregulated foods and drugs to mobilize the public and the Congress to push the Reagan Administration back toward more regulation rather than less.

March 9, 1981

The Health Research Group (HRG) of Public Citizen, Inc. and the Center for Science in the Public Interest (CSPI) announced today that they have filed suit against the Food and Drug Administration (FDA) to stop the illegal marketing of 23 color additives used in food, drugs or cosmetics. These 23 additives have never been shown to be safe, as required by law, and were, until January 31, 1981, on a provisional list which allowed them to be marketed while further tests were being carried out to determine their safety.

Three of these color additives, FD&C Blue No. 2, FD&C Green No. 3 and FD&C Yellow No. 6, were provisionally listed for use in food. Examples of the types of foods containing these dyes are candy and beverages (Blue 2 and Green 3) and beverages, candy, baked goods and gelatin (Yellow 6).

Three other dyes, FD&C Yellow No. 5, FD&C Blue No. 1 and FD&C Red No. 3, have been approved for use in food and drugs but are provisionally listed for use in cosmetics because of unresolved questions about their safety. The other additives are only used in drugs and/or cosmetics.

The FDA failed to issue a regulation to further extend the provisional listing for these drugs beyond January 31, 1981, in response to President Reagan's freeze on government regulations. Therefore, these additives can no longer legally be marketed. HRG and CSPI have fought in the past to have these additives removed from the market and are now suing to force the FDA to follow its own regulations and remove these products. (Other petroleum-derivative food dyes which are still on the market are permanently listed. They include Red 3, Red 40, Blue 1, Yellow 5, Citrus Red 2 and Orange B.)

Fourteen petroleum derivative (often called coal-tar) food dyes have already been banned over the years--usually when evidence of toxicity was belatedly "discovered." These include Red 2, Violet 1, Red 4 and others.

\*      \*      \*      \*      \*      \*

HRG Director, Dr. Sidney M. Wolfe, said today that "The FDA, although well aware that these dyes are being illegally used, has told the food, drug and cosmetic industry that FDA 'does not intend to take any regulatory action' [See History, p. 2].

"Instead of protecting the millions of Americans who consume foods, drugs and cosmetics by banning these 23 dyes of unproven safety, the FDA is protecting the profits of the industries which make and use the dyes by urging them to break the law and continue their illegal marketing of these products."

HEALTH RESEARCH GROUP • 2000 P STREET, N.W., WASHINGTON, D.C. 20036 • (202) 872-0370

## History

The Color Additive Amendments of 1960 require that a new color additive (for foods, drugs or cosmetics) be proven safe through long-term animal tests before it can be marketed. These amendments provided for a 2 1/2 year grace period during which time additives already on the market could continue to be sold while tests were carried out. These additives were, thus, said to be on a provisional list. Congress allowed the FDA the freedom to extend this 2 1/2 year period if more time was needed for tests to be carried out. Between 1963 and 1980 the FDA used this power to grant approximately 15 extensions of marketing deadlines to these provisional additives.

Yet another extension was set to take effect on January 31, 1981 when the last extension expired. This extension would have given each color additive one to three more years of provisional status. On January 29, President Reagan asked that all pending regulation be delayed, and as a result, these 23 color additives lost their legal right to be marketed when their provisional listing expired on January 31.

On February 2, 1981, Mark Novitch, Acting FDA Commissioner, informed the manufacturers' trade associations "that FDA does not intend to take any regulatory action against any food, drug and cosmetic product which bears or contains . . . any of the colors which as of January 31, 1981, were on the provisional list."[1]

The Health Research Group and the Center for Science in the Public Interest feel that the FDA has made a mockery of the Food, Drug and Cosmetic Act and the regulatory process first by allowing these additives to remain on the market without adequate assurance of their safety for 18 years past the congressional deadline, and second by failing to halt the sale of these additives now that their provisional listing has expired.

The three dyes that were provisional for use in food are all coal-tar dyes. Many dyes of this type are known or suspected carcinogens and some, such as Red No. 2, have been removed from the market. The fact that these dyes were provisionally listed means that the FDA has not yet been able to resolve questions about their safety.

---

1   Letter from Dr. Mark Novitch, Acting FDA Commissioner, to Cosmetic, Toiletry and Fragrance Association, Inc., Feb. 2, 1981. Same letter sent to Pharmaceutical Manufacturers Association and Certified Color Manufacturers Association.

Blue 2 (Indigotine)

In 1979, 88,000 pounds of Blue 2 were used in the United States, mostly in candy and beverages.

A 1966 study[2] found that rats fed 2% or 5% Blue 2 in their diet for two years showed growth inhibition. This same study found cancer in 11 of 24 rats receiving 5% Blue 2 as opposed to 6 of 24 control animals with no exposure. Among females, 6 of 12 animals had breast cancer as opposed to 3 of 12 controls. Although these differences were not statistically significant, they certainly raise questions as to the carcinogenicity of Blue 2.

An ongoing study of Blue 2 has shown that rats fed this additive have lowered plasma glucose levels. No data on carcinogenicity has been made available from this study.[3]

Green 3 (Fast Green FCF)

In 1979, 6000 pounds of Green 3 were used in candy and beverages.

The only oral feeding tests of Green 3 that have been completed to date showed no evidence of carcinogenicity[4] although reviewers for the International Agency for Research on Cancer do not consider these tests to have been adequately done.[5]

A current study involving the feeding of Green 3 to rats has shown a greater cancer incidence among treated animals after one year, although the difference is not yet statistically significant.[6]

---

2   Hansen et al., Tox. Appl. Phar., **8**,:29-36, 1966.

3   FDA Memo from Dr. Michael Bolger to Dr. Joan Schwing, April 8, 1980.

4   Hansen, et al., Fd. Cos. Tox., **4**:389-410, 1966.

5   IARC Monographs, **16**:192, 1978.

6   FDA Memo from Ronlad Biskup to Dr. Joan Schwing, February 1, 1980.

Yellow 6 (Sunset Yellow FCF)

In 1977, almost 1.3 million pounds of Yellow 6 were used in gelatin, beverages, candy and baked goods.

Although there is no evidence to suggest that Yellow 6 is carcinogenic, there are other unresolved questions as to its safety. Yellow 6 is very similar to Yellow 5 which is known to cause allergic reactions, such as rashes and asthma, in many people.

An ongoing study in which Yellow 6 is being fed to mice has shown a trend towards a relationship between dose and mortality. 7/

Other Additives

The other additives involved in the HRG and CSPI lawsuit are: FD&C Blue No. 1, D&C Green No. 5, D&C Orange No. 5, D&C Orange No. 17, FD&C Red No. 3, D&C Red No. 6, D&C Red No. 7, D&C Red No. 8, D&C Red No. 9, D&C Red No. 19, D&C Red No. 21, D&C Red No. 22, D&C Red No. 27, D&C Red No. 28, D&C Red No. 30, D&C Red No. 33, D&C Red No. 36, D&C Red No. 37, FD&C Yellow No. 5, D&C Yellow No. 6.

---

7     FDA Memo from Neil Sass, Ph.D., to Dr. Joan Schwing, March 28, 1980.

UNITED STATES DISTRICT COURT
FOR THE DISTRICT OF COLUMBIA

WENDE McILWAIN,

    3605 Underwood
    Chevy Chase, Maryland 20015
    (301) 652-3695

NANCY HENDREE SIMPSON,

    2900 Woodley Road, NW
    Washington, D.C. 20008
    (202) 232-6346

PUBLIC CITIZEN,

    1346 Connecticut Ave., NW
    Washington, D.C. 20036
    (202) 872-0320

and

CENTER FOR SCIENCE IN THE PUBLIC INTEREST,

    1755 S Street NW
    Washington, D.C. 20009
    (202) 332-9110

        Plaintiffs,

v.

DR. MARK NOVITCH, ACTING COMMISSIONER,

    Food and Drug Administration
    200 C Street SW
    Washington, D.C. 20204

        Defendant.

## COMPLAINT FOR DECLARATORY AND INJUNCTIVE RELIEF
(Unlawful Agency Action)

1. This action seeks a declaratory judgment that the Food and Drug Administration (FDA) has no authority to approve the sale of 23 color additives which have not been proven safe, and an order directing the FDA to take appropriate action to prohibit their shipment in interstate commerce.

### Jurisdiction

2. This court has jurisdiction over this action pursuant to 28 U.S.C.

sections 1331, 1337, and 1361.

## Parties

3. Plaintiffs Wende McIlwain and Nancy Hendree Simpson are individuals who regularly use food, drug and cosmetic products which contain the 23 color additives that have not been proven to be safe.

4. Plaintiff Public Citizen is a non-profit, public interest organization supported by annual contributions from approximately 40,000 persons. Its Health Research Group (HRG) is engaged in research and advocacy to promote safe food, drugs and cosmetics, and in 1977, brought a lawsuit to enjoin FDA from extending the period during which color additives that had not been proven safe could be sold (Health Research Group v. Califano, Civ. No. 77-0293). Public Citizen brings this action on behalf of its contributors and employees who use food, drugs and cosmetics containing the 23 color additives.

5. Plaintiff Center for Science in the Public Interest (CSPI), a non-profit, consumer-research and advocacy organization which has long opposed the use of color additives that have not been proven safe, brings this action on behalf of its approximately 25,000 members.

6. Defendant Dr. Mark Novitch is sued in his capacity as Acting Commissioner of the FDA.

## Statutory Framework and Facts Giving Rise To Plaintiffs' Claims For Relief

7. The Color Additives Amendments to the Food, Drug and Cosmetic Act, enacted by Congress on July 12, 1960, 21 U.S.C. section 376, establish a comprehensive system for regulating the use of color additives in foods, drugs and cosmetics.

8. The Color Additive Amendments prohibit the sale of color additives until the additive has been proven "safe" for its intended use, at which time the FDA may "permanently list" an additive for use in foods, drugs or cosmetics,

except that the FDA may place additives on a "provisional list" pursuant to the requirements of the transitional provisions to the Amendments, 79 Stat. 399, sections 201-203, 21 U.S.C. section 376 (note)("the transitional provisions").

9. The transitional provisions provide that the FDA may place all color additives which were commercially established on July 12, 1960 on the "provisional list" for 2-½ years following enactment of the statute, therby allowing the additives to be marketed while tests of their safety were conducted.

10. The transitional provisions also provide that the FDA may extend the "closing date" beyond the 2-½ year interim period, if an extension is both "consistent with the public health" and "with the objectives of carrying to completion in good faith, as soon as reasonably practicable, the scientific investigations necessary for making a determination as to [permanently] listing [the color] additive." Section 203 (a), 21U.S.C. section 376 (note).

11. The 23 color additives involved in this lawsuit, listed at 45 Fed. Reg. 75228-9 (Nov. 4, 1980) and 21 C.F.R. section 81.27(d), are synthetically produced coal tar dyes which serve no purpose other than adding color to foods, drugs and cosmetics. None of the 23 color additives have been proven to be safe.

12. During the 2-½ years between July 12, 1960, when the Color Additive Amendments were enacted, and January 12, 1963, the 23 color additives were provisionally listed pursuant to section 203(a) of the transitional provisions.

13. Between January 12, 1963 and January 31, 1977, the FDA granted approximately 14 extensions of the closing dates for these additives, thereby allowing the 23 color additives to remain on the market for 14 additional years even though they had not been proven to be safe.

14. On January 31, 1977, despite HRG's objections, the FDA extended the closing date for the 23 color additives for an additional four years to January 31, 1981. 42 Fed. Reg. 6992, 7000 (Feb. 4, 1977). However, recognizing that Congress "did not anticipate in 1960 that color additives would be provision-

ally listed in 1976" and declaring its commitment to "close the books" on the provisional listing, the FDA adopted a strict standard applicable to all future requests for extensions of the January 31, 1981 closing date:

> In the unlikely event that unforeseen and unavoidable circumstances arise to make compliance with the the requirements of the final regulation virtually impossible, the Commissioner will consider requests for brief extensions of the closing dates. The Commissioner cautions, jowever, that such requests will be considered only if "extraordinary circumstances" exist and maximum effort has been given to meeting the deadlines.

42 Fed. Reg. 6992, 6998 (Feb. 4, 1977); 21 C.F.R. section 81.27(d).

15. HRG challenged the extension of the closing date to January 31, 1981 in <u>Health Research Group v. Califano</u>, Civ. No. 77-0293 (D.D.C. 1977). Although the Court dismissed the case in a Memorandum dated September 23, 1977, the FDA entered into a binding agreement with HRG providing, <u>inter alia</u>, that:

> In considering any request [for a postponement of the January 31, 1981 closing date] the Food and Drug Administration will evaluate the additive on an individual basis, and if any request is granted, it will, prior to January 31, 1981, publish in the Federal Register a specific justification for both the extension and period of time allowed.

16. On November 14, 1980, the FDA proposed regulation in response to petitions from three manufacturers' trade associations which would again extend the provisional listing of the 23 color additives, this time for 1-3 years. 45 Fed. Reg. 75226-9 (Nov, 14, 1980).

17. HRG and CSPI filed timely comments with the FDA objecting to the proposed extensions.

18. The FDA has not issued a final regulation extending the closing date for the 23 color additives beyond January 31, 1981.

19. Since February 1, 1981, the 23 color additives have not been on either the FDA's provisional lis or its permanent list.

20. By letters dated February 2, 1981, defendant informed the three trade associations which had petitioned the FDA for extensions of the January 31,

1981 closing date that although the FDA had intended to promulgate regulations extending the closing date of the provisional listing, it had not done so because of President Reagan's January 29, 1981 memorandum to Executive agencies urging the postponement of final rules for 60 days. Defendant's letter also advised the trade associations "that FDA does not intend to take regulatory action against any food, drug and cosmetic product which bears or contains.... any of the colors which as of January 31, 1981, were on the provisional list."

### Causes of Action

21. As a result of the actions of the FDA, food, drugs and cosmetics containing the 23 provisionally listed color additives continue to be sold in interstate commerce although the color additives have not been proven safe as required by the 1963 Color Additive Amendments.

22. By advising the manufacturers' trade associations that the FDA will not take regulatory action against the foods, drugs and cosmetics that contain the 23 color additives, defendant violated both the Color Additive Amendments, 21 U.S.C. section 376, and the FDA's own regulations, 21 C.F.R. section 81.27.

23. The FDA has no authority to extend the provisional listing of the 23 color additives because it has failed to comply with the transitional provisions of the Color Additive Amendments, 21 U.S.C. section 376 (note), the agency's own standards, 42 Fed. Reg. 6998 (Feb. 4, 1977), and the Stipulation agreed to in Health Research Group v. Califano, supra.

24. The FDA has abused its discretion and acted arbitrarily and capriciously and contrary to the law in violation os section 10(e) of the Administrative Procedure Act, 5 U.S.C. section 706, by allowing the sale of color additives which have not been proven safe pursuant to the Color Additive Amendments.

25. Unless this Court orders the FDA to rescind its authorization of the continued sale of the 23 color additives and to advise the manufacturers accordingly, the individual plaintiffs and those represented by the organization-

al plaintiffs will be subjected to continued risk to their health by exposure to the 23 color additives which have not been proven to be safe, as required by the Color Additive Amendments.

WHEREFORE, plaintiffs pray that this Court enter an Order:

(1) declaring that the continued sale of the 23 color additives is in violation of the law;

(2) declaring that FDA has no authority to extend the provisional listing of any of the 23 color additives beyond January 31, 1981;

(3) directing FDA to take all appropriate steps to remove the 23 color additives from the market;

(4) awarding plaintiffs their costs, disbursements, and a reasonable attorneys' fee; and

(5) granting plaintiffs such other and further relief as may be just and proper.

Respectfully submitted,

Katherine A. Meyer

William B. Schultz

Alan B. Morrison

Public Citizen Litigation Group
Suite 700, 2000 P St. NW
Washington, D.C. 20036

March 9, 1981    Attorneys for Plaintiffs

# TAKING STEPS TOWARD SAFER FOOD

Section Five

## TAKING STEPS TOWARD SAFER FOOD

Having journeyed this far, you just could be most eager to become part of the solution for these preventable problems. Two roles are important here. One is that of a more discriminating food buyer; the other is that of a more active public citizen. The two roles go hand in hand and together lead to greater effectiveness. Moreover, if you have young children, they will benefit very much from being around such informed shopping behavior and community activity.

Readings in this section reflect these two roles, starting with consumer self-improvement advice. One of the most popular government publications is "Nutrition and Your Health; Dietary Guidelines for Americans." Millions of copies of this informative pamphlet were distributed in 1981. Nonetheless, the U.S. Department of Agriculture, responding to some meat and dairy industry criticisms of the pamphlet, announced that the government would discontinue its distribution. So because of their value and scarcity, it is doubly important that these dietary guidelines be reprinted here. A complementary publication is issued by the Public Health Service called "Health Style: a Self Test," which quizzes you about how well you are doing to stay in good health. Find out how you score.

Eating more nutritiously and wisely is also a way of minimizing your exposure to contaminants and chemicals in food. Dr. Michael Jacobson, director of the Center for Science in the Public Interest, says:

"By eating healthful, natural foods — like whole grains, legumes, fresh vegetables and fruit, low fat dairy products and lean poultry and meat — you can avoid most questionable food additives, the filth that too often gets into hot dogs and other processed foods, and the dangerous environmental pollutants that accumulate in animal fats. Of course, some environmental pollutants are virtually ubiquitous, so even the most carefully chosen diet will not be totally free of unwanted guests."

Eating fresh foods from your own garden and fruit trees serves similar objectives. (Thirty-three million American households, according to a Gallup Poll survey, grow some of their food in home or community gardens.)

Knowing more about good food can easily pique your curiosity about the entire U.S. food system, from the fields to your table. Here the researchers from Rodale Press in Emmaus, Pennsylvania 18049, have some fascinating material for you. Called "The Cornucopia Papers," they guide your interest toward the soil, the rapid loss of agricultural lands through erosion and development, the long distribution systems from farm to market, how much food states import from one another (compared to a generation or two ago) and much more about the food crunch and price spiral that will come if present conditions do not change.

The "Cornucopia Project" is designed for Americans who want change for a more self-sustaining food growing system. Letters from people across the country are pouring into Rodale Press for additional free information about the Project, such as the one by a mother from Orangevale, California, who wrote: "I'd like to know what I *can* do. With the help of the Cornucopia Project, I hope to do my share — for my children and future grandchildren."

Meanwhile, the food and soft drink lobby is looking to its own future horizons. It is demanding from Congress action to cut away and weaken the nation's food health and safety laws. It hopes to do this while America sleeps, with very little publicity. Legions of food and soft drink lobbyists are swarming all over Capitol Hill arguing their case before the legislators, in offices, corridors and swank restaurants. The legislators are wondering if weakening the food safety laws would create a storm of protest back in their Districts. Well, answering that is up to you, your friends and neighbors. Making Congress aware won't take all that many people, just enough informed and active citizens — the second role.

First, you should know that you are not alone in your concern. Sounding the alert on behalf of the food safety laws are several consumer groups that have prepared useful materials to help you catch the attention of your members of Congress. The article "Coalition Vows to Defend Food Safety Laws" illustrates this effort. Included in this section are the Coalition for Safe Food's suggested courses of action for you to consider. A two-page primer on "How to Learn Federal Secrets" shows you the easy use of the Freedom of Information law, with a sample letter requesting specific information from the Food and Drug Administration (or any other agency). If you have difficulty obtaining information you believe the government should send you on food issues, contact the Freedom of Information Clearinghouse, P.O. Box 19367, Washington, DC 20036.

When you have the information the next step is making your Representatives aware of your concern. Short, pointed articles describing how a bill becomes a law, how to form a telephone tree, how to write your Representative, how to conduct a public candidates' forum, and how to profile your Representative can save you time, trial and error. Also, remember that your members of Congress speak frequently to business chambers and clubs; they should be as willing to speak to a consumers' gathering that you organize. Such a meeting can enhance their sensitivity to your recommendations.

Concluding this section are additional reference materials: 1) summaries of several General Accounting Office (GAO) reports on food safety — single copies of GAO reports are free and can be obtained by writing to the General Accounting Office, Washington, DC 20548; 2) a list of the major food safety laws; 3) a list of some free information sources; 4) a list of leading food safety consumer groups that would be delighted to hear from you; 5) a list of administration officials and members of Congress with leading responsibilities for the food safety laws; and 6) a list of some major food corporations and their addresses. Don't hesitate to write to them.

In conclusion, if these selected materials make you look at the food you buy more critically, an important purpose will be served. But, becoming part of the "clean up our food movement" would be even more significant. Preventing problems is always better than trying to cope with them on a daily basis. As the eminent British mathematician and philosopher Alfred North Whitehead wisely said: "Duty arises from our potential control over the course of events." These materials, by showing the potential control that an individual can have over the course of events, should nourish a heightened sense of duty. Duty is a many-splendored side of freedom. You live in a country where you can exercise both. Whether you and your family and fellow citizens are going to be free from the poisons flowing into the food supply will depend on the assertion of duties. That means, the rest is up to each one of us — to make the difference.

# Nutrition and Your Health

Dietary Guidelines for Americans

 **Eat a Variety of Foods**

 **Maintain Ideal Weight**

 **Avoid Too Much Fat, Saturated Fat, and Cholesterol**

 **Eat Foods with Adequate Starch and Fiber**

 **Avoid Too Much Sugar**

 **Avoid Too Much Sodium**

 **If You Drink Alcohol, Do So in Moderation**

U.S. Department of Agriculture
U.S. Department of Health and Human Services

# Nutrition and Your Health

Dietary Guidelines for Americans

What should you eat to stay healthy? Hardly a day goes by without someone trying to answer that question. Newspapers, magazines, books, radio, and television give us a lot of advice about what foods we should or should not eat. Unfortunately, much of this advice is confusing.

Some of this confusion exists because we don't know enough about nutrition to identify an "ideal diet" for each individual. People differ — and their food needs vary depending on age, sex, body size, physical activity, and other conditions such as pregnancy or illness.

In those chronic conditions where diet may be important — heart attacks, high blood pressure, strokes, dental caries, diabetes, and some forms of cancer — the roles of specific nutrients have not been defined.

Research does seek to find more precise nutritional requirements and to show better the connections between diet and certain chronic diseases.

But today, what advice should you follow in choosing and preparing the best foods for you and your family?

The guidelines below are suggested for most Americans. They do not apply to people who need special diets because of diseases or conditions that interfere with normal nutrition. These people may require special instruction from trained dietitians, in consultation with their own physicians.

These guidelines are intended for people who are already healthy. No guidelines can guarantee health or well-being. Health depends on many things, including heredity, lifestyle, personality traits, mental health and attitudes, and environment, in addition to diet.

Food alone cannot make you healthy. But good eating habits based on moderation and variety can help keep you healthy and even improve your health.

## DIETARY GUIDELINES FOR AMERICANS

- **Eat a variety of foods**
- **Maintain ideal weight**
- **Avoid too much fat, saturated fat, and cholesterol**
- **Eat foods with adequate starch and fiber**
- **Avoid too much sugar**
- **Avoid too much sodium**
- **If you drink alcohol, do so in moderation**

February 1980

# Eat a Variety of Foods

You need about 40 different nutrients to stay healthy. These include vitamins and minerals, as well as amino acids (from proteins), essential fatty acids (from vegetable oils and animal fats), and sources of energy (calories from carbohydrates, proteins, and fats). These nutrients are in the foods you normally eat.

Most foods contain more than one nutrient. Milk, for example, provides proteins, fats, sugars, riboflavin and other B-vitamins, vitamin A, calcium, and phosphorus—among other nutrients.

No single food item supplies all the essential nutrients in the amounts that you need. Milk, for instance, contains very little iron or vitamin C. You should, therefore, eat a variety of foods to assure an adequate diet.

The greater the variety, the less likely you are to develop either a deficiency or an excess of any single nutrient. Variety also reduces your likelihood of being exposed to excessive amounts of contaminants in any single food item.

One way to assure variety and, with it, a well-balanced diet is to select foods each day

# Maintain Ideal Weight

If you are too fat, your chances of developing some chronic disorders are increased. Obesity is associated with high blood pressure, increased levels of blood fats (triglycerides) and cholesterol, and the most common type of diabetes. All of these, in turn, are associated with increased risks of heart attacks and strokes. Thus, you should try to maintain "ideal" weight.

But, how do you determine what the ideal weight is for you?

There is no absolute answer. The table on the following page shows "acceptable" ranges for most adults. If you have been obese since childhood, you may find it difficult to reach or to maintain your weight within the acceptable range. For most people, their weight should not be more than it was when they were young adults (20 or 25 years old).

It is not well understood why some people can eat much more than others and still maintain normal weight. However, one thing is definite: to lose weight, you must take in fewer calories than you burn. This means that you must either select foods containing fewer calories or you must increase your activity — or both.

**Suggested Body Weights**
Range of Acceptable Weight

| Height (feet-inches) | Men (Pounds) | Women (Pounds) |
|---|---|---|
| 4'10" | | 92-119 |
| 4'11" | | 94-122 |
| 5'0" | | 96-125 |
| 5'1" | | 99-128 |
| 5'2" | 112-141 | 102-131 |
| 5'3" | 115-144 | 105-134 |
| 5'4" | 118-148 | 108-138 |
| 5'5" | 121-152 | 111-142 |
| 5'6" | 124-156 | 114-146 |
| 5'7" | 128-161 | 118-150 |
| 5'8" | 132-166 | 122-154 |
| 5'9" | 136-170 | 126-158 |
| 5'10" | 140-174 | 130-163 |
| 5'11" | 144-179 | 134-168 |
| 6'0" | 148-184 | 138-173 |
| 6'1" | 152-189 | |
| 6'2" | 156-194 | |
| 6'3" | 160-199 | |
| 6'4" | 164-204 | |

NOTE: Height without shoes; weight without clothes.
SOURCE: HEW conference on obesity, 1973.

**TO IMPROVE EATING HABITS**
- **Eat slowly**
- **Prepare smaller portions**
- **Avoid "seconds"**

If you need to lose weight, do so gradually. Steady loss of 1 to 2 pounds a week — until you reach your goal — is relatively safe and more likely to be maintained. Long-term success depends upon acquiring new and better habits of eating and exercise. That is perhaps why "crash" diets usually fail in the long run.

Do not try to lose weight too rapidly. Avoid crash diets that are severely restricted in the variety of foods they allow. Diets containing fewer than 800 calories may be hazardous. Some people have developed kidney stones, disturbing psychological changes, and other complications while following such diets. A few people have died suddenly and without warning.

from each of several major groups: for example, fruits and vegetables; cereals, breads, and grains; meats, poultry, eggs, and fish; dry peas and beans, such as soybeans, kidney beans, lima beans, and black-eyed peas, which are good vegetable sources of protein; and milk, cheese, and yogurt.

Fruits and vegetables are excellent sources of vitamins, especially vitamins C and A. Whole grain and enriched breads, cereals, and grain products provide B-vitamins, iron, and energy. Meats supply protein, fat, iron and other minerals, as well as several vitamins, including thiamine and vitamin $B_{12}$. Dairy products are major sources of calcium and other nutrients.

**TO ASSURE YOURSELF AN ADEQUATE DIET**
**Eat a variety of foods daily, including selections of**
- **Fruits**
- **Vegetables**
- **Whole grain and enriched breads, cereals, and grain products**
- **Milk, cheese, and yogurt**
- **Meats, poultry, fish, eggs**
- **Legumes (dry peas and beans)**

There are no known advantages to consuming excess amounts of any nutrient. You will rarely need to take vitamin or mineral supplements if you eat a wide variety of foods. There are a few important exceptions to this general statement:

- *Women in their childbearing years* may need to take iron supplements to replace the iron they lose with menstrual bleeding. Women who are no longer menstruating should not take iron supplements routinely.

- *Women who are pregnant or who are breastfeeding* need more of many nutrients, especially iron, folic acid, vitamin A, calcium, and sources of energy (calories from carbohydrates, proteins, and fats). Detailed advice should come from their physicians or from dietitians.

- *Elderly or very inactive people* may eat relatively little food. Thus, they should pay special attention to avoiding foods that are high in calories but low in other essential nutrients — for example, fat, oils, alcohol, and sugars.

Infants also have special nutritional needs. Healthy full-term infants should be breastfed unless there are special problems. The nutrients in human breast milk tend to be digested and absorbed more easily than those in cow's milk. In addition, breast milk may serve to transfer immunity to some diseases from the mother to the infant.

Normally, most babies do not need solid foods until they are 3 to 6 months old. At that time, other foods can be introduced gradually. Prolonged breast or bottlefeeding — without solid foods or supplemental iron — can result in iron deficiency.

You should not add salt or sugar to the baby's foods. Infants do not need these "encouragements" — if they are really hungry. The foods themselves contain enough salt and sugar; extra is not necessary.

**TO ASSURE YOUR BABY AN ADEQUATE DIET**
- **Breastfeed unless there are special problems**
- **Delay other foods until baby is 3 to 6 months old**
- **Do not add salt or sugar to baby's food**

**TO LOSE WEIGHT**
- **Increase physical activity**
- **Eat less fat and fatty foods**
- **Eat less sugar and sweets**
- **Avoid too much alcohol**

Gradual increase of everyday physical activities like walking or climbing stairs can be very helpful. The chart below gives the calories used per hour in different activities.

**Approximate Energy Expenditure by a 150 Pound Person in Various Activities**

| Activity | Calories per hour |
| --- | --- |
| Lying down or sleeping | 80 |
| Sitting | 100 |
| Driving an automobile | 120 |
| Standing | 140 |
| Domestic work | 180 |
| Walking, 2-½ mph | 210 |
| Bicycling, 5-½ mph | 210 |
| Gardening | 220 |
| Golf; lawn mowing, power mower | 250 |
| Bowling | 270 |
| Walking, 3-¾ mph | 300 |
| Swimming, ¼ mph | 300 |
| Square dancing, volleyball; roller skating | 350 |
| Wood chopping or sawing | 400 |
| Tennis | 420 |
| Skiing, 10 mph | 600 |
| Squash and handball | 600 |
| Bicycling, 13 mph | 660 |
| Running, 10 mph | 900 |

SOURCE: Based on material prepared by Robert E. Johnson, M.D., Ph.D., and colleagues, University of Illinois.

A pound of body fat contains 3500 calories. To lose 1 pound of fat, you will need to burn 3500 calories more than you consume. If you burn 500 calories more a day than you consume, you will lose 1 pound of fat a week. Thus, if you normally burn 1700 calories a day, you can theoretically expect to lose a pound of fat each week if you adhere to a 1200-calorie-per-day diet.

Do not attempt to reduce your weight below the acceptable range. Severe weight loss may be associated with nutrient deficiencies, menstrual irregularities, infertility, hair loss, skin changes, cold intolerance, severe constipation, psychiatric disturbances, and other complications.

If you lose weight suddenly or for unknown reasons, see a physician. Unexplained weight loss may be an early clue to an unsuspected underlying disorder.

# Avoid Too Much Fat, Saturated Fat, and Cholesterol

If you have a high blood cholesterol level, you have a greater chance of having a heart attack. Other factors can also increase your risk of heart attack — high blood pressure and cigarette smoking, for example — but high blood cholesterol is clearly a major dietary risk indicator.

Populations like ours with diets high in saturated fats and cholesterol tend to have high blood cholesterol levels. Individuals within these populations usually have greater risks of having heart attacks than people eating low-fat, low-cholesterol diets.

Eating extra saturated fat and cholesterol will increase blood cholesterol levels in most people. However, there are wide variations among people — related to heredity and the way each person's body uses cholesterol.

Some people can consume diets high in saturated fats and cholesterol and still keep normal blood cholesterol levels. Other people, unfortunately, have high blood cholesterol levels even if they eat low-fat, low-cholesterol diets.

*There is controversy about what* recommendations are appropriate for healthy Americans. But for the U.S. population *as a whole,* reduction in our current intake of total fat, saturated fat, and cholesterol is sensible. This suggestion is especially appropriate for people who have high blood pressure or who smoke.

The recommendations are not meant to prohibit the use of any specific food item or to prevent you from eating a variety of foods. For example, eggs and organ meats (such as liver) contain cholesterol, but they also contain many essential vitamins and minerals, as well as protein. Such items can be eaten in moderation, as long as your overall cholesterol intake is not excessive. If you prefer whole milk to skim milk, you can reduce your intake of fats from foods other than milk.

### TO AVOID TOO MUCH FAT, SATURATED FAT, AND CHOLESTEROL
- Choose lean meat, fish, poultry, dry beans and peas as your protein sources
- Moderate your use of eggs and organ meats (such as liver)
- Limit your intake of butter, cream, hydrogenated margarines, shortenings and coconut oil, and foods made from such products
- Trim excess fat off meats
- Broil, bake, or boil rather than fry
- Read labels carefully to determine both amount and types of fat contained in foods

# 4

## Eat Foods with Adequate Starch and Fiber

The major sources of energy in the average U.S. diet are carbohydrates and fats. (Proteins and alcohol also supply energy, but to a lesser extent.) If you limit your fat intake, you should increase your calories from carbohydrates to supply your body's energy needs.

In trying to reduce your weight to "ideal" levels, carbohydrates have an advantage over fats: carbohydrates contain less than half the number of calories per ounce than fats.

*Complex* carbohydrate foods are better than *simple* carbohydrates in this regard. Simple carbohydrates — such as sugars — provide calories but little else in the way of nutrients. Complex carbohydrate foods — such as beans, peas, nuts, seeds, fruits and vegetables, and whole grain breads, cereals, and products — contain many essential nutrients in addition to calories.

Increasing your consumption of certain complex carbohydrates can also help increase dietary fiber. The average American diet is relatively low in fiber.

Eating more foods high in fiber tends to reduce the symptoms of chronic constipation, diverticulosis, and some types of "irritable bowel." There is also concern that low fiber diets might increase the risk of developing cancer of the colon, but whether this is true is not yet known.

To make sure you get enough fiber in your diet, you should eat fruits and vegetables, whole grain breads and cereals. There is no reason to add fiber to foods that do not already contain it.

**TO EAT MORE COMPLEX CARBOHYDRATES DAILY**
- Substitute starches for fats and sugars
- Select foods which are good sources of fiber and starch, such as whole grain breads and cereals, fruits and vegetables, beans, peas, and nuts

# Avoid Too Much Sugar

The major health hazard from eating too much sugar is tooth decay (dental caries). The risk of caries is not simply a matter of how much sugar you eat. The risk increases the more frequently you eat sugar and sweets, especially if you eat between meals, and if you eat foods that stick to the teeth. For example, frequent snacks of sticky candy, or dates, or daylong use of soft drinks may be more harmful than adding sugar to your morning cup of coffee — at least as far as your teeth are concerned.

Obviously, there is more to healthy teeth than avoiding sugars. Careful dental hygiene and exposure to adequate amounts of fluoride in the water are especially important.

Contrary to widespread opinion, too much sugar in your diet does not seem to cause diabetes. The most common type of diabetes is seen in obese adults, and avoiding sugar, without correcting the overweight, will not solve the problem. There is also no convincing evidence that sugar causes heart attacks or blood vessel diseases.

Estimates indicate that Americans use on the average more than 130 pounds of sugars and sweeteners a year. This means the risk of tooth decay is increased not only by the sugar in the sugar bowl but by the sugars and syrups in jams, jellies, candies, cookies, soft drinks, cakes, and pies, as well as sugars found in products such as breakfast cereals, catsup, flavored milks, and ice cream. Frequently, the ingredient label will provide a clue to the amount of sugars in a product.

### TO AVOID EXCESSIVE SUGARS
- **Use less of all sugars, including white sugar, brown sugar, raw sugar, honey, and syrups**
- **Eat less of foods containing these sugars, such as candy, soft drinks, ice cream, cakes, cookies**
- **Select fresh fruits or fruits canned without sugar or light syrup rather than heavy syrup**
- **Read food labels for clues on sugar content — if the names sucrose, glucose, maltose, dextrose, lactose, fructose, or syrups appear first, then there is a large amount of sugar**
- **Remember, how often you eat sugar is as important as how much sugar you eat**

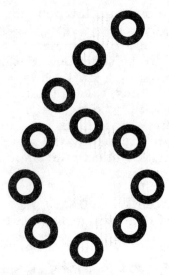

# Avoid Too Much Sodium

Table salt contains sodium and chloride — both are essential elements.

Sodium is also present in many beverages and foods that we eat, especially in certain processed foods, condiments, sauces, pickled foods, salty snacks, and sandwich meats. Baking soda, baking powder, monosodium glutamate (MSG), soft drinks, and even many medications (many antacids, for instance) contain sodium.

It is not surprising that adults in the United States take in much more sodium than they need.

The major hazard of excessive sodium is for persons who have high blood pressure. Not everyone is equally susceptible. In the United States, approximately 17 percent of adults have high blood pressure. Sodium intake is but one of the factors known to affect blood pressure. Obesity, in particular, seems to play a major role.

In populations with low-sodium intakes, high blood pressure is rare. In contrast, in populations with high-sodium intakes, high blood pressure is common. If people with high blood pressure severely restrict their sodium intakes, their blood pressures will *usually* fall — although not always to normal levels.

At present, there is no good way to predict who will develop high blood pressure, though certain groups, such as blacks, have a higher incidence. Low-sodium diets might help some of these people avoid high blood pressure if they could be identified before they develop the condition.

Since most Americans eat more sodium than is needed, consider reducing your sodium intake. Use less table salt. Eat sparingly those foods to which large amounts of sodium have been added. Remember that up to half of sodium intake may be "hidden," either as part of the naturally occurring food or, more often, as part of a preservative or flavoring agent that has been added.

**TO AVOID TOO MUCH SODIUM**
- **Learn to enjoy the unsalted flavors of foods**
- **Cook with only small amounts of added salt**
- **Add little or no salt to food at the table**
- **Limit your intake of salty foods, such as potato chips, pretzels, salted nuts and popcorn, condiments (soy sauce, steak sauce, garlic salt), cheese, pickled foods, cured meats**
- **Read food labels carefully to determine the amounts of sodium in processed foods and snack items**

# If You Drink Alcohol, Do So in Moderation

Alcoholic beverages tend to be high in calories and low in other nutrients. Even moderate drinkers may need to drink less if they wish to achieve ideal weight.

On the other hand, heavy drinkers may lose their appetites for foods containing essential nutrients. Vitamin and mineral deficiencies occur commonly in heavy drinkers—in part, because of poor intake, but also because alcohol alters the absorption and use of some essential nutrients.

Sustained or excessive alcohol consumption by pregnant women has caused birth defects. Pregnant women should limit alcohol intake to 2 ounces or less on any single day.

Heavy drinking may also cause a variety of serious conditions, such as cirrhosis of the liver and some neurological disorders. Cancer of the throat and neck is much more common in people who drink and smoke than in people who don't.

One or two drinks daily appear to cause no harm in adults. If you drink you should do so in moderation.

- **Remember, if you drink alcohol, do so in moderation**

**For further reading on diet and its relationship to good health, send for:**

*FOOD* — A publication on food and nutrition by the U.S. Department of Agriculture. Home and Garden Bulletin No. 228. Science and Education Administration. Stock No. 001-000-03881-8.

*HEALTHY PEOPLE* — The Surgeon General's Report on Health Promotion and Disease Prevention. Public Health Service, U.S. Department of Health, Education, and Welfare. Stock No. 017-001-00416-2.

The above publications are for sale from:

Superintendent of Documents
U.S. Government Printing Office
Washington, D.C. 20402

**For assistance with your food and nutrition questions, contact the dietitian, home economist, or nutritionist in the following groups:**

Public Health Department
County Extension office
State or local medical society
Hospital outpatient clinic
Local Dietetic Association office
Local Heart Association office
Local Diabetes Association office
Local Health Center or Clinic

*NOTE: These recommendations are intended only for populations with food habits similar to people in the United States.*

Home and Garden Bulletin No. 232
Previously issued as an unnumbered publication.

# HEALTH STYLE *a self test*

U.S. DEPARTMENT
OF HEALTH AND HUMAN SERVICES
Public Health Service

Office of Disease Prevention and Health Promotion

Office of Health Information, Health Promotion
and Physical Fitness and Sports Medicine

DHHS Publication No. (PHS) 81-50155

U.S. DEPARTMENT OF HEALTH AND HUMAN SERVICES • Public Health Service

## How This Booklet Can Help You

All of us want good health. But, many of us do not know how to be as healthy as possible. Good health is not a matter of luck or fate. You have to work at it.

Good health depends on a combination of things . . . the environment in which you live and work . . . the personal traits you have inherited . . . the care you receive from doctors and hospitals . . . and the personal behaviors or habits that you perform daily, usually without much thought. All of these work together to affect your health. Many of us rely too much on doctors to keep us healthy, and we often fail to see the importance of actions we can take ourselves to look and feel healthy. You may be surprised to know that by taking action individually and collectively, you can begin to change parts of your world which may be harmful to your health.

Every day you are exposed to potential risks to good health. Pollution in the air you breathe and unsafe highways are two examples. These are risks that you, as an individual, can't do much about. Improving the quality of the environment usually requires the effort of concerned citizens working together for a healthier community.

There are, however, risks that you can control: risks stemming from your personal behaviors and habits. These behaviors are known as your lifestyle. Health experts now describe lifestyle as one of the most important factors affecting health. In fact, it is estimated that as many as seven of the ten leading causes of death in the United States could be reduced through common sense changes in lifestyle.

That's what the brief test contained in this booklet is all about. The few minutes you take to complete it may actually help you add years to your life! How? Well to start, it will enable you to identify aspects of your present lifestyle that are risky to your health. Then it will encourage you to take steps to eliminate or minimize the risks you identify. All in all, it will help you begin to change your present lifestyle into a new HEALTHSTYLE. If you do, it's possible that you may feel better, look better, and live longer too.

## Before You Take the Test

This is not a pass-fail test. Its purpose is simply to tell you how well you are doing to stay healthy. The behaviors covered in the test are recommended for most Americans. Some of them may not apply to persons with certain chronic diseases or handicaps. Such persons may require special instructions from their physician or other health professional.

You will find that the test has six sections: smoking, alcohol and drugs, nutrition, exercise and fitness, stress control, and safety. Complete one section at a time by circling the number corresponding to the answer that best describes your behavior (2 for "Almost Always", 1 for "Sometimes", and 0 for "Almost Never"). Then add the numbers you have circled to determine your score for that section. Write the score on the line provided at the end of each section. The highest score you can get for each section is 10.

# A Test for Better Health

### Cigarette Smoking

|  | Almost Always | Sometimes | Almost Never |
|---|---|---|---|

If you never smoke, enter a score of 10 for this section and go to the next section on *Alcohol and Drugs*.

1. I avoid smoking cigarettes. — 2　1　0
2. I smoke only low tar and nicotine cigarettes *or* I smoke a pipe or cigars. — 2　1　0

Smoking Score: _____

### Alcohol and Drugs

|  | Almost Always | Sometimes | Almost Never |
|---|---|---|---|

1. I avoid drinking alcoholic beverages *or* I drink no more than 1 or 2 drinks a day. — 4　1　0
2. I avoid using alcohol or other drugs (especially illegal drugs) as a way of handling stressful situations or the problems in my life. — 2　1　0
3. I am careful not to drink alcohol when taking certain medicines (for example, medicine for sleeping, pain, colds, and allergies), or when pregnant. — 2　1　0
4. I read and follow the label directions when using prescribed and over-the-counter drugs. — 2　1　0

Alcohol and Drugs Score: _____

### Eating Habits

|  | Almost Always | Sometimes | Almost Never |
|---|---|---|---|

1. I eat a variety of foods each day, such as fruits and vegetables, whole grain breads and cereals, lean meats, dairy products, dry peas and beans, and nuts and seeds. — 4　1　0
2. I limit the amount of fat, saturated fat, and cholesterol I eat (including fat on meats, eggs, butter, cream, shortenings, and organ meats such as liver). — 2　1　0
3. I limit the amount of salt I eat by cooking with only small amounts, not adding salt at the table, and avoiding salty snacks. — 2　1　0
4. I avoid eating too much sugar (especially frequent snacks of sticky candy or soft drinks). — 2　1　0

Eating Habits Score: _____

### Exercise/Fitness

|  | Almost Always | Sometimes | Almost Never |
|---|---|---|---|

1. I maintain a desired weight, avoiding overweight and underweight. — 3　1　0
2. I do vigorous exercises for 15-30 minutes at least 3 times a week (examples include running, swimming, brisk walking). — 3　1　0
3. I do exercises that enhance my muscle tone for 15-30 minutes at least 3 times a week (examples include yoga and calisthenics). — 2　1　0
4. I use part of my leisure time participating in individual, family, or team activities that increase my level of fitness (such as gardening, bowling, golf, and baseball). — 2　1　0

Exercise/Fitness Score: _____

### Stress Control

| | Almost Always | Sometimes | Almost Never |
|---|---|---|---|
| 1. I have a job or do other work that I enjoy. | 2 | 1 | 0 |
| 2. I find it easy to relax and express my feelings freely. | 2 | 1 | 0 |
| 3. I recognize early, and prepare for, events or situations likely to be stressful for me. | 2 | 1 | 0 |
| 4. I have close friends, relatives, or others whom I can talk to about personal matters and call on for help when needed. | 2 | 1 | 0 |
| 5. I participate in group activities (such as church and community organizations) or hobbies that I enjoy. | 2 | 1 | 0 |

Stress Control Score: _____

### Safety

| | Almost Always | Sometimes | Almost Never |
|---|---|---|---|
| 1. I wear a seat belt while riding in a car. | 2 | 1 | 0 |
| 2. I avoid driving while under the influence of alcohol and other drugs. | 2 | 1 | 0 |
| 3. I obey traffic rules and the speed limit when driving. | 2 | 1 | 0 |
| 4. I am careful when using potentially harmful products or substances (such as household cleaners, poisons, and electrical devices). | 2 | 1 | 0 |
| 5. I avoid smoking in bed. | 2 | 1 | 0 |

Safety Score: _____

## Your HEALTHSTYLE Scores

After you have figured your scores for each of the six sections, circle the number in each column that matches your score for that section of the test.

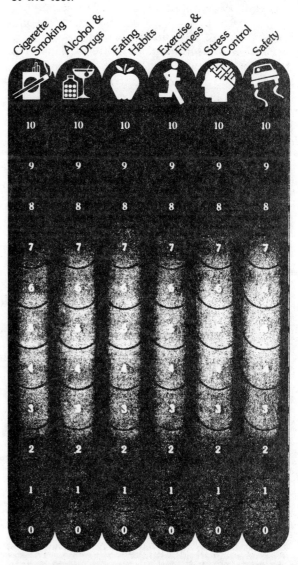

Remember, there is no total score for this test. Consider each section separately. You are trying to identify aspects of your lifestyle that you can improve in order to be healthier and to reduce the risk of illness. So let's see what your scores reveal.

## What Your Scores Mean to YOU

### Scores of 9 and 10

Excellent! Your answers show that you are aware of the importance of this area to your health. More importantly, you are putting your knowledge to work for you by practicing good health habits. As long as you continue to do so, this area should not pose a serious health risk. It's likely that you are setting an example for your family and friends to follow. Since you got a very high score on this part of the test, you may want to consider other areas where your scores indicate room for improvement.

### Scores of 6 to 8

Your health practices in this area are good, but there is room for improvement. Look again at the items you answered with a "Sometimes" or "Almost Never". What changes can you make to improve your score? Even a small change can often help you achieve better health.

### Scores of 3 to 5

Your health risks are showing! Would you like more information about the risks you are facing and about why it is important for you to change these behaviors. Perhaps you need help in deciding how to successfully make the changes you desire. In either case, help is available. See the last page of this booklet.

### Scores of 0 to 2

Obviously, you were concerned enough about your health to take the test, but your answers show that you may be taking serious and unnecessary risks with your health. Perhaps you are not aware of the risks and what to do about them. You can easily get the information and help you need to improve, if you wish. A source of contact appears on the last page. The next step is up to you.

## YOU Can Start Right Now!

In the test you just completed were numerous suggestions to help you reduce your risk of disease and premature death. Here are some of the most significant:

*Avoid cigarettes.* Cigarette smoking is the single most important preventable cause of illness and early death. It is especially risky for pregnant women and their unborn babies. Persons who stop smoking reduce their risk of getting heart disease and cancer. So if you're a cigarette smoker, think twice about lighting that next cigarette. If you choose to continue smoking, try decreasing the number of cigarettes you smoke and switching to a low tar and nicotine brand.

*Follow sensible drinking habits.* Alcohol produces changes in mood and behavior. Most people who drink are able to control their intake of alcohol and to avoid undesired, and often harmful, effects. Heavy, regular use of alcohol can lead to cirrhosis of the liver, a leading cause of death. Also, statistics clearly show that mixing drinking and driving is often the cause of fatal or crippling accidents. So if you drink, do it wisely and in moderation.

*Use care in taking drugs.* Today's greater use of drugs—both legal and illegal—is one of our most serious health risks. Even some drugs prescribed by your doctor can be dangerous if taken when drinking alcohol or before driving. Excessive or continued use of tranquilizers (or "pep pills") can cause physical and mental problems. Using or experimenting with illicit drugs such as marijuana, heroin, cocaine, and PCP may lead to a number of damaging effects or even death.

*Eat sensibly.* Overweight individuals are at greater risk for diabetes, gall bladder disease, and high blood pressure. So it makes good sense to maintain proper weight. But good eating habits also mean holding down the amount of fat (especially saturated fat), cholesterol, sugar and salt in your diet. If you must snack, try nibbling on fresh fruits and vegetables. You'll feel better—and look better, too.

*Exercise regularly.* Almost everyone can benefit from exercise—and there's some form of exercise almost everyone can do. (If you have any doubt, check first with your doctor.) Usually, as little as 15-30 minutes of vigorous exercise three times a week will help you have a healthier heart, eliminate excess weight, tone up sagging muscles, and sleep better. Think how much difference all these improvements could make in the way you feel!

*Learn to handle stress.* Stress is a normal part of living; everyone faces it to some degree. The causes of stress can be good or bad, desirable or undesirable (such as a promotion on the job or the loss of a spouse). Properly handled, stress need not be a problem. But unhealthy responses to stress—such as driving too fast or erratically, drinking too much, or prolonged anger or grief—can cause a variety of physical and mental problems. Even on a very busy day, find a few minutes to slow down and relax. Talking over a problem with someone you trust can often help you find a satisfactory solution. Learn to distinguish between things that are "worth fighting about" and things that are less important.

*Be safety conscious.* Think "safety first" at home, at work, at school, at play, and on the highway. Buckle seat belts and obey traffic rules. Keep poisons and weapons out of the reach of children, and keep emergency numbers by your telephone. When the unexpected happens, you'll be prepared.

## Where Do You Go From Here?

Start by asking yourself a few frank questions:
*Am I really doing all I can to be as healthy as possible? What steps can I take to feel better? Am I willing to begin now?* If you scored low in one or more sections of the test, decide what changes you want to make for improvement. You might pick that aspect of your lifestyle where you feel you have the best chance for success and tackle that one first. Once you have improved your score there, go on to other areas.

If you already have tried to change your health habits (to stop smoking or exercise regularly, for example) don't be discouraged if you haven't yet succeeded. The difficulty you have encountered may be due to influences you've never really thought about—such as advertising—or to a lack of support and encouragement. Understanding these influences is an important step toward changing the way they affect you.

*There's Help Available.* In addition to personal actions you can take on your own, there are community programs and groups (such as the YMCA or the local chapter of the American Heart Association) that can assist you and your family to make the changes you want to make. If you want to know more about these groups or about health risks contact your local health department. There's a lot you can do to stay healthy or to improve your health—and there are organizations that can help you. Start a new HEALTHSTYLE today!

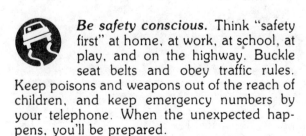

☆ U.S. Government Printing Office: O-1981-344-829

# THE CORNUCOPIA PAPERS

by Robert Rodale

Stocking Up on Future Food

The Cornucopia Project

Have You Ever Asked These Questions About Your Food?

Plan Your Food Future Now

"No $7.00 Loaf of Bread for Me"

The Cornucopia Project, Rodale Press, 33 E. Minor St., Emmaus, PA 18049

*Fifth in a series.*

# 24,000 tons of broccoli.

To many people, broccoli is a vegetable with a funny name.

To New Yorkers, it's a vegetable with a price tag that's no joke.

And for you, wherever you live, the story of how broccoli got to be so expensive provides one often-forgotten reason why your food bills are so high. It may even help you to lower them.

Last year, New York area residents bought about 24,000 tons of broccoli — almost all of it from the West Coast, 2,700 miles away.

It cost nearly $6 million to move that one vegetable across the country. More than 950,000 gallons of fossil fuel were consumed in the shipping process. And, besides the impact on price of this transcontinental journey, consider what must have happened to the broccoli's "fresh-picked" flavor along the route.

*Yet, this trip wasn't necessary:*

Broccoli prefers cool weather. Except for southern Florida, it could have been grown successfully — and delivered to New Yorkers less expensively — from virtually anywhere else in the continental U.S. Including New York's own backyard gardens.

A special case? Not at all. There are literally dozens of fruits and vegetables that could be — and should be — grown closer to the people who will consume them.

Sometimes, of course, long-distance food shipments are unavoidable. Oranges won't grow in North Dakota; apples don't thrive in southern Florida.

But those few exceptions can't explain why, today, for every two dollars we spend in the U.S. to grow food, we spend another dollar to move it around. Not just to New York. To Massachusetts, which imports more than 80 percent of its food. And to every other section of the country.

It's a crazy-quilt pattern of unplanned food-growing and unrealistic distribution which cannot continue much longer. Our economy will be unable to survive the expense. Our energy resources will be unable to supply the demand.

Clearly, the nation's food system must be re-examined, and then reorganized. In fundamental ways. And without delay.

That's why Rodale Press recently launched *The Cornucopia Project*, which, as far as we know, is the first attempt by a non-governmental organization to make a systematic study of America's food system.

Because of our deep interest in health and nutrition, we're trying to help people — as individuals and in small groups — to understand more about what's happening today on the nation's farms, in its supermarkets, at its dinner tables. And then to do something about it.

The first step in this venture is to assemble information about every aspect of the nation's food system — and how it affects you. Just a small sample of the information we've already gathered is represented by what you've been reading in this message.

If you'd like more — without charge — please write:
Robert Rodale,
Rodale Press,
33 E. Minor St.,
Emmaus,
PA 18049.

**RODALE PRESS, INC.**
Emmaus, PA 18049

Rodale publishes: *Prevention, Rodale's New Shelter, Organic Gardening, Bicycling, Executive Fitness Newsletter, New Farm,* and *Theatre Crafts*, as well as hardcover and paperback books under the Rodale Books imprint.

*Sixth in a series.*

# The losing of America.

The United States will lose 26 square miles of its land today.

It will lose another 26 square miles tomorrow, and every day this year.

But not to a foreign power.

We are giving up our land to the ravenous demands of an unrealistic food system. And, before you turn the page, telling youself that land loss is someone else's problem — too remote, too big for you — *it isn't*.

You *can* help to face this dilemma. And help yourself, too.

If you know these facts:

Each time an Iowa farmer grows one pound of corn for you and us, he uses up five to six pounds of topsoil. And in the wheatlands, for each pound harvested, up to 20 pounds of soil disappear — never again to be available for growing food.

In 1980—and again in 1981—three million acres of farmlands will be lost to erosion. Five billion tons of topsoil are displaced annually — enough to cover all five boroughs of New York City with a 13½-foot-deep layer of soil.

And while this is going on, each year, another three million acres of agricultural land are lost to development: new homes, factories, and other structures.

Let's concede those three million acres to the requirements of "progress."

But why is it that, in the areas still being farmed, we're eating up more land than food?

Because today's food producers are "burning up" the soil with chemicals in order to maintain high yields. Because, when they over-use chemicals, the soil becomes more vulnerable to wind and water erosion. Because the ways of commerce are now overruling the wisdom of crop rotation. And because modern tractors don't permit the strip cropping and contour plowing that are most effective for preventing erosion.

*Man*-made blights, these. Symptoms of a food system that is dangerously out of touch with our new world of limited resources. A food system that is headed for disaster, unless we do something about it.

That's not easy to believe. Not with our bumper harvests almost every year, and our grain-clogged warehouses.

But the danger signals are inescapable: We have already lost half of our land. Each day, the U.S. population increases by 5,000 people while the farmlands shrink. And prices soar, because the food system needs more and more expensive energy to operate.

Is there still time? We believe there is.

America *can* produce food at less cost to the consumer, and to the environment. The nation's citizens *don't* have to pay for food today with fertile soil they'll need tomorrow.

That's why we launched *The Cornucopia Project* — a concentrated effort to collect information about the ways the U.S. food system must be restructured to prevent a food crisis.

The information we've already assembled could be of great importance to you, personally, and we'd be pleased to share it. No cost or obligation, of course. Just write:
Robert Rodale,
Rodale Press,
33 E. Minor St.,
Emmaus, PA
18049.

**RODALE PRESS, INC.**
Emmaus, PA 18049

Rodale publishes: *Prevention, Rodale's New Shelter, Organic Gardening, Bicycling, Executive Fitness Newsletter, New Farm,* and *Theatre Crafts,* as well as hardcover and paperback books under the Rodale Books imprint.

*Seventh in a series.*

# Join our conspiracy.

*A loaf of bread costs $7.66.*

*Tomatoes, three large ones, $5.24.*

*A ten-ounce jar of instant coffee comes to $45.06.*

Is this merely the shopping list of a paranoid housewife trapped in a frantic nightmare? Not at all.

Based on a recent report to the White House, those are some of the food prices you may be paying at the end of the century.

But if you believe only inflation is to blame, think about this:

Each day of the coming decades, there will be 5,000 new people to feed in the U.S. And the food they'll need will have to be grown on less and less land, because six million acres of agricultural land are being lost annually —half to developers, half to the ravages of erosion and irresponsible agricultural practices.

Soon, with mathematical certainty, the growing number of consumers will converge with a declining farmland base — and America can then expect periodic food crises.

Prices will skyrocket. Shortages will be a fact of life. And our food system will cease to be a major source of America's strength.

An overly pessimistic scenario?

Hard to take seriously as you shop in the supermarkets packed with an amazing variety and quantity of food?

Look closer. You will see sobering evidence of the system's irrationality.

We need energy. Yet we waste immense amounts of it, year after year, by continuing to grow fruits, vegetables, and other foods thousands of miles away from the people who will consume them.

We need farmers. Yet, along with our farmlands, our farmers are disappearing. Back in the 1930's, 40 percent of all Americans lived and worked on farms; now, only three percent.

And, of the 32,000 food-manufacturing firms in this country, 75 percent of all their profits is reaped by just 50 of them. With each passing year, power over our food system passes into fewer hands.

Observing this, some people think they detect a conspiracy to control production, distribution, and prices.

We see, instead, too many people ignoring the danger signals. So, we propose a "people's conspiracy" which we call *The Cornucopia Project.*

Quite simply, *The Cornucopia Project* aims to help people — as individuals and in small groups—to understand more about what's happening today on the nation's farms, in its supermarkets, at its dinner tables. And then to do something about it.

To this end, we've already assembled information which we will be glad to share with anyone who is ready to join our conspiracy.

There are no dues. No membership obligations. We're just looking for people who want to stop the waste of soil, food, and energy that is threatening America.

If that's you, we have facts that can be important to you. Please write: Robert Rodale, Rodale Press, 33 E. Minor St., Emmaus, PA 18049.

**RODALE PRESS, INC.**
Emmaus, PA 18049

Rodale publishes: *Prevention, Rodale's New Shelter, Organic Gardening, Bicycling, Executive Fitness Newsletter, The New Farm,* and *Theatre Crafts,* as well as hardcover and paperback books under the Rodale imprint.

# Stocking Up on Future Food

BELIEVE ME, I know that you've got enough worries without me thrusting one more in your direction. What with inflation, recession, energy shortages, and general uncertainty about the future, you might be justified in asking me to spare you thinking about a problem that hasn't even happened yet. But I'm going to suggest you do just that, and my purpose is entirely constructive. Burying our heads in the sand and trying not to think about troubles that are clearly down the road is exactly what got us into the present mess. The signs were very clear ten years ago that oil couldn't gush forth at low prices forever, that cars had gotten too big and that homes were using too much energy for heating. We were eating free lunches that weren't really free, but paid little or no attention and hoped for the best. Looking back now, it's clear to me the biggest mistake our society made during the 1970's was not so much our wasteful ways, but our failure to look ahead to see where our profligate life-style was taking us. Let's not make that mistake again.

The future of our food supply is what we have to start thinking about now. The question is no longer *if* a food crunch is going to follow on the heels of the energy crisis, but *when*. Optimists say we've got until the year 2000 or even 2010 until the number of hungry mouths converges with our declining agricultural land base. I'm an optimist, but I can also read statistics. Furthermore, I've learned a lot lately about how a problem like energy shortage can mix in with another problem, like growing enough food, to make bad news arrive much earlier than expected. The bottom line, according to my reckoning, is that the real crunch is only a few years ahead and spot outbreaks of the problem are here now. It's also clear that getting enough food at reasonable prices has, for some people, been a problem for years.

We have trouble seeing the big picture, though. American farmers have produced surpluses of food for a hundred years or more, and are still doing it now. How could it be that surpluses will turn suddenly into shortages? Well, just think back a few years when there was a global glut of oil, and the price for crude at the well was $5 a barrel. Things can change quickly, and there doesn't have to be an absolute shortage of something people want badly to cause prices to drive through the ceiling. Just a hint that there are limits to supply, and that total needs are beginning to bump against supply limits, is enough to change the picture quickly.

The food problem may even be tougher to cope with than oil and energy. Not everyone has to drive a car, and many jobs can be accomplished with little energy use. But everyone has to eat every day. The number of eaters is increasing at the rate of 6.2 million a month globally, and 159,000 a month here in the U.S. And our land base is declining. Every year we lose over 3 million acres of agricultural land to development. Five billion tons of topsoil wash or blow away. There are other problems. Water for irrigation is being pumped up from sub-surface storage areas (in some areas) faster than it's being replaced by rainfall, and the end of that water supply is in sight. Pollution caused by overuse of chemicals and such environmental factors as acid rain is raising doubt about long-term productivity of many farming areas.

One of the worst aspects of the problem, in my humble opinion, is that the whole structure of our agriculture and our food processing and distribution system has evolved the wrong way. Instead of producing as much food as possible close to where people are, and recycling mineral-rich and nitrogen-rich wastes back to the land, everything is spread out over vast distances. Nitrogen is shipped thousands of miles, while the same nutrient could be captured from the air right on the farm if leguminous crops were used in the rotation. I could cite several other examples of fertilizer waste, were space available. And we should not forget that phosphate supplies in the U.S. are limited. The price for phosphate has already shot up, and may soon go still higher.

Long-distance food shipping is another part of the problem. The reports I get are that beef-feeders in the Midwest can't make money because the cost of moving meat to the big markets in the East is so high. Vegetables and fruits are also moved thousands of miles. That used to make economic sense when gas prices were low, but now (and for our future) it's just not right.

So much for stating the problem. Let's begin to focus on the solution.

Curbing population growth is important. Fortunately, people seem to be getting that message and some progress is being made. But growth trends are in place, and it appears certain that both world population and U.S. population will increase substantially before leveling off.

Looking at our food system, we need to switch it away from being a mining operation. We are mining the soil of its fertility, and are draining oil and gas fields of their wealth to get the energy to do that. Mines always run out and become depleted. We can't afford to let our food system just "run out" someday.

Widespread application of organic farming techniques is the way to build permanence into our farms. Organic farming feeds the soil instead of mining it. Humus is built up, creating an intensely alive soil environment that liberates minerals and other nutrients gradually in ways that allow a soil to remain fertile for many thousands of years. There is also recycling of nutrients. Rotations which include legumes both capture nitrogen from the air and help to prevent erosion. Pollution of soil, water, air and food is prevented because toxic chemicals are not used.

To suggest organic farming as a way to keep food prices down might strike some people as strange, since in today's market organically grown food usually costs more. That is a strictly temporary situation. Conventional farmers are actually burning up their farms to produce today's low-priced commodities. They are consuming their capital—the fertility of their land—in order to try to stay solvent financially. That is a giant hidden cost which doesn't show up in your food bills today, but which will have to be paid soon. We are getting close to the danger point now. Organic farmers refuse to sell their soil along with the food they move to market. So their current prices are somewhat higher.

Just how important organic methods can be to our food future is brought out forcefully in the report and recommendations on organic farming just issued by the special organic farming study team of the U.S. Department of Agriculture. I told you about that project last year, explaining that it was the first time a group of American scientists, economists and statisticians attempted to take an objective look at organic methods.

Generally speaking, their preliminary conclusions are very encouraging. "The challenge to agriculture in this decade will be to develop farming systems that can produce the necessary quantity and quality of food and fiber without adversely affecting our soil resources and the environment," Anson R. Bertrand, Director of the Science and Education Administration, says in the report's preface. Here is his next sentence, which is extremely important: "It is likely that organic technology and management expertise will play an important role in meeting that challenge."

I suggest that you get a copy of the report and read it. You may be able to get one from the office of your Congressman, or from the USDA itself. Although the report specifically addresses itself to farming, not organic gardening, I think you'll find ideas and suggestions that will expand your perception of methods you can use. More importantly, it will give you information about some basic changes needed in our food system.

The study team's report dispels the notion that a switch to organic methods would contract the food supply drastically. A total conversion now would reduce production, true, but there would still be enough food for America's needs and some for export. The most constructive parts of the report suggest excellent directions for research and education projects that could make organic farming much more efficient than it is now, without sacrificing the respect for preservation of resources that is so characteristic of organic farmers. We have to realize that all the achievements of organic farming to date have been accomplished largely outside the farm research and education establishment.

One conclusion of the report is that "a total or even a significant shift to organic farming would require overall major changes in the overall structure of agriculture." I agree, and have been implying in this article that such changes are needed badly to avoid a food crunch that could be much more difficult for us to cope with than the current energy crunch. That brings me to my second suggestion for action.

Rodale Press has long recognized the need for change in agricultural structure. In fact, we think the whole food system should be changed in ways that will reduce the distance food and food resources are shipped, that will encourage the eating of more fresh food (which cuts energy needs), and generally converts our food system from its current "gas-guzzling" condition to a leaner, more efficient model that can keep running for long into the future. Those changes are also needed, as the USDA team points out, to make organic farming practical for a larger number of farmers.

The question really is not if structural changes of that type can be made. The only doubt is *when* they will be imposed on us and on our food system by such pressures as rising energy costs, growing population, declining soil fertility, the need for a secure store of food close to where people live, and high prices for fertilizer ingredients, such as phosphorus. If we keep going blindly on our present course — the way we did with automobile design and use in the early 1970's — the switch to a regional and more natural food production system will come to us in one great big bang, probably having the same kind of impact on our economy and our lives as the stock market crash of 1929. Nobody wants that.

I feel that if we are going to head off that kind of collapse we need to use entirely new methods to guide ourselves toward change. The challenge is so great. We need the courage to look at what, up to now, has been seen as the strongest part of the American resource system — our food production apparatus — and decide that it needs a drastic revision to be able to serve us well in the future. That is not an easy task. Certainly it is too massive and important a job to be left just to our elected officials, or to farm scientists, farmers or food companies. The rethinking has to start within our own heads.

The first step in that effort is the creation of a list of questions that need to be asked about the strengths and weaknesses of our present food system. Then those questions must be circulated publicly so that discussion and debate can begin. But that discussion must start from a solid understanding of just how vulnerable our present system is to disruption. Without the facts about soil erosion, potential water shortages, rising fertilizer prices, increasing costs of transportation, and the need to recycle nutrients, a public discusson of the issues probably wouldn't even start, and certainly wouldn't get very far.

Rodale Press has started a project to get those facts, and to present them to you. We call it The Cornucopia Project because looking beyond the bare bones of those facts I can see emerging the outline of a more natural and *sustainable* food system that will be far more productive and efficient than the system we are using today. It is my conviction that America's vast natural resources can yet be turned into a veritable horn of plenty for centuries to come. But we need to unhook ourselves from the wasteful and destructive habits that have been so characteristic of our past.

In the next issue of ORGANIC GARDENING I will begin a series of editorials that will show exactly how we are consuming the heritage of generations of Americans yet unborn in order to feed ourselves today. And I will start asking some questions about future food that you need to think about and answer. But this will be more than just another series of articles that you read in a magazine. I am hoping that they will be the start of a process of thinking that could lead to coordinated action. For that reason, we are going to be asking you to respond, and especially to start planning for ways that a better food system can be created.

The road ahead may seem rocky and uncertain. But we are going to have to travel it anyway. Let's think now about smoothing out some of the many bumps, and get on with the journey.

# The Cornucopia Project

I DO NOT think that a conspiracy exists—in the usual sense, at any rate—to monopolize the production, processing and sale of food in the United States, and to profit enormously when food begins to become scarce. The food business is too vast and diverse for any group to be able to get together and decide that they are going to try to corner the market, or restrict production in some way. However, the food system—large as it is—is starting to behave as if a cartel to control the price of food will soon be very easy to form.

Every year there are fewer people producing food, on larger and larger farms. Every day the population of the U.S. grows by 5,000 people, and available agricultural land shrinks by almost 26 square miles. Since our food system needs large amounts of energy to operate, the rising cost and declining availability of energy has meant food price increases, with the strong possibility that food costs will go up at an even higher rate in the future. And the processing and distribution of food is becoming concentrated in the hands of a few large companies.

The problem of a potential food shortage is more immediate in some states than others, but could eventually affect our whole country. Massachusetts is particularly vulnerable. It now imports 40 to 50 percent of its food from California. But in 20 years California will have no food available to send to other states because its own population is growing and by then will need all that can be grown there. Massachusetts is also concerned about a rail or truck strike because only a 14-day supply of food can be stored within the state.

The Cornucopia Project is an effort by Rodale Press to encourage thinking and action now, while there is still time to act. The primary goal of the project is to supply people like yourself with facts about the strengths and weaknesses of the American food system—facts which are crucial to the improvement of that system but which have never been gathered in one place for easy use. I feel certain that if you know more about what is likely to happen, you can take action that will improve your own food supply, save you money, and help change the whole system in ways that will prevent an explosion in food costs that would make the energy crunch look like a minor problem in comparison.

One of the things the project has examined is the structure of our food system. This has changed markedly in the last 50 years, and now a relatively small number of people have a great deal of control over the production and processing of our food supply. If these changes continue for just a few more years, an OPEC of food will exist right here, with one major difference. The sheiks of food we will have to bow down to will be our own countrymen. If you don't want that to happen, the first thing to do is to try to get a handle on what that word *structure* means.

Agriculture Secretary Bob Bergland defines structure as "how farming is organized, who controls it, and where it is heading." I think we need to extend the meaning of the word even further—to the whole food system. Farming is the foundation of the system, of course, but there are many other aspects to consider as well.

Your own garden has a very compact structure. You are close to it, and know what is going on. Usually, you own the soil on which your garden grows, so you can control the entire process of operating it. There is no business or institutional structure standing between you and your garden—at least, in most cases. True, you may buy seed from a seed company, but if you want to save your own from one season to another that is possible to do. Fertilizer can be made from the organic wastes of your home and your yard, so you can be in total control of the building of your garden's soil fertility. And the closeness of your garden to you removes the need to ship food long distances. I think you need to realize how tight and efficient a garden's structure is in order to understand what happens when the production of food moves farther away from you.

As soon as you begin to buy food, the structure of the food production system becomes less clear and more complex, and your control over the producing, processing and shipping of that food is diluted. Fifty years ago the food structure was still very compact here in the United States. When we bought vegetables, for example, they usually came from a farm just outside town. We knew the farmer, and could see his crops growing anytime we wanted. He brought food to town in a pickup truck that had to travel only a few miles.

The same close structure was characteristic of the production and distribution of almost all foods. We used to get meat from local farmers, and flour from mills only a few miles from our home. Potatoes, fruit, cheese—they all came from nearby farms. There was no big capital structure between us and our food. The system was more primitive than what we have now, but it worked, and we usually knew what was going on.

Many factors have combined over the past few decades to change that situation, and to move the food system farther away from us, and actually out of our sight. I think the biggest factor was the tremendous flow of energy into our society. Cheap and abundant gasoline made it possible to move goods and people almost anywhere at low cost. So the growing and processing of food moved farther away, where we couldn't see what was happening. True, we got some advantages. Most people liked what was happening because they got more convenience. Centralized farms and processing plants turned out food that came in neat packages, and which could be prepared quickly—or even eaten without preparation. We also could get fresh fruit and vegetables at any time of the year, no matter where we lived. But we lost something important: control.

That system has achieved many successes, and is in many ways the best in the world. No other society as large as ours can claim to have the abundance and variety of food we enjoy today. And the price of food here in the United States is still relatively low compared to what it costs in other countries. But is this structure rock solid and permanent — something that can be continued in its present form for years, decades and centuries to come? Are we sure that we can continue to get good food at reasonable prices, or does the current nature of the food system create unnecessary vulnerability?

The answer is that the structure is not solid, nor does it form the basis for continued abundance of the kind we have known up to now. If energy and fertilizer costs keep going up for a few more years at their current rates, the system may just stop working. At the very least, it will stop working as a source of moderately priced food, and will produce food priced out of reach of many people.

The concentration of food-system control in the hands of large operations is another aspect of structure that needs to be looked at very carefully. The trend toward bigness starts with farming itself and works its way through the whole system.

Back in the 1930's, 40 percent of all Americans lived and worked on farms. Today farming is the occupation of only three percent. There are 32,000 food-manufacturing firms in this country, but 75 percent of the net profits go to just 50 of them. With each passing year, power over our food system passes into fewer and fewer hands.

The decline in the size of our land base usable for producing food needs close study as well. The United States is losing agricultural land at the amazing rate of nearly six million acres every year. About half of that lost land is covered over by the building of highways, factories, homes and other structures, and the other half is eroded away by rain and wind. That is a giant, irreplaceable loss.

My belief is that there is not a conspiracy to cause the destruction of that much land, or to give control of the food system to a few businessmen, or to contrive energy shortages to increase food prices. But I also believe that many important people are sitting on their hands and not doing much to prevent such happenings, figuring that their organizations will profit mightily when these factors cause a sellers' market in food to occur. That conspiracy could be somewhat passive, in the sense that the participants don't have a clear plan to make it occur. But the fact that saving our land for the production of food or decentralizing the control of our food system is not a high priority is clear evidence that there is at least a conspiracy of silence and inaction.

Dan Morgan, writing in the July issue of *Atlantic*, goes even further than I. In his article, "The Politics of Grain," he advocates that our government and the elements of the food trade get together and fix the prices for our grain sold in foreign markets to increase our power to use food as a weapon in world trade. He says that our food structure is well along the way toward being ready for that kind of action. Here are his words:

"There are, in fact, striking parallels between the present grain market and the world oil market of the late 1960's, before the OPEC cartel reached maturity.... Now, a decade later, in the world grain economy, a handful of companies that resemble the oil giants in their control of communication, processing plants, depots, transportation, and financial facilities dominate the system that markets the American surplus."

What can we do about this situation? Through The Cornucopia Project we are searching for ways to solve the serious problems facing our food system. I believe there are solutions, and that we can create a food system that avoids the problems of our present one, works better, and provides for our needs without wiping out our farmland or our savings.

You can participate in The Cornucopia Project in two ways. One is by reading the articles in this series. My hope is that I will be able to present enough information to you in these pages, over the next few months, to give you a broad view of the food challenge and what might be done about it. But I also want you to know that The Cornucopia Project exists as a real information-gathering and service organization. Medard Gabel, an experienced investigator of the global food situation, heads the project office. He and his staff of researchers and planners will be able to serve other organizations and people who are in need of specialized information, or who want to become active in developing and implementing the project's goals. Together we can insure a safe and nutritious food supply not only for ourselves, but also for our children and grandchildren.

## SOURCES OF MORE INFORMATION

If you want to become active in the effort to prevent us from being trapped by the growing weaknesses of our own food system, these publications will be of value to you.

*Future Food*, by Colin Tudge, published by Harmony Books, New York. This is a paperback book, selling for $9.95, that outlines a way to plan a diet that will help you and at the same time help to "rationalize" the food system, to use the author's word.

*Report and Recommendations on Organic Farming* of the U.S. Department of Agriculture. Available at no charge from: Office of Governmental Public Affairs, U.S. Department of Agriculture, Washington, D.C. 20250.

Reprinted from the September, 1980, issue of ORGANIC GARDENING.

# Have You Ever Asked These Questions About Your Food?

AS I SAID in the first two articles in this series, the coming food crunch is likely to be a much more serious problem than the energy crisis we have been living with for the past few years. Energy problems are beginning to have a ripple effect through the food system, which is just one small part of the problem. Every day there are more mouths to feed, both here and abroad. At the same time our soil is wasting away at a rate that is becoming frightening. It is possible to extend current trends into the foreseeable future and see an America with little topsoil left, and many millions of people unable to pay the high prices asked for food of decent quality.

There is a bright side to this situation, though. Tremendous potential for increased production of food remains—if we use more of the idle land around our homes and cities for intensive production of vegetables, fruit, beans and many other types of food for people. (Horticulture is far more productive of food than is agriculture, per unit of land area.) Also, by using land close to where we live, the expensive and wasteful practice of shipping food long distances will be reduced.

The Cornucopia Project of Rodale Press is an effort to collect information about the ways the U.S. food system needs to be restructured to prevent a food crisis. We think that if people have a chance to look at the big picture that is formed by all the elements that constitute our food system, they will be able to see both the size of the problem and a road leading to a solution. As long as we just think about our own little corner of the system, without trying to see how everything fits together, the present decline of our food production base will not only continue, but will accelerate.

The best way to know how vulnerable you are to future food-cost inflation, and also to know what you can do about it, is to ask yourself more probing questions about your food. These are the traditional questions people usually ask themselves:
"What am I hungry for?"
"What is nourishing for me?"
"What will it cost?"

Those are very important questions, and I don't mean to downgrade them. But go deeper! Try to ask yourself additional things that will tell you how dependent you are on a food system that may not be able to function in the old way much longer. And think of the information you will need to make your food system more reliable and more self-sustaining.

We may be able to help. Here are some probing questions about food that have been drawn up by researchers at The Cornucopia Project. If they seem to you to be easy and you can answer them well, then you have an important depth of knowledge that can be a useful tool in working for greater food security. If you draw a blank on most of these questions, then you need to get to work quickly to expand your knowledge.

### WHERE DOES YOUR FOOD COME FROM?

Think of this question in an item-by-item way, as well as generally. How much comes from your garden, and how much do you buy? Do you have any idea where the lettuce you buy is grown, or where your meat was produced? What about cheese, bread, flour, potatoes and other staples? Try to get a handle on where they originate and whether you could get any of them from sources closer to your home.

### DO YOU KNOW HOW MUCH YOUR FAMILY EATS IN A YEAR?

The importance of food becomes more apparent when we get an idea of the quantity used over a period of time. The average American eats 1,451 pounds of food a year. That breaks down to 371 pounds of fruits and vegetables, 143 pounds of cereals, 239 pounds of meat and fish, 353 pounds of dairy products, plus 345 pounds of everything else. Multiply that by 230 million people in the U.S. and you can see how important is the system that keeps food coming to us regularly.

### DO YOU KNOW HOW MUCH YOU PRODUCE IN YOUR GARDEN?

Have you kept track of what you've grown, how much, and what its value is? Last year, according to the Gallup survey sponsored by Gardens For All, 33 million American households (42 percent) grew some of their own food in home or community gardens. The total value of produce from all our gardens in 1979 was $13 billion—an average yield of $386 for each garden. With an average cost of only $19, the total saving was $367 for every family growing food in a typical garden of only 600 square feet—or a plot approximately 25 feet on each side.

Record-keeping is the basis of productive planning. If you know how much food you are growing, you can tell how effective your methods are, and what impact fertilizer, soil improvement, weather and improved tillage have. Good records don't have to be complex or time-consuming to collect. And you'll find that they become very useful, over a period of years, to tell what is happening to your garden's fertility.

### DO YOU KNOW HOW MUCH YOU *COULD* PRODUCE?

What would happen to your annual production rate if you enlarged your garden, put in more time, or grew varieties that produce more in a smaller space? What if you used home food systems in addition to gardening, such as small-scale fish production, making tofu and other foods from soybeans, grinding your own flour, and so forth? What percentage of your total food needs could you handle that way, and what would be the impact on your budget?

## HOW LARGE IS YOUR FOOD RESERVE?

Suppose your normal sources of food were cut off for any reason. The cause could be a natural disaster, an economic crash, or panic buying for a variety of reasons. How long could you keep going on the food you have on hand? We now see the importance of a national security reserve of oil (which has not yet been created). A food reserve is even more basic, and should be localized to be most effective. The worldwide store of grain is down to about a 50-day supply. How do you rate in food security by comparison?

## DO YOU KNOW WHERE THE NEAREST FARM IS, AND WHAT FOOD YOU COULD GET FROM IT?

Today, many farmers buy almost all their food in stores. Farming has become specialized, with even family farms producing only one or two crops, which are processed into food at distant locations. That single-cropping and specialization is a big cause of the vulnerability of our food system. Do you have any balanced, diversified farms near you? Can you buy local produce at roadside stands?

## HOW MUCH FOOD COULD YOUR COMMUNITY PRODUCE?

Garden potential counts in trying to answer that question, but so do a host of small-farming options. Much vacant land near towns and cities could be used for intensive food production if people saw the need, and if economic factors favored regional food production. An inventory of local food-production potential is an important first step toward planning a food system which would operate with less reliance on high-cost energy, and which could protect regions against inflation and shortages. Are there potential farm/garden lands in your area that are currently idle?

## DO YOU KNOW HOW MUCH SOIL IS LOST IN THE PRODUCTION OF YOUR FOOD?

Each pound of corn grown in Iowa is matched by the erosion of six pounds of topsoil. Wheat can be even more destructive. Twenty pounds of topsoil are lost for every pound of wheat produced in eastern Washington State. Most other farm crops cause unacceptable losses of soil as well. When you eat, do you ever think about how much irreplaceable soil is being lost in the process of bringing that food to your table? Are there ways that these losses can be reduced, or even reversed?

## WHAT IS THE ENERGY COST OF YOUR FOOD?

Very little energy is needed to produce a potato in your garden or on a nearby farm. But a box of instant potato flakes requires nearly 10 times as much energy as to produce the same amount of raw potatoes. Some foods are as much like a gas-guzzler as any overstuffed old-style car, while others are as lean and efficient as a front-wheel drive subcompact. Do you know which category different foods are in?

## WHO IS IN CONTROL OF OUR FOOD SYSTEM?

As I pointed out last month, you are in almost total control of the food you grow in your garden. But food that you buy can be under the control of organizations or even countries you know little about. Consumers who know little about the control of their food supply can't do much to change things. Why do some items suddenly jump in price, while others hold steady or even drop? Who decides what new products will appear in stores, or what old ones will disappear?

## DO YOU KNOW HOW GOVERNMENT FOOD POLICY AFFECTS YOU?

Many government programs influence farming, shipping, storage and processing of food. But most consumers have no idea what these policies are or what they are doing to the cost and availability of food. Use of drugs to fatten animals faster can reduce effectiveness of drugs you may need badly when sick, but government policy has yet to stop that dangerous practice. The 1981 farm bill could help improve the food system, but few nonfarmers will try to influence it. How can one individual affect government policy? Do you want to learn more about food policy before a food emergency happens?

## DOES WHAT YOU CHOOSE TO EAT MAKE A DIFFERENCE?

Think for a moment about the implication of a large number of individual changes in food preferences. Every vote counts at election time, and so do purchases at the food store. Food companies are now planning to introduce and sell more natural and less-processed products, but most companies don't see those foods becoming mainstream items soon. However, if more people switched to more fresh and healthful foods, the impact would be felt up and down the commercial food chain, and policy changes would result. How can we get companies to produce and market more fresh and healthful foods?

## COULD YOU PLAN TO MAKE YOURSELF MORE SELF-RELIANT?

There are simple steps everyone could take to create more food security. Building a bigger reserve of food is one. Another is to produce more yourself, and do more home processing. That could mean equipping your home with more storage space and tools for processing. Learning to use more whole grains and similar staple foods might be another useful step. Can you make a list of all the things *you* could do to increase your personal food independence and form them into a plan for action?

## HOW MUCH SELF-RELIANCE IN FOOD IS POSSIBLE FOR THE AVERAGE PERSON?

Getting an answer to that question is one of the goals of The Cornucopia Project. In fact, the process by which we try to get an answer illustrates very well how The Cornucopia Project can work. The first step is for you to think through your food situation and see how dependent you are on our food system, as well as how much self-reliance you could achieve. Then tell us, and we will begin to form a picture of the potential for self-reliance nationally. Of course, the purpose of The Cornucopia Project is not limited to promoting home food production, but that could be an important aspect of national food security. You may be able to benefit by knowing more about how much food you can produce, and we can use information about that potential to show how important individual action can be.

## WHERE IS THE U.S. FOOD SYSTEM GOING?

We are all aware of how farms, food stores, and the type of food available have changed. If that kind of change continues into the future, what kind of food will we have, and what will the system of producing and delivering food be like? Is this the kind of food system we want? Think about those questions in terms of your own desire for food, and your community's situation as well.

## WHERE *COULD* THE FOOD SYSTEM BE HEADED?

Are we as individuals trapped in a system of producing food over which we have no control, or is there still time for us to learn what is happening and exert pressure for change? What options for learning and for influence do we have? Is it possible for us to create a plan that would show how the abundant resources of this country could be managed to produce large amounts of good food at reasonable cost and without burning up our soil and fuel resources?

I have only one more question to ask of you. It is this: What do you think of these questions? In attempting to answer that, keep in mind that sometimes asking the right questions can be more important than getting the "right" answers. A good series of probing questions can start people thinking about a taken-for-granted thing like food in new ways. There is more to food than breakfast, lunch and dinner. There is life in our food, and we need to start asking how secure is the source of that life, who controls it, and what we can do to protect it for the future. If we keep asking the right questions, both of ourselves and of the organizations and people who produce food, we can create a consciousness that will lead to constructive action.

You may be able to think of other, better questions that should be asked. And you may feel that some of my questions are more important than others. Our plan is to use your response to this suggested list of questions to create a Cornucopia Agenda of questions that could form the basis for a new kind of analysis of the merits and possibilities for improvement of the U.S. food system. We need your thinking, and your ideas, to make that agenda as effective as possible.

Reprinted from the October, 1980, issue of ORGANIC GARDENING.

# Plan Your Food Future Now

NEXT TIME YOU eat, look for a moment at your fork. Think how useful it is, and also how symbolic. A fork is the main tool you use to get food into your mouth. The *idea* of a fork also symbolizes choice. A road forks in front of you. Choose one of the forking roads, and you get to an entirely different destination than if you take the other.

Eating is often a forking experience, in more ways than one. There are choices to be made. Usually the choices are simple, and have no permanent impact on us. "Tonight I'll eat fish, and tomorrow chicken." Eventually, all those menu choices even out, and add up to nothing significant.

I'm convinced, though, that right now all of us face another kind of food choice—a forking in our food road—that is entirely different. It's a choice that we've not had to make before. Also, we don't have much time to make up our minds. We are moving quickly and the road is dividing before us. We must go one way or another, and this is not one of those "meat today and fish tomorrow" choices. It is a choice that will decide where our food comes from, how good it is, and especially how much we must pay for it all the rest of our lives. If we don't make up our minds quickly and take the right road, we are likely to spend the rest of our lives being mere passengers instead of navigators. And the journey could be hazardous indeed.

There are hills and valleys in the road ahead of us, and we can't see over all of them. The road maps we have are only of partial usefulness because our journey for food is into the future. We can only guess at what detours and washouts may crop up in the roads ahead. But there is a useful tool of the futurist that we can work with. It is the scenario—a description of a probable future based on attempts to understand how decisions made today change the nature of the "place" we are going to reach at some point in time ahead. The writer of scenarios knows where we have been, and tries to show how course corrections now and in the future change the nature of the destination we will arrive at some day. The scenarios are not the same because the decisions that are made along the way to each destination are different. There is no guarantee of total accuracy, of course, because no one can be sure what course-correcting decisions society will make in the future. Nor can we predict acts of God or other unforeseen events. But I believe scenarios are extremely useful "seat-of-the-pants" navigation tools into our future.

For simplicity, let's decide that our future food fork has only two prongs, and write simple scenarios for each. One prong will be the conventional food scenario, which assumes that the food system will move ahead as it has in the past. Farming will be based on widespread and growing use of chemical fertilizers, synthetic pesticides, and other inputs depending heavily on fossil fuel. Soil erosion will persist at its current rapid rate. The number of farms will continue to decline, as remaining farms become larger. Food produced on farms will be shipped long distances to processing plants and markets. Fewer people will work on farms and in the food system as a whole. Consumers will go to fewer and fewer stores for their food purchases, and will be encouraged to buy types of food that are highly processed for quick use. In other words, the industrialized and high-tech trends of our food past are projected into the future.

My suggestion of what will happen if this scenario is followed is based on the thinking and planning that has gone into The Cornucopia Project, which I have discussed in some depth in my three previous editorials. A very similar projection of our food future is also made in the Global 2000 Report to the President, just released by the Council on Environmental Quality and the U.S. Department of State. I am also drawing to some extent on the information in a study of the future of U.S. agriculture completed recently by The Conservation Foundation.

Because space here is limited, let's just focus on one issue—the cost of food.

Our Cornucopia Project studies suggest that food will become terribly expensive in the future as the amount of good land available for production declines and the number of mouths to feed increases. The Conservation Foundation experts were "guardedly optimistic" that the amount of food available a couple of decades ahead would be adequate. But at a press conference they concluded that "people would have to spend a larger portion of their income for food in the future." The Global 2000 report puts a number on that cost increase for the period just past the year 2000—the beginning of the 21st century. "In order to meet projected demand, a 100 percent increase in the real price of food will be required," the U.S. government report concludes.

Even without the extra impact of inflation, that kind of food-cost increase would be devastating. Imagine a quick shopping trip in 2001. Some friends drop in for the weekend and you need some extra salad ingredients, fruit and bread. You cash your paycheck at the store, and have to pay exactly double—in proportion to today's rates—for whatever food you need. That scenario doesn't even consider the impact of inflation—just the rise in price caused by the fact that food is in short supply and more people want to buy it.

Let's factor in inflation as well. I heard a government spokesman say recently that we'll be lucky to have 10 percent inflation for quite a while, and I agree. So we'll use 10 percent as our inflation rate out to the year 2000, and I've started from a base of average retail food costs during 1980.

You pick up a couple heads of California *iceberg* lettuce. Two for $9.20. Tomatoes are next. Three large ones cost $5.24. Then green onions—$3.78 a bunch. A loaf of bread costs $7.66. A two-pound bunch of bananas sells for $6.56. Finally, a ten-ounce jar of instant coffee comes to $45.06.

Of course, I realize that your wages have gone up some too in those 20 years, but by now we should have learned that prices go up faster than wages, in almost all instances. You also need to keep in mind that food is not going to be the only necessity of life that will rise in price. Forecasts of energy, transportation and housing costs also indicate that people are going to have to "pay a larger percentage of their income" for them. Where all these "larger percentages" are going to come from, I surely don't know.

## THE CORNUCOPIA ALTERNATIVE

Fossil fuels are expensive now, and they're going to cost plenty more in future years. The conventional food system relies heavily on fossil fuels for the production of fertilizers and pesticides, for powering all the equipment used, and for shipping food long distances to processing plants and markets. Any scenario that is going to lead us toward reasonable food prices in the future will have to cut fuel use in the agriculture system.

What I call the Cornucopian approach to food production does that. Preserving the soil is a vital first step. Our soil is now eroding at a terrible rate, and only the large-scale use of synthetic fertilizers is propping up production. Such practices as proper plowing, crop rotation, and the building up of organic matter and nutrients in the soil will permit intensive use of our land for food production, while at the same time preventing erosion and eliminating the need for chemical fertilizers and pesticides. But planning for these changes will take imagination and effort — starting right now. Good land is the only foundation for low-cost food production. If our land continues to disappear, any hope we have of avoiding super-luxury prices for food will disappear with it.

Other elements of a Cornucopian scenario will focus on how that land is used. To head off disastrous price increases, people will have to figure out ways to produce more food close to home. Transporting food long distances is an idea whose time has passed. Soon we simply will not be able to afford the fuel needed to move meat, vegetables, fruit and other foods thousands of miles.

I hope people see the need quickly for a return to regionalized food production. Right now, we need clear thinking about the Cornucopian scenario, so future towns can be built in ways that will allow them to be food self-reliant. More space for gardening must be planned into housing developments. Space for small, high-intensity farms should be allowed near towns. Ways to recycle garbage and other organic wastes back to the land — to preserve vital fertilizer nutrients — must also be provided for in future building plans.

Food that is eaten fresh, close to where it is grown, can be provided to consumers with much less capital investment than highly processed food. Our nation is going to need all the capital it can muster for energy and reconstruction of our industrial plants to make them energy-efficient. The food system can be transformed in Cornucopian ways with very little capital — provided people return to a diet of largely fresh foods, eaten in season. We need dozens of scenarios that will create plans for that kind of diet re-education for different regions of this country.

Processing and storage of food will have to be done at home, or in the local community, at least to a large extent. The financial advantages of home food processing and storage systems are great now, but they'll become overwhelming soon. As food prices continue their upward march, only those people who can put their homes or neighborhoods to full use as food system centers will be able to eat well. Education is needed now to teach everyone how to use home and neighborhood resources to provide food security. Along with a revival of what I call home food systems will come a flowering of small-scale shops and processing plants that will prepare the foods produced in each region.

That is just a brief outline of some of the options for change that could be put into a Cornucopian scenario. You probably have other ideas — particularly steps that could make a low-cost and sustainable food system work well in your area. I almost forgot to mention sustainability. That's the major reason that the conventional food system scenario is going to lead to such fantastically high food prices. The base of production — both in land and energy sources — is being consumed as food is produced. We are eating up now the heritage of productive capacity that will be needed even more in a few short years. No wonder prices are going to go up so much!

Another very important point that I should add has even more to do with economics. It also concerns management goals and business opportunities. The very high prices that are going to be paid for food after the year 2000 (and probably before) will not go to Arabs and other foreigners but to American food companies (unless they are bought by Arabs). America is now and is likely to remain a world center of food production, and the profits to be earned on food are likely to be large indeed.

However, any food company that continues to rely largely on old, energy-expensive ways to manage is likely to find itself soon having severe problems. In fact, that is already happening to at least a few long-haul shippers of food. Conversely, any food company that is based on production and sale of food that fits the Cornucopian model is likely to benefit from the new realities of the food marketplace. By having the courage and wisdom to begin changing now, such companies can help assure a supply of good food for the future, and continued prosperity for themselves.

One goal of Cornucopia Project thinking is to help develop a food system where the opportunity to participate is wide open to everyone. We can't each have our own oil well, but each and every one of us can either grow a garden, sprout seeds, bake bread, or do any one of hundreds of other home food production projects. And many exciting new opportunities for commercial food production will open up as well.

Reprinted from the November, 1980, issue of ORGANIC GARDENING.

# "No $7.00 Loaf of Bread for Me"

ORGANIC GARDENING READERS by the thousands have written us for more information about The Cornucopia Project. Comments of many reveal a strong desire to find out more about the workings of the U.S. food system. Large numbers of people also say they want to work to make that system more sustainable, and less vulnerable to food-price inflation and outright disruption.

"I am particularly interested in facts the public needs to know to understand the relationship between energy availability and our food supplies," writes K. E. Hardy of Pocatello, Idaho. "For instance, how much of the food grown in California and Florida leaves those states for out-of-state markets? What total amounts are imported by each state? How many barrels of oil does it take per month or per year to serve those out-of-state markets?"

No one has a clear answer to those questions yet, but one function of The Cornucopia Project is to get them.

Here is a letter from a person whose name I will withhold:

"I work on one of the few small farms still existing in western Massachusetts. It is really disillusioning to become aware of the terrible farming practices and waste built into our current farming system. Even on just the farm I have worked the past three seasons I have seen a rapid decline in care and quality since I began there. There are strong pressures on the owners to use more and more chemicals and to plant more crops (often left to rot in a field or cooler) to compete with large-scale growers. The character and quality of the produce, land and farm itself have really changed."

Many of the letters were written in a tone of urgency, bordering on panic.

"I'd like to know what I *can do*," wrote Dehaine Couchman-Turner of Orangevale, California. "With the help of The Cornucopia Project I hope to do my share—for my children and future grandchildren."

I expect this kind of response to continue and to increase. There is no escaping the problems that our food system faces. Even in the few months since Medard Gabel began planning The Cornucopia Project and I started writing this series of editorials, inflation has pushed food cost higher at a frightening rate—and at a time of year when food costs normally go down. Grocery prices went up 2.3 percent in August, the biggest jump in five years. We have little time to waste, if we are going to head off even more serious food-cost inflation that could create devastating hardship.

Right now, our big need is to understand exactly what The Cornucopia Project is, and to begin making decisions about which parts of the food problem need working on first. Until now we have deliberately not sharpened the focus of The Cornucopia Project because we wanted you to help do that job. You must participate in deciding what The Cornucopia Project is and how it will work, or there isn't any chance at all that it will accomplish anything. The American food system is so large and complex that the thinking of one small group cannot fully explain its problems. No effort based in one office, or one region, can hope to make a significant impact on inflationary price trends, or build real strength into the system. Only millions of people, working in a large number of different ways, can begin to stop the waste of soil, food and energy that is going to cause the price of food to shoot toward the stratosphere.

Here are some clues, though, that may help you plan what The Cornucopia Project can accomplish—and what it can mean to you. Think first about education. We all have so much to learn about our food system. So many vitally important things are happening, including many that are draining away the natural resource heritage of future generations. We eat much of our food in blindness, not thinking that each leaf of lettuce and slice of bread is paid for with a pinch of lost topsoil, or a few drops of irreplaceable oil. Many new lessons about food must be learned, and taught to others.

Another help is to remember what we've been through with energy, and to plan how to avoid the same thing happening to food. There is a good possibility that the food system now is in the same situation that our energy system was about the year 1970. The looming energy problem was then obvious to any planner who looked at how much oil there was in the world, who owned it, how unstable various political regions were, and what might happen to prices if a shortage psychology could be created. The problem could have been minimized and maybe even prevented if clear-headed energy education had taken place 10 years ago, and a broad-scale plan for more production, conservation and storage of reserves worked out. But nothing was done, and the energy crisis hit with full force.

As far as we at ORGANIC GARDENING and The Cornucopia Project have been able to find out, almost no food system planning is now going on in the U.S. I know of only one city that has a plan—Knoxville, Tennessee. We have talked to some state governments that are thinking about creating a plan, but little real action has taken place. One federal agency is beginning to work on a plan for a portion of the U.S. History seems to have taught us nothing. Our vulnerability to a food crisis increases day by day, yet we have hardly begun to try to think and act our way out of it.

Don't panic, though. I'm convinced that you, acting on your own behalf and as a good citizen, have much more power to plan a way out of the food problem than you did to participate in a solution to the energy crisis. For one thing, you now know that a food problem is going to happen. You've been warned! Much more important, though, is the decentralized nature of food production. To make it big in energy you need an oil or gas well, or a coal mine. Those energy sources are literally as good as gold.

But with food, the gold is the soil. In fact, soil is better than gold, because you can't eat the yellow metal. And many millions of Americans are very close indeed to a patch of land that can produce much more food than it is now yielding. That is why we chose the word cornucopia to be the name of this project. A cornucopia is ready to begin pouring out food all over America, if only we will bring the production, storage and processing of food closer to where people are, and help them to use more of their own time and energy to make the earth bloom.

How do you get started? I think a notebook is essential. Everyone working to think through and act on the food problem should get a regular, 3-ring notebook and a pack of lined paper to record their personal ideas and other Cornucopian efforts. I've done that myself, and already have thought of plenty of information about my own food flows that can be collected and put to good use.

All the basic literature on The Cornucopia Project is designed to fit in a 3-ring notebook. The latest edition of the introductory leaflet which thousands of people have sent for is punched for easy insertion, and whatever other data-collection papers that you or others suggest for use can easily be inserted. Of course, a notebook is not absolutely essential. Don't stop thinking about The Cornucopia Project if you fail to get a notebook. But keeping your information in one place sure makes the process of planning easier and more effective. Easy reference is essential. Eventually, you may have a series of notebooks.

As I wrote in the October issue editorial, one of our first steps is to decide what questions we think should be asked about the U.S. food system. There hasn't yet been time for us to get your feedback on the questions I suggested, but we'll have heard from you by the time you get to read this. My goal now is to complete the Cornucopia Agenda by February, and print a notebook sheet then listing the questions that are central to the project.

You can get a head start by simply copying out of the October issue those questions you feel are important, and adding your own. Write them on the blank, lined pages in your notebook, perhaps putting each question on a separate sheet. Leave plenty of space for whatever answers or other information about each question that you'll want to retain.

All of us need much more information about the flows of food into and out of our localities. Finding how to reduce those flows—creating more local production and use of food—is the part of The Cornucopia Project that is likely to yield the quickest and also most permanent benefits. The best place to start working to reduce the travel distance of your food is in your own home.

I suggest we all start using our notebooks to keep a food diary. Let's aim to record our use of food week by week, month by month. The cost of food should be entered, but that's only the beginning. I think you'll find it very useful to note how much food comes fresh from your garden, how much you use from your stock of home-preserved food, how much you buy, and also how much you waste.

I'm not suggesting that you do this kind of note-keeping the rest of your life. Do it long enough to get a base of data about your purchases and production of food that you can use for planning. (If you want to keep on doing it permanently, that's O.K., of course.) Once you get a fairly good picture of your family's food flow, you can make your analysis more sophisticated.

Note how much food you eat fresh. That's important! A major goal of The Cornucopia Project is to encourage more use of fresh food, thereby reducing the energy cost that goes into processing. Also keep records that will help you appreciate the value of food you process yourself. You might even put a dollar value on that category of food, to help justify investments in more equipment to increase the amount of food you process at home.

Where does the food come from that you buy? An important part of this planning process is to get a base of data that you can use to shorten the food chain you are feeding on. Can you eat more food that is grown in your own state or county? Only by keeping records will you be able to know enough to judge how effective is your plan to do that.

I think you'll be amazed at how useful your food diary will be. You can use the information in it to cut costs and reduce waste. But its value can extend far beyond you and your home and family. One possibility is that The Cornucopia Project will find a way to begin collecting some data from your diaries to form a picture of food flow efficiency over a wide area.

You can also use the information in your diary as an educational tool. If you can prove with written records that you can produce a great deal of your own food and can eat good food at much lower cost than the average person, that information will encourage other people to do the same thing. Many who have written us about The Cornucopia Project are members of food co-ops or other organizations with a food or environmental interest. Such organizations can prove their effectiveness much more easily when members keep records.

Pushing back the day of the $7.00 loaf of bread is the bottom line of this project. If we don't become active now and start planning for a different future, as sure as anything the day will soon come when the cost of food will be more frightening than the numbers that flick into place on the gas pump when you fill your tank.

Not long ago, I happened to be paging through a copy of Henry David Thoreau's famous book, *Walden*, and was pleased to see that he was a faithful recorder of information about his food. In that classic little book are many pages about the things he ate, how much he grew, what he sold or exchanged, and especially what his food cost. The bottom line of his food budget for eight months of solitary life on the shores of Walden pond came to the amazingly low figure of $8.70. How times have changed! And what does the future portend?

Reprinted from the December, 1980, issue of ORGANIC GARDENING.

## CSPI NEWS

# Coalition Vows to Defend Food Safety Laws

Resistance is building among consumer, medical, labor, and health organizations to recently introduced, industry-sponsored legislation that would radically overhaul the nation's food safety laws.

The Coalition for Safe Food, organized by the Center for Science in the Public Interest, the Community Nutrition Institute, and Congress Watch, recently charged that Senate bill 1442 "has the potential to rewrite 75 years of law governing the safety of the American food supply." The coalition characterized the bill, backed by Sen. Orrin Hatch (R.-Utah), as "an industry wish list" that contains "not one provision benefiting the health of the public."

At least 34 organizations have already co-signed a letter that is to be distributed to all members of the House and Senate. The coalition is also developing a media campaign and organizing grass-roots lobbying before the measure is considered by Hatch's congressional panel.

### Major Rewriting

According to the coalition, the bill is anything but the minor modification of the 1934 Food, Drug, and Cosmetic (F,D&C) Act that industry claims. Consumerists had originally thought that industry would be satisfied to see the controversial Delaney "anti-cancer" Clause repealed (*Nutrition Action*, July, '81). But the scope of S. 1442 is far greater.

Bruce Silverglade, Legal Affairs director for CSPI, said that the bill sabotages the F,D&C Act in three principal ways. The industry legislation would:

• weaken the current definition of "safe," characterizing it as the absence of significant harm to public health, rather than no harm to public health;

• gut the Delaney Clause which now prohibits the approval of a human or animal carcinogen as a food additive;

• hasten approval of new food substances by the Food and Drug Administration (FDA), while simultaneously making it harder for the agency to get harmful substances off the market by forcing the FDA to first consider economic factors relating to industry profitability.

The legislation appears haunted by the ghosts of earlier food safety battles—particularly the controversies over saccharin and nitrites. For example, without actually striking the language of Delaney—"that no additive shall be deemed safe if it is found to induce cancer"—the Hatch bill effectively repeals the Clause. If the FDA found that a new food additive "does not present a significant risk to human health" (as opposed to animal health), then the Delaney Clause is rendered inoperative.

### Risk-Benefit Analysis

Furthermore, the proposed bill would require the FDA to approve human carcinogens for use in foods if the substance's economic or other benefits outweigh the risks. And finally, the FDA would be given authority to permit a "phase-out" of any additive found dangerous to the public health in order to prevent economic penalties to industry. At present, the Delaney Clause stipulates that the FDA must immediately withdraw the substance from the marketplace (though, in fact, "immediate" stretches out over many months).

Myron Zeitz, an attorney for the Community Nutrition Institute, described the industry action this way: "Delaney is being repealed, and it is being repealed worse than if industry had just stricken the Clause from the F,D&C Act."

One of the worst aspects of the bill, according to William Schultz, an attorney with the Public Citizen Litigation Group, is the way new procedural steps would entangle the FDA in long and costly fights to remove harmful substances from the food supply. Current review processes already take two to three years, hardly expeditious to Schultz' thinking. By creating new advisory panels, requiring frequent risk assessment documents, and permitting more opportunities for industry appeals, Schultz predicted that the review process would stretch to eight or nine years. All the while, the contested substance could remain in food.

Moreover, the Hatch bill gives the FDA less time and more responsibility to evaluate applications for marketing new substances. In the specific case of packaging materials, manufacturers would only have to submit summaries of experiments proving that the "food contact substance" was safe. Unless the

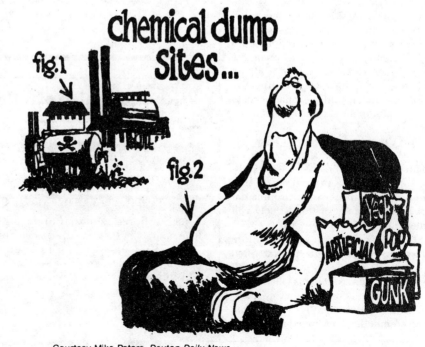

Courtesy Mike Peters, *Dayton Daily News*

FDA could show reason to question the studies within 90 days, the material would automatically be permitted in the food supply.

## CSPI Alternatives

In contrast to the industry bill, the Center for Science in the Public Interest is calling for the food safety laws to be strengthened. It cites several inadequacies in the current F,D&C Act to buttress its charges:

• Additives in use prior to 1958 are virtually exempt from food safety laws, and should be tested in accordance with contemporary standards;

• Some "possibly unsafe" color additives stuck on a "provisional list" for more than 20 years should be tested in a reasonable period of time;

• Food additives should be licensed for ten years, and not approved permanently, since improvements in testing procedures could uncover dangers overlooked in previous tests;

• Additives should be tested for their effect on behavior since they may have significant impact on the central nervous system;

• Testing regimens for food additives should be broadened to consider interactions between commonly used food ingredients, alcohol, and drugs.

As of now, hearings on S. 1442 are scheduled for mid-November. Sen. Hatch had postponed the hearings from September so that the administration would have time to prepare its position on the issues. At presstime, an aide to Hatch said that it was not known whether the administration would offer amendments to S. 1442, a policy statement, or its own bill.

FDA Commissioner Arthur Hayes recently told an American Frozen Food Institute convention that a joint position of the FDA, U.S. Department of Agriculture, and the U.S. Environmental Protection Agency on food safety will emerge by the end of the year.

But Silverglade, of the coalition, promised opposition to the Hatch bill when the issue is finally joined. "We will not stand by and watch this bill increase health risks to consumers solely to advance the financial interests of some food companies," Silverglade said. ∎

## SO, THE FDA HAS EVIDENCE TO WARRANT BANNING A FOOD ADDITIVE, THEN WHAT? UNDER S. 1442, THE CASE WILL DRAG ON FOR AT LEAST EIGHT YEARS. HERE'S WHY:

**Year One**
**Year Two**
**Year Three**

**Risk Assessment**

• Requires FDA to speculate how much disease will result from use of an additive and to permit its use if the threat to the public health is "insignificant."

**Year Four**

**Advisory Committees:**

• Would allow food manufacturers to force FDA to refer findings that an additive is dangerous to panel of non-government scientists.

**Year Five**
(Additives placed on the Interim List have remained there for more than ten years)

**Interim Status (Further Testing):**

• Requires that GRAS substances stay on the market pending additional research *after* FDA and advisory committee determine that the safety of the additive is questionable.

**Year Six**

**Risk/Benefit Analysis:**

• Must be performed before restricting the use of additives which have a substantial history of use and which have no readily available substitutes.

• Would require FDA to consider factors relating to industry profitability before banning an additive.

**Year Seven**

**Phase-outs:**

• Even if all other steps are met, additives would not be immediately removed from the market but phased out gradually.

**Year Eight**

**Judicial Review of Science:**

• Final agency actions could be appealed to Federal Court, further delaying effective dates and enforcement of restrictions.

## Key Senators in Food Safety Battle...

The Coalition for Safe Food urges individuals and organizations to write their legislators about the pending revision of the food safety laws before hearings, which are scheduled for the middle of November. Key senators include:

• Lowell Weicker (R.-Conn.) and Robert Strafford (R.-Vt.), two advocates of progressive food policies whose votes on the Republican-dominated Labor and Human Resources Committee are crucial to defeating S. 1442.

• Edward Kennedy (D.-Mass.), the ranking minority member of the committee, whom consumer groups hope will take the lead among Democrats opposing the bill.

Address all correspondence: *Senator, U.S. Senate, Washington, D.C. 20510.*

For more information about the coalition, write:

**Bruce Silverglade**
**CSPI**
**1755 S St. NW**
**Washington, D.C. 20009**

## Suggested Courses of Action

**1. Circulate the suggested model letters** to your members. Request that your members revise these letters in their own words and send them to members of the Senate Labor and Human Resources Committee and to the Senate Agricultural Committee, which will be considering S. 1442 in the near future. The enclosed postcard may be duplicated and sent in lieu of a letter. However, personal letters are more effective.

**2. Circulate copies of the model petition** at meetings of your organization, in classrooms, at supermarkets, or wherever you may find people who wish to express their support for strong food safety laws. Completed petitions should also be sent to members of these committees.

**3. Write letters** to the editors of your local newspapers expressing support for strong food safety laws.

**4. Consider alternative means of communication.** Call or send a telegram to your senator's office.

**5. Organize groups** of interested persons to meet with your Congressional representatives or their aides. A visit can be arranged for the local or Washington office.

**6. Get others involved.** Circulate fact sheets to other interested groups and have them co-sign letters to members of these committees expressing their support for strong food safety laws.

### Committeee on Labor and Human Resources
Room 4230, Dirksen Senate Office Building
Washington, DC 20510

Orrin G. Hatch, Utah, Chairman

| | |
|---|---|
| Robert T. Stafford, Vt. | Edward M. Kennedy, Mass. |
| Dan Quayle, Ind. | Jennings Randolph, W. Va. |
| Paula Hawkins, Fla. | Harrison A. Williams, Jr., N.J. |
| Don Nickles, Okla. | Claiborne Pell, R.I. |
| Lowell P. Weicker, Jr., Conn. | Thomas F. Eagleton, Mo. |
| Gordon J. Humphrey, N.H. | Donald W. Riegle, Jr., Mich. |
| Jeremiah Denton, Ala. | Howard M. Metzenbaum, Ohio |
| John P. East, N.C. | |

### Committee on Agriculture, Nutrition, and Forestry
Room 322, Russell Senate Office Building
Washington, D.C. 20510

Jesse Helms, N.C., Chairman

| | |
|---|---|
| Robert Dole, Kan. | Walter D. Huddleston, Ky. |
| S.I. Hayakawa, Cal. | Patrick J. Leahy, Vt. |
| Richard G. Lugar, Ind. | Edward Zorinsky, Neb. |
| Thad Cochran, Miss. | John Melcher, Mont. |
| Rudy Boschwitz, Minn. | David Pryor, Ark. |
| Roger W. Jepsen, Iowa | David L. Boren, Okla. |
| Paula Hawkins, Fla. | Alan J. Dixon, Ill. |
| Mark Andrews, N.D. | Howell Heflin, Ala. |

## Model Letter

Dear Senator:

I am writing to express my strong opposition to S. 1442, the Food Safety Amendments of 1981. I am particularly opposed to how this bill changes the definition of a "safe" food additive and permits potentially carcinogenic and other hazardous substances to enter our nation's food supply.

S. 1442 represents the latest attempt by the food industry to force the Food and Drug Administration to consider factors relating to industry profitability before proposing that unsafe food additives be banned. Such proposals undermine the safety of our food supply and could make a mockery out of the law by requiring the FDA to meet a series of burdensome procedural requirements which may delay the removal of an additive from the market for more than nine years.

One of these procedural requirements calls for the FDA to conduct a risk/benefit analysis to determine whether an additive should be banned. If this analysis is applied to food safety questions consumers are bound to lose, as the benefits we receive in terms of improved health will not be measurable for 20 or 30 years to come, while industry's lost profits will be measured today. We will open a Pandora's box, in which industry's cries of excess costs cannot be countered with proof of immediate health benefits.

Strong regulation of food additives is essential to the public health. I urge you to support bills which would strengthen, not weaken, the food safety laws.

Your opposition to S. 1442 and similar industry-backed proposals will be essential to maintaining these health and safety protections.

Please let me know what action you plan to take to preserve the integrity and strength of our food safety laws, and consequently the health of today's and future generations of Americans.

Thank you very much for your concern. I await your reply.

Sincerely,

## Model Postcard

Dear Senator _____:

The Center for Science in the Public Interest has informed me that Congress is considering severely weakening the "DELANEY CLAUSE" of the Food, Drug and Cosmetic Act. This clause now protects Americans from cancer-causing additives.

Please pay close attention to CSPI's warnings that food safety laws should be *strengthened,* not weakened — and do your utmost to protect the health of our nation's citizens.

Sincerely,

## Model Letter

Dear Senator:

We are writing to you to express our strong opposition to S. 1442, titled "The Food Safety Amendments of 1981." We are alarmed that S. 1442 may be the vehicle for revision of the food safety laws in the 97th Congress.

The bill would seriously weaken our food safety laws by relaxing the definition of "safe" food, effectively repealing the anti-cancer Delaney Clause and undermining the regulatory framework designed to protect the public health. Because it constitutes a one-sided "wish list" for the food industry, S. 1442 would ultimately damage public confidence in the safety of American food and contribute to an increase in human disease.

We believe that this result would be an intolerable one for the consumer, for responsible members of the food industry, including those Americans who are employed in the food sector, and for the government which is charged with protecting the public health through enforcement of the Food, Drug and Cosmetic Act, the Meat Inspection Act and other laws which make a safe food supply a reality in this country.

It is cruelly ironic that in the year we are celebrating the 75th anniversary of the Food, Drug and Cosmetic Act and the Meat Inspection Act, some segments of the food industry are working to sabotage the foundation for American health — a safe food supply. We urge you to oppose S. 1442 and help prevent the havoc and danger which it would produce. The American people will not tolerate this assault upon the food safety laws.

Thank you for your consideration of our views.

Sincerely,

# PETITION

## SUPPORT STRONG FOOD SAFETY LAWS

We oppose any weakening of our nation's food safety laws. Food additives which may cause cancer or other diseases must be removed from our food supply. New food additives must be thoroughly scrutinized before they are allowed on the market. S. 1442 and H.R. 4014 would allow additives which have *been shown to cause cancer* to be used in food. These bills must not be allowed to become law.

**NAME**                           **ADDRESS**

1.
2.
3.
4.
5.
6.
7.
8.
9.
10.
11.
12.
13.
14.
15.
16.
17.
18.
19.
20.

# FOI: How To Learn Federal Secrets

*by James Greene*

Big Brother is being watched! Yes, a full 5 years before its time, *1984* is being repudiated. *1984* was the title of George Orwell's novel that described a future time when every facet of a person's daily life would be controlled and monitored. Citizens were warned that Big Brother (the government) was watching them at all times. And, through technology, it was true. Big Brother even looked in (via two-way television set) on the citizen's diligence in performing the required morning exercises.

But this is 1979, and the opposite appears to be coming true. For today, it is Government that is being watched—watched to the extent that every bureaucrat senses that someone is looking over his shoulder on every memo that he writes.

Thanks to the Freedom of Information Act, someone is watching as that memo is written—or, more correctly, someone other than the intended recipient may be reading that memo. The Freedom of Information Act was first passed in 1966 and greatly strengthened in 1974. It has opened thousands of Government file cabinets to public scrutiny.

This act set up FOI offices in all Government agencies, and was designed to make the release of records the rule rather than the exception. It also required the Federal Government to justify the denial of any record. In addition, it gave individuals the right to go to court, if necessary, to obtain records.

The 1974 amendments required Federal agencies to publish rules stating how FOI requests would be serviced, established uniform fees for search and reproduction of paperwork, and set a 10-day time period in which agencies must decide whether to release or deny the record. These amendments also modified some of the original exemptions and encouraged agencies to release portions of a record after nonreleasable information was deleted.

The exceptions or grounds for denial include: internal operating rules of agencies; information protected by statutes, including trade secrets and financial data from private businesses obtained by FDA in carrying out its statutory obligations; trade secrets obtained under explicit or implicit pledges of confidentiality; intra-Agency and inter-Agency memorandums and letters that precede adoption of an official position; personel and medical files whose disclosure would constitute an invasion of privacy under law; and information that is part of an investigation for law enforcement purposes.

Agencies are required to submit an annual FOI Report to Congress under the amendments. These reports list, among other details, the number of requests denied, the names of persons responsible for those decisions, and how many denials were appealed. FOI reports also cite any disciplinary action taken against agency officials who withheld releasable information, and copies of fee schedules and agency rules relating to FOI.

The public benefits as news media people and public interest groups have gained access to previously denied materials. The law makes it possible for Government officials to be held more accountable for their actions.

Individual citizens may benefit, too. All they have to know is what to ask for. In most cases, their requests will be honored.

However, a funny thing happened on the way to unlocking the file cabinet. John Q. Public and his representatives found that in many cases they are surrounded by other information seekers. At some agencies—such as FDA—most of the information seekers are from businesses that are regulated by the agency. And, as might be expected, much of the information being sought is about rival companies.

For instance, FDA handled more than 32,000 FOI requests in 1978. Just 11 percent of those were from private individuals, the news media, and public interest groups. On the other hand 78 percent—or some 26,000 requests—came from industries or from FOI service companies that specialize in providing information to business and industry. Another 8 percent of the requests were from lawyers, many representing businesses, while 2 percent came from hospitals and health organizations.

Most FOI requests to FDA seek information on drugs, foods, and medical devices, or administrative records, compliance manuals, and other guidelines used by the Agency to regulate manufacturers. In addition, many companies want to know how their competitors fared in FDA plant inspections. So, they request inspection records, which detail violations of FDA Good Manufacturing Practice Regulations. These reports also list insanitary conditions found by FDA inspectors and any seizure or recall of defective products.

In 1978, only one other Government organization received more FOI requests than FDA. The Department of Defense, which includes 14 separate components, received nearly 56,500 requests. The Department of HEW was second in total requests with over 51,000, which includes the 32,286 requests handled by FDA. The Social Security Administration was a distant second within HEW with a total of about 7,000 FOI requests. Other agencies that regulate consumer products received nowhere near the number of FOI requests as those made to FDA. The Consumer Product Safety Commission, for example, received only 6,800 requests, while the Environmental Protection Agency totaled about 2,000.

FDA's workload under the Freedom of Information Act continues to grow each year. In 1974, before the Agency issued final FOI regulations, there were only about 13,000 requests. That number is expected to nearly triple this year. FDA also increased its FOI Headquarters staff in Rockville, Md., from 20 to 28 employees and placed an FOI officer in each of its six bureaus. In addition, the Agency has designated an individual in each of its 33 regional and district offices to coordinate and answer requests sent from headquarters.

## SAMPLE REQUEST LETTER

Dr. Donald Kennedy
Commissioner
Food and Drug Administration
5600 Fishers Lane
Rockville, Md. 20857

Re: Freedom of Information Act Request

Dear Dr. Kennedy:

    Under the provisions of the Freedom of Information Act, 5 U.S.C. 552, I am requesting access to (identify the records as clearly and specifically as possible).

    If there are any fees for searching for, or copying, the records I have requested, please inform me before you fill the request. (Or: . . . please supply the records without informing me if the fees do not exceed $_____.)

(Optional) I am requesting this information (state the reason for your request if you think it will assist you in obtaining the information.)

(Optional) As you know, the act permits you to reduce or waive fees when the release of the information is considered as "primarily benefiting the public". I believe that this request fits that category and I therefore ask that you waive any fees.

    If all or any part of this request is denied, please cite the specific exemption(s) which you think justifies your refusal to release the information, and inform me of the appeal procedures available to me under the law.

    I would appreciate your handling this request as quickly as possible, and I look forward to hearing from you within 10 days, as the law stipulates.

                                Sincerely,

                                  John Q. Public
                                Anytown, U.S.A.   Zip Code

Consumers who want information from FDA under the FOI Act can use this format to assure their requests are processed quickly and efficiently.

---

The additional personnel and equipment needed to process these requests, as well as day-to-day operating expenses, cost the Agency about $2.4 million last year. The Agency, in turn, received and turned over to the U.S. Treasury about $158,000 in fees charged to requesters as provided by the '74 amendments, which established a system of nominal fees for search, reproduction of paperwork, and postage. All or part of these fees can be waived by FOI if the information requested is considered primarily beneficial to the public. The Agency also is considering raising the fees to more nearly reflect the actual costs incurred in processing the requests.

Despite its high volume of FOI requests, FDA has one of the lowest denial rates in Government. In 1978, FDA denied less than 2 percent of all requests on the grounds that the information requested dealt with trade secrets of companies and opened investigatory files. Denial rates for other regulatory agencies such as the Environmental Protection Agency, the Department of Energy, the Federal Trade Commission, and the Occupational Safety and Health Administration ranged from 7 to 15 percent.

The majority of FOI requests are mailed directly to FDA Headquarters in Rockville. Once an FOI request is received by FOI personnel it is entered into a daily log and given a control number. This number and pertinent information about the request, including the date received, the type of information desired, and the action taken by the Agency, are fed into a computer. A daily printout of this information is also made. These measures allow the Agency to keep track of the current status of each request.

Of the requests received by FDA from or on behalf of businesses, nearly a quarter seek research data and other scientific information on new drugs that have been submitted to FDA for evaluation. A drug company, for example, might want data on the safety and effectiveness of a new drug which has been developed by another drug firm and clinically tested.

The FOI office processes the request and sends it to New Drug Evaluation in the Bureau of Drugs. From there it is sent to the proper division within the Bureau for evaluation. A request for safety and effectiveness data on a new drug to treat certain infections, for example, would be forwarded to the Division of Anti-Infective Drug Products.

This type of information is available to the public upon request to FDA, providing the drug has been approved by the Agency. If it still is being evaluated for approval the data is not releasable. All denials are reviewed by the FOI office and then by the FDA General Counsel's office.

If a denial is upheld then the Agency officially informs the requester in writing that the information is not releasable and cites the reasons. All denials can be appealed to the Department of Health, Education, and Welfare.

Despite extensive use of the FOI law for corporate prying and the added expense to Government agencies for processing requests, few people have advocated its repeal. The act has achieved its primary goal of opening the majority of Government files to public scrutiny. It allows citizens the opportunity to double check—in most cases—the "inner workings" of Big Brother.

*James Greene is a staff writer with FDA's Office of Public Affairs.*

### The long and winding road
# How a bill becomes a law...

The light on the clock blinks as the buzzers begin to ring. Representatives leave their business in mid-air as they move from their offices to vote on a bill.

The bill — after much nurturing and politicking — has almost reached its final stage. It is the rare piece of legislation that gets this far. If it does manage to pass one chamber of Congress it still has to survive the rigors of the other chamber, as well as the conference committee, and presidential signature. In the 96th Congress, for instance, more than 14,000 bills were introduced; only 736 of these ever became law — with 123 of these being private laws affecting only a handful of constituents. The remaining bills died a number of deaths, raising the question, "What characteristics must a bill have to become a law?"

**1** Introduction of a bill in the House and Senate is a simple procedure. In the House, members just drop their bills into the "hopper," a mahogany box next to the clerk's desk at the front of the chamber. In the Senate, members generally submit their proposals and accompanying statements to clerks in the Senate chamber, or they may introduce their bills from the floor. House and Senate bills are printed and made available to all members.

One measure of a bill's chance of passage is the number of sponsors and cosponsors it has. A Representative will often collect signatures of other members before introducing a bill. The signatures show committee members the extent of the bill's general support.

**2** After a bill is introduced it is referred to the appropriate committee, where it must pass its first test. It is carefully scrutinized for its merit, political viability, degree of controversy, effect on previously enacted legislation, presidential and party approval, and public awareness or support. During the 96th Congress only 1,492 of the originally introduced bills made it through House or Senate committees. (Not all of the other bills disappeared, since some became amendments to other bills voted out of committee.)

The key figure in the committee is the chairperson. Although the power of this position has been tempered by recent congressional reforms, the chairperson still makes final decisions on whether to keep a bill in full committee or send it to subcommittee, to schedule public hearings, schedule a mark-up, or ignore it altogether. These decisions are usually based upon the chairperson's interest in a bill and upon the bargaining power of the bill's promoters. If the bill's proponents are successful, public hearings will be scheduled, moving the bill one step closer to the floor.

For example, in the spring of 1980 — after years of prompting by consumer groups — the trucking deregulation measure was finally given committee attention. "When Sen. Howard Cannon, chairman of the Science, Commerce, and Transportation Committee, came out in favor of a strong bill its chance for passage increased 100 percent," said Nancy Drabble, Director of Congress Watch. "Prior to his approval many Commerce Committee members were not very sympathetic to deregulation, primarily because they were worried about the effect of the legislation on small town service. Cannon's stand helped make up a lot of minds."

The purpose of hearings is to present committee members with supporting and opposing arguments for a bill. Hearings can also become an arena for political battles, drawing media attention to a bill. At this stage, lobbyists and constituents play a vital role in shaping legislation, by influencing the selection of people who testify at a hearing. The schedule of witnesses is often biased towards the opinions of committee staff. Constituents can urge their Representative to select someone who will testify in their interests.

**3** A bill is then scheduled for committee mark-up. (Not all bills that have weathered hearings reach this stage.) During the mark-up members discuss, negotiate, and eventually vote on the final language of the bill. Sometimes, a compromise acceptable to a majority of the committee members is crafted through amendments to the bill. At this stage there is a lot of backroom politicking by staff members, outside lobbyists, and the Representatives themselves. A strong amendment can create major changes in the intent of the original bill. An entire mark-up often is spread out over several weeks. Finally, there is a vote to report the bill for floor consideration or to table it — postponing it indefinitely.

The House places great weight on the recommendations of the committee, regarding its members as authorities because of the time spent in hearings and mark-up. If a bill is reported out of committee with strong bipartisan support, it often means smooth sailing on the floor of the House. If not, the same controversies that are unresolved in committee are likely to be raised during floor debate.

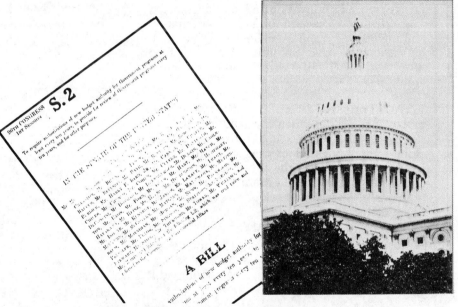

After the committee votes to report a bill to the floor of the House it must publish a Committee Report. The Report states the purpose of the bill and the rationale behind it. Committee members who oppose the bill often file dissenting views in the Report. Non-committee members often rely on this report as the major source of information on the bill. Any citizen can obtain the report by writing to his or her Representative.

**4** When a bill is reported out of committee it does not move immediately to the floor. First it is assigned to one of a variety of calendars. The *Private Calendar* is for bills concerning individual constituent problems. The *Consent Calendar* is for noncontroversial measures. The *Union Calendar* is for revenue and appropriation bills. And the *House Calendar* is for all other major legislation.

In addition, a few bills can bypass the calendar altogether. Under the suspension calendar the House by a two-thirds vote can suspend the traditional procedures to consider noncontroversial measures. These bills go directly to the floor. "In the closing days of the session," noted Drabble, "everything from the seed patent bill to a controversial auto safety measure may be thrown on the suspension calendar to force a vote. They are not always noncontroversial measures."

**5** Once a bill reaches the floor, each party has a floor manager from the committee that reported the bill. They control the time allotted to each member for asking questions or making statements, and decide to accept or oppose amendments. Amendments, besides affecting the substance of the bill, can be used to stall its consideration.

**6** The final vote is the moment of reckoning. Most votes are taken in one of two ways. In the *voice votes* members call out aye or nay. In this case the *Congressional Record* does not list how individual members voted. This type of vote can be used when members want to dodge reprisals from their constituents on particularly controversial measures such as the pay raise.

A *recorded vote* is taken on one member's request if supported by 44 other members. In this case, Representatives use electronic cards to record both their name and how they are voting.

**7** Before a bill goes to the President for signature into law it must pass both chambers in identical form. If the House and the Senate pass different bills they must iron out the differences in a conference committee. When the final bill is voted out of conference it goes to each chamber for another vote. At this point, it must be approved or rejected as a whole.

**8** The last hurdle before a bill becomes a law is presidential approval. He can sign it or let it stand for ten days while Congress is in session, and it becomes a law.

The simple act of introducing a bill sets off a complex and variable series of events that may or may not result in its final passage in Congress. Most bills follow a path in which the steps are fairly predictable, governed by rules and convention. The outcome of the process, on the other hand, is almost always unpredictable. ∎

*For more information on how a bill becomes a law, write your Representative for a free copy of* How Our Laws Are Made.

# Forming a Telephone Tree

The telephone tree can be one of an activist's most useful tools. The purpose of the tree is to get information out quickly. It can provide the impetus for its members to act on issues they may have learned of through the mails. Or when activists can't wait for the mail because of the unpredictability of the legislative process, the tree can be used to spread vital facts.

If you already have a phone contact system for alerting your members about meetings or actions, consider using it to generate letters, telegrams, or phone calls to your Representative in Washington. If you don't, talk about creating a telephone tree at your next meeting.

A tree usually consists of a lead coordinator (the "trunk"), several branch coordinators ("branches"), and callers ("leaves"). An effective tree should have one lead coordinator, 5-7 "branches", and each branch should have 5-7 "leaves".

1. ACTIVATING THE TREE: The tree moves into action when the lead coordinator receives information that must be passed along and phones all of her or his branches. Each branch phones the message to each of his or her leaves, and all members of the tree then write, telegram or telephone the message to their elected officials.

2. WRITE DOWN THE MESSAGE: It is important that all along the line people write the message down and read it back to the person calling them so that it gets communicated accurately. If the message deals with legislation, it should include the bill name and number, the Representative to be contacted, his or her address and phone number, and the request or statement to be made. Always suggest that the calls, letters, or telegrams be dispatched as soon as possible unless you are following a specific timetable.

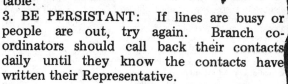

3. BE PERSISTANT: If lines are busy or people are out, try again. Branch coordinators should call back their contacts daily until they know the contacts have written their Representative.

4. REVERSE FLOW OF CALLS: Everyone should be encouraged to call their coordinator if they have any question about issues or the groundrules of using the tree. Branch coordinators can then get feedback from the lead coordinators, and the lead coordinators from the branches.

5. KEEP THE TREE IN GEAR: To be effective, a strong, well-organized telephone tree must be maintained all year round. Such maintainance is quite a challenge given the unpredictable nature of politics. There can be a lull in activity for several months and then a sudden burst of legislative alerts within a 4 week period. The network must be ready to respond at any time to alert activists to action, but also be prepared for periods of rest.

6. ADDING TO THE TREE: Continually seek to expand the leaves, branches and trunk of your tree. Each time an individual expresses a good deal of interest in your work, ask if they want to become a part. Make sure, however, that they will respond to your calls. A tree of only mildly interested people will cause an enormous amount of work and minimal results.

*For a more detailed pamphlet on telephone trees, contact Congress Watch.

# How to write your Representative

Writing to your Representative can have an impact — there are letters that electrify, that counter information, that change votes. And on important issues, many Members have their clerks count their mail the way geologists read seismographs.

Depending on the issue and the timing, alternative forms of communication — e.g. telephone calls, telegrams, postcards — may be more appropriate. But in general, the best way to get the attention, and the vote, of your Representative is through a well-written letter. Below are some suggestions derived from an article by Representative Morris Udall, on how to write such a letter. For more detailed suggestions, ask Rep. Udall for a copy of "The Right to Write."

Address the letter properly: "Hon.          , House of Representatives, Washington, D. C. 20515" or "Senator          , U. S. Senate, Washington, D. C. 20510"

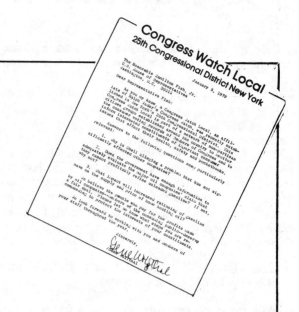

Identify the bill or issue you are writing about. Some 20,000 bills are introduced in each Congress, so it is important to be specific. When writing about a bill try to use the bill number or describe it by popular title (e.g., "public financing" or "clean air").

Write your letter in time to make a difference. Try and write your Representative while the bill in question is still in committee or has not yet been to the floor for a vote. Then your message will be timely and have a greater effect.

Be reasonably brief. Most Representatives get over one hundred pieces of mail a day. Your opinions and arguments stand a better chance of being read if they are stated as concisely as possible.

Give your reasons for taking a stand. Do not simply say "Vote for HR 1" or "I oppose decontrol of oil prices". Explain the concrete effects you believe a certain vote will have and use facts and figures to back up your argument.

Be Constructive. If you believe the bill is the wrong approach to a given problem, outline what you feel the correct approach would be.

Personalize your letters. Representatives will usually be aware of what various lobbying groups are saying, but they also need to hear about *your* experiences and observations; what the bill will do to and for you. They may also not be fully aware of new conditions and developments in your state because they spend such a major portion of their time in Washington. A sincere, well thought-out letter can help fill in some of the gaps for them.

Ask some specific questions. You will get a more focused response if you ask for a specific answer or opinion.

*These "how-to" articles are reprinted from* The Congress Watcher, *a bi-monthly publication of Public Citizen's Congress Watch. For a free copy write Comps, 215 Pennsylvania Ave., SE, Washington, DC 20003.*

# How To Do A Candidates Forum

One of the most valuable ways to educate citizens about current national issues this season is to host a "Candidate's Forum." These are face-to-face meetings between candidates for elected office in your area. (Another excellent tool—preparing a profile of candidates—will be discussed in the next issue.)

The first step in planning the forum is to assess the whys, wheres, whens, etc. The League of Women Voters' *Voters Service/Citizens Information Handbook* includes an excellent list of suggestions that can be used to help plan a candidates forum.* Key issues for consideration include:

- Can you get the needed help from volunteers to do the work involved?

- Will your community turn out for a candidates meeting?

- Can you get the cooperation of the political parties?

- Can you get the cooperation of the candidates? Will date and place affect their response?

- Are there other organizations that will cooperate with or assist you?

- Will the media cover your meeting? Do you want the radio/TV to cover the meeting live?

- Do you want to do a candidate meeting before the primary? Before the general election? Both?

If you've answered these questions and are still inclined to have a candidates forum, the next step is to begin planning. It can be useful to reach out to as many supportive groups as possible--e.g., labor, senior citizen, neighborhood organizations--to be involved in the planning process. At the meeting, share responsibilities and decisionmaking, and divide activities. At least three committees can be formed: *logistics*, in charge of determining place, time, date, equipment needed; *program*, in charge of the issues, content, candidates chosen, invitations, format, time limitations, moderator and audience participation; and *publicity*, in charge of getting the word out to organizations, press, schools, libraries, etc.

Once the responsibilities are assigned, the logistics committee must determine a place and time, and the program committee must invite the candidates. The most efficient way of doing this is to present candidates with several dates and have them give preferences as to which is the most convenient for them. Make sure to clear dates with all candidates and parties, (this may encourage candidate participation) and do your best to reserve a facility as far in advance as possible. Timing for the event can be critical in gaining an acceptance from candidates and in ensuring citizen interest. An event no more than one to two weeks before the election is usually preferable.

When you have received answers from your candidates, you can assess what to do next. If you've had *yes* responses, you can proceed-- if not, don't give up. Talk with the political parties in your district, influential people in the community, the press--see if you can find persuasive people to assist in getting a yes answer. A news story concerning the negative responses may help.

At the forum, all candidates present their views on consumer issues in short introductory speeches, and answer questions from a panel of three to four experts on consumer issues. Panel questions should focus on major issue areas such as health care, energy, taxes and food. Each panelist can ask the candidates one question to which they can respond. (It's helpful to limit the time for responses e.g., two minutes for each.) If you have only two or three candidates present, you might want to give them the opportunity to respond briefly to each other, to instill a debate flavor. The panelist may follow up on the response briefly.

Ample time should be left for general questions from the audience.

*For more detailed information, write for: League of Women Voters Education Fund. *Making a Difference: A Voters Service/Citizens Information Handbook, 1978.*

# Profiling your representative

One way to gain influence with your Representative is by holding a candidate's forum, another is by doing a profile.

Profiles are not difficult to write. In fact, much of the material needed is readily available; all you have to do is compile it.

The most crucial information to include is factual.

**1.** Get a brief appraisal of the background and voting record of the Representative. An excellent text to consult is the *Almanac of American Poltics 1980*. Since there are over 1,000 votes taken during each Congress, it is important to single out the crucial votes affecting the public interest. A list of key consumer votes is published annually by Congress Watch.

More information about the Representative's voting record should be available from either his/her Washington, D.C. or district office. In addition, the office will usually have press clippings by or about the Representative. Besides evaluating the willingness of the office to disseminate information about the Representative in your profile, you may want to include an evaluation of the quality of the material—particularly the newsletters (e.g., are they self-serving? objective? informative?).

For information on the finances of the Representative, call or write the Federal Election Commission, 1325 K St., N.W., Washington, D.C. 20463 (202) 523—4068. This information is also on file at each state capitol. Minimally, a profile would want to list the major contributors (e.g., the "political action committees (PACs) that gave over $500). A more complete analysis might also link major special interest group contributions and votes on key issues (e.g., compare oil industry contributions with votes on the windfall profits tax).

A free copy of the *Financial Disclosure Reports of Members of the U.S. House* from January 1978 to December 1978 is available from the Office of the Clerk of the House, Room H—105, The Capitol, Washington, D.C. 20515. The local library may also have a copy on hand. The recent *Financial Disclosure Reports* are not yet bound but can be obtained for the cost of copying from the Office of Records and Registration, 1036 Longworth House Office Building, Washington, D.C. 20515. The reports include salaries, outside income, holdings and honoraria of the Representative.

The January issue of the *Congressional Quarterly* (Vol. 38, No. 2, p. 99) lists voting participation percentages for each Representative. The list will tell you what percentage of the time the Representative was present for votes. *Congressional Quarterly* (Vol. 38, No. 3), also compiles statistics on how often a Representative voted with her/his party on partisan issues: this can be a useful indication of the outlook of the Member.

Many groups rate the Representative on key votes and publish percentage ratings. It is important to include ratings by a variety of groups to give a more accurate picture of how he/she functions as a legislator.

**2.** For insights into the Representative's performance in Washington, you should interview congressional aides. You can also consult the staff of your local newspaper covering the Representative. Often seasoned observers of the political scene, reporters can either give you the information you are seeking or suggest names of other knowledgeable people.

You also might interview key local activists and community leaders for their impressions of the Representative's performance. An interview with the Member should be sought. It should be entered with a set of well-prepared questions and either a tape-recorder or a friend to help you take notes. A list of sample questions is included in the brochure entitled *How to Write a Profile*, available at Congress Watch.

**3.** Once the profile is researched, written, and edited, it is ready for publication and distribution. Send copies to the local media requesting a press conference. Also distribute copies to citizen groups, unions, and community organizations, churches and any other local institutions that might be interested.

—Allouette Kluge

*For a detailed manual on* How to Write a Profile, *write Congress Watch, 215 Pennsylvania Ave., S.E., Washington, D.C. 20003.*

**Graphics by Linda Cahill**

# A Partial Listing of Summaries of Reports Dealing with Food-Related Issues Available Without Charge from the General Accounting Office Washington, D.C. 20548

**Problems in Preventing the Marketing of Raw Meat and Poultry Containing Potentially Harmful Residues #HRD-79-10**  4/17/79

GAO estimates that 14 percent by dressed weight of the meat and poultry sampled by the Department of Agriculture between 1974 and 1976 contained illegal and potentially harmful residues of animal drugs, pesticides, or environmental contaminants. Many of these substances are known to cause or are suspected of causing cancer, birth defects, or other toxic effects.

Action taken by the Food and Drug Administration, the Environmental Protection Agency, and the Department of Agriculture to protect consumers from illegal and potentially harmful residues have not been effective because:

—the extent of public exposure to illegal residues has not been accurately measured.
—Contaminated meat and poultry are generally marketed before the violation is discovered and some cannot be recalled.
—Efforts to prevent future shipments of meat and poultry containing illegal residues have been inadequate.

**Food Salvage Industry Should Be Prevented From Selling Unfit and Misbranded Food to the Public #HRD-79-32**  2/14/79

Potentially adulterated food in dirty, rusty, swollen, and severely dented cans or torn packages is being sold to the public and to health care facilities. Product labels are often missing or incomplete. These conditions in the food industry are the same as those GAO reported in 1975.

For the most part, recommendations made in GAO's earlier report have not been carried out. The Food and Drug Administration and the Department of Agriculture believe that regulation of this industry is a relatively low priority because there is not a significant health risk to consumers of salvaged food products.

**Department of Agriculture's Beef Grading: Accuracy and Uniformity Need to Be Improved #CED-78-141**  7/21/81

The Department of Agriculture's beef grading program serves as a basis for price quotations among feeders, packers, suppliers, retailers, and others along the marketing chain. GAO found that grading was not always accurate or consistent from one section of the country to another. The Department needs to: increase research efforts to develop instruments to accurately measure beef carcass characteristics; establish a grading accuracy standard; improve its management of the program; and resolve questions about the adequacy and usefulness of the current beef grade standards.

**Improving Sanitation and Federal Inspection at Slaughter Plants: How to Get Better Results for the Inspection Dollar #CED-78-141**  7/30/81

Excerpted in this publication.

**Need For More Effective Legislation of Direct Additives To Food #HRD-80-90**  8/14/80

The Federal Food, Drug and Cosmetic Act requires that the safety of direct food additives be based on scientific evidence and that the evidence be reviewed and approved by the Food and Drug Administration. However, the Act exempts from review and approval substances generally recognized as safe by "experts" or approved for use before 1958 and allows the safety determination of some of those substances to be based on experience drawn from common use in food. The safety of several of these exempted substances, including saccharin, cyclamates, and nitrite, has been questioned.

GAO recommends that the Congress amend the law to eliminate the exemptions and that the Secretary of Health and Human Services publish regulations establishing criteria and guidelines for assessing the safety of additives. Regulations listing substances affirmed as generally recognized as safe should be revised to indicate the kinds of evidence that support their safety.

**Further Federal Action Needed to Detect and Control Environmental Contamination of Food #CED-81-19**                                                                                                                         12/31/80

The Food Safety and Quality Service and the Food and Drug Administration took much longer than necessary to identify the presence and source of chemical contamination involving polychlorinated biphenyls in the Western States in 1979. PCBs leaked into the drainage system of a hog slaughter plant and eventually were processed into grease and animal feed. These products and the animls fed on them were ultimately distributed to 19 states, Canada, and Japan, in the form of contaminated chickens, eggs, and even dessert products. These agencies and the Environmental Protection Agency have taken several actions to prevent similar incidents or to deal more effectively with such incidents if they occur in the future.

The agencies have implemented many but not all of GAO's previous recommendations concerning the control of chemicals in the nation's food supply and in the environment. Additional action, which the Office of Technology Assessment has suggested, is needed to clearly define which agency will assume the leadership role in future contamination situations.

**Food Waste: An Opportunity To Improve Resource Use #CED-77-118**                                      9/16/77

About one-fifth of all food produced for human consumption is lost annually in the United States. In the world context of rising population, uncertain weather, and concern with the availability of resources, every opportunity should be taken to improve food system management in this country. More attention should be directed at the causes of food waste, new management technology for reducing loss, and improvement of consumer understanding.

# PRINCIPAL LAWS GOVERNING FOOD SAFETY

Below are the principal federal laws governing the safety of the food supply. Under these laws there are regulations and standards that are much more detailed. The federal agencies, your member of Congress, or some local law libraries should be able to provide access to these laws and regulations or, within reason, with copies of the sections you want. *It should be noted* that what the laws say should be done is very often quite far from what the agency or department is able or willing to implement. Also note that there are state and local laws governing food safety and inspection right down to the retail level of restaurants and food stores.

A number of these federal agencies could use still more legislative authority. This is especially true of the FDA. Under current law, food processors have no obligation to notify FDA when products subject to their control have become adulterated, or when they recall products because they believe the products may be in violation of the law. Food processors are not required to permit FDA to inspect food processing records that may bear on whether the products are adulterated or mislabeled. FDA does not have subpoena power, the authority to compel the production of documents or the testimony of witnesses, and it has no embargo or detainment authority. Many weeks may pass between the time of an inspection of a food processing facility and the filing in court of papers to effect a seizure or obtain an injunction concerning violative products; in the meantime, the firm holding the violative products may ship them into the stream of commerce.

1. The Federal Food, Drug and Cosmetic Act, enacted in 1938 and amended in 1958 and 1960, was passed to protect the public from potential cancer-causing chemical food additives. Under the law all food dyes are supposed to be proven safe before they are used. Also under the law, the FDA has the authority to preserve the safety of the nation's foods by prosecution, seizure, injunction and recalls. The FDA also regulates the use of animal drugs and intentional additives in food and sets tolerances or action levels for contaminants that are unavoidably present in food or feeds.

2. The pesticide tolerance setting functions, governed by the Federal Food, Drug and Cosmetic Act, were transferred in 1970 to the newly created Environmental Protection Agency. The pesticide registration system, administered under the Federal Insecticide, Fungicide, and Rodenticide Act of 1947, currently being reviewed, was moved from the U.S. Department of Agriculture to the Environmental Protection Agency in the same year.

3. The Federal Meat Inspection Act (1906), the Poultry Products Inspection Act (1957), and the Wholesome Meat Act (1967) require that livestock and poultry slaughtered at plants that do business in interstate or foreign commerce be federally inspected. Federal inspection is also required at slaughter plants that do intrastate business in states not having their own inspection program. These laws require that federally inspected meat and poultry plants operate in a sanitary manner and that the products they sell be *wholesome, unadulterated,* and *properly labeled.* Plant managers are primarily responsible for meeting these requirements. The Food Safety and Inspection Service of the U.S. Department of Agriculture is responsible for meat and poultry inspection activites. The FSIS devises ways to ensure that harmful chemical residues are not present in meat, poultry, and egg products.

4. The Safe Drinking Water Act of 1974, as amended, requires the Environmental Protection Agency to establish federal standards for drinking water, protect underground sources of drinking water, and establish a joint federal-state system for assuring compliance with the resultant regulations. The authorization of the SDWA expires in September 1982, and will be reviewed at that time.

5. The Agricultural Marketing Act of 1946 provides for the development and promulgation of grade standards, and the inspection and certification of transportation and processing facilities for fish and shellfish and all processed fish products. The authority for these regulations was transferred to the U.S. Department of Commerce in 1971.

# Conclusions and Recommendations of
## *"Cancer-Causing Chemicals in Food"* —
A Report by the House Subcommittee on Oversight and Investigations,
Committee on Interstate and Foreign Commerce,
December 1978

The Subcommittee on Oversight and Investigations concludes that the federal programs designed to protect the American public from toxic chemicals in food, which are administered jointly by the United States Department of Agriculture (USDA), the Environmental Protection Agency (EPA), and the Food and Drug Administration (FDA), are ineffective. In some cases, existing laws need amending. However, in many more instances, the EPA, FDA and USDA are failing to carry out the vital responsibilities invested in them to protect the public from dangerous chemical residues in food. These failures are briefly summarized here:

**EPA:** The federal chemical residue monitoring system hinges on a strong pesticide tolerance setting program. However, the Environmental Protection Agency's system for setting tolerances (safe, legal limits of chemical residues that may be found in specific foods) is outdated, ineffectual, and showing few signs of improvement. EPA's Office of Pesticide Programs is veering away from the health-oriented language of the Federal Food, Drug, and Cosmetic Act, which administers the tolerance setting program, toward the "risk-benefit" balancing language of the Federal Insecticide, Fungicide and Rodenticide Act, under which pesticides are registered for use. In so doing, the EPA is putting the public at greater risk. Many tolerances remain in effect for pesticides that are known to be suspect carcinogens. Scores of other tolerances are for chemicals that have never been tested for carcinogenicity or other equally serious effects.

**USDA:** The United States Department of Agriculture, which is responsible for monitoring meat and poultry for residues of chemicals such as pesticides and animal drugs, is doing a poor job of finding residue violations and of preventing (in conjunction with the Food and Drug Administration) the marketing of contaminated meat. The Department's two residue monitoring programs (on the farm and in the slaughterhouse) are seriously ineffective. The "in slaughterhouse" monitoring program tests few animals and doesn't look for many harmful chemicals known to occur in meat and poultry. The "retest" program, which enjoins growers suspected of marketing violative livestock to submit tissue samples for laboratory analysis, is easily avoided by farmers. Consequently, much of the meat and poultry consumed in this country may be contaminated. The USDA inspection stamp goes on all produce that has passed inspection for *visible* signs of health and cleanliness. The stamp is no guarantee that meat is free of chemical contaminants.

**FDA:** The Food and Drug Administration's role in the regulation of toxic substances in food includes monitoring agricultural produce (other than meat and poultry) for residues of pesticides and environmental contaminants, investigating and prosecuting cases of residue violations in meat and poultry reported to FDA by USDA, and sharing responsibility, with USDA, for keeping contaminated produce out of interstate commerce. FDA's chemical monitoring program has dangerous shortcomings. Many chemicals occurring in food that are known to be suspect carcinogens or may potentially cause birth defects and genetic mutations are not monitored. FDA investigates few of the residue violations reported to it by USDA and rarely prosecutes violators. Combined USDA-FDA programs to remove contaminated meat from the marketplace almost never result in meat or poultry recalls.

General summaries of Subcommittee recommendations to the three agencies are as follows:

**EPA should** overhaul its tolerance setting program. Existing pesticide tolerances should be reassessed. Those lacking safety test and residue data should be revoked unless registrants produce required data before January of 1981.

Tolerances for substances suspected of causing cancer, reproductive problems, birth defects and genetic mutations should be revoked, no later than July of 1981. EPA should discontinue its practice of issuing "interim tolerances" for pesticides which have not been proven safe, or which are suspected of being safety hazards, immediately.

Tolerance exemptions that have been granted over the years should be reassessed. Many tolerance-exempt pesticide ingredients are suspect carcinogens. The practice of exempting all "inert" ingredients from the tolerance-setting requirement should be curtailed by January of 1980 since some inerts are more highly toxic than active pesticides.

Tolerance-setting methodology should be brought up to date with current scientific knowledge. EPA's method of computing the "Food Factor" should be based on demographic information about food consumption.

EPA should develop a system for pesticide safety testing which removes testing from the manufacturer's own labs and places it in the hands of independent, impartial laboratories. Such a system could include Federal licensing of independent laboratories and Federal monitoring of these laboratories to assure objectivity.

Positions devoted entirely to reassessment and overhaul of the pesticide tolerance setting system should be created and filled within EPA's Office of Pesticide Programs.

**USDA should** allocate additional resources and personnel to its chemical residue monitoring programs. Rapid test procedures for chemicals known to occur as residues in meat and poultry should be developed on a high priority basis. The percentage of animals randomly tested for chemical residues in slaughterhouses should be increased, and where "multi-residue" test methods do not exist, USDA should use "single-residue" methods selectively.

A national identification system for livestock should be developed so that growers will not be able to evade USDA's "pretest" chemical residue monitoring program.

**FDA should** expand its residue monitoring program (for foods other than meat and poultry) substantially. Chemical detection methods should be improved to include more substances and require less time to perform. Where "multi-residue" tests are lacking, "single-residue" methods should be used on a selective basis.

Chemical residue monitoring should be based on a knowledge of where particular residues are likely to occur. Residue violation findings should be backed up by strong efforts to remove contaminated produce from the market.

FDA should improve its "market basket" survey, based on knowledge of consumption habits and chemical usage.

FDA should continue its efforts to integrate state-federal chemical monitoring efforts.

The Subcommittee also recommends that the U.S. Congress take action to improve the federal system for monitoring toxic pesticides in food. Recommendations to Congress are summarized here as follows:

**Congress should** consider passing legislation making it unlawful to use potentially carcinogenic, teratogenic or mutagenic pesticides on agricultural commodities, unless such substances degrade and leave no harmful metabolites in food.

Congress should consider livestock quarantine legislation, giving the United States Department of Agriculture the authority to detain from slaughter all animals suspected of containing violative chemical residues, pending laboratory test results.

Congress should give the Food and Drug Administration the authority to impose civil penalties on growers who send livestock contaminated with residues of animal drugs or pesticides to slaughter.

Overall, the Subcommittee recommends that all three agencies responsible for protecting the public against chemically contaminated food, the EPA, FDA and USDA, strive for increasingly effective programs, improved research and development, greater inter-agency coordination, and a heightened awareness of the need to protect the public from unnecessary exposure to harmful chemicals in food.

# Key Government Officials in Food Safety

**John Block,** as *Secretary of U.S. Department of Agriculture,* has final authority over all USDA issues. Block was formerly the Director of Agriculture in Illinois, where he owns a 3,000-acre farm. Nutrition and anti-hunger lobbyists fear Block will be insensitive to the needs of their constituencies, at least partly due to his lack of expertise. In terms of international trade, he sees food, at least in part, as a "weapon, but the best way to use that is to tie countries to us. That way they'll be far more reluctant to upset us."

**Richard Lyng,** *Assistant Secretary of USDA,* has been called a "street-smart professional" and an "excellent administrator-type." He was previously deputy director and director of the California State Department of Agriculture. He is also a former president of the American Meat Institute, a group strongly opposed to government regulation of the meat industry. Lyng is a strong opponent of the Delaney Clause, maintaining that if the U.S. pursues a "zero-risk" food policy, it will end up with "zero-food."

**C.W. McMillan** is *Assistant Secretary of Agriculture* in charge of Marketing and Inspection Services. For 22 years, McMillan was the executive vice-president of the National Cattleman's Association which, as one staffer of the House Agriculture Committee noted, "tends to look upon federal regulations on meat the way the National Rifle Association looks upon gun control." His responsibilities include supervision of the Food Safety and Inspection Service (FSIS), and he is also chief administrator of the Agricultural Marketing Service, which administers standardization, grading, voluntary and mandatory inspection, marketing orders, regulatory, and related programs.

**Arthur Hull Hayes, Jr.,** *Commissioner of the Food and Drug Administration,* is chief policy maker in FDA matters concerning food, drugs, radiological health, medical devices, and toxicological research. He is recognized as one of the foremost clinical pharmacologists in the U.S. He is also an authority on hypertension (high blood pressure), a condition aggravated by diets high in sodium content; nonetheless, Hayes has said he favors only voluntary labeling of salt content in food products. He has suggested he would back changes in the Delaney Clause to allow for the use of cost/benefit analysis.

**Anne Gorsuch** is *Administrator of the Environmental Protection Agency,* a position she calls the "hottest spot" in domestic politics. Nicknamed by some as the "Dragon Lady" for her reputation of refusal to compromise or negotiate, she has generally followed an anti-regulatory line. She has proposed enormous budget cuts for 1983 that many maintain will cripple EPA's ability to enforce the nation's environmental laws . Enforcement actions have been running at about a rate of one-eighth of that of the total pursued by previous administrations. Regarding water safety, Rep. Toby Moffett wrote to her: "Your lack of action suggests that the Reagan Administration in general and EPA in particular have no real interest in assisting the further protection of. . . groundwater resources."

**Sen. Orrin G. Hatch** (R-Utah) is Chairman of the Senate Committee on Labor and Human Resources. He has introduced S. 1442, the so-called "Food Safety Amendments of 1981," which are supported by the food and soft drink industry. A staunch conservative, he was elected to the Senate in 1976, and faces re-election in 1982.

**Sen. Edward M. Kennedy** (D-Mass.) is the Ranking Minority Member of the Labor and Human Resources Committee. Elected in 1962, his seat is up in 1982. He has a reputation for being one of the hardest working members of the Senate.

**Sen. Jesse Helms** (R-N.C.) is Chairman of the Senate Committee on Agriculture, Nutrition, and Forestry. Generally a proponent of strong right wing views, he believes in almost entirely unregulated free enterprise, though he supports subsidies to the agribusiness and nuclear industries. First elected in 1972, he will be up for re-election in 1984.

**Rep. Albert Gore, Jr.** (D-Tenn.) was elected in 1976 in the Fourth Congressional District of Tennessee. He has recently introduced a Food Safety Bill which, while not as extreme as S. 1442, still redefines "safe" and in general leaves much to be desired.

**Rep. Henry A. Waxman** (D-Cal.) is Chairman of the House Health and Environment Subcommittee. Elected from the 24th District in 1974, he is considered a strong consumer advocate.

**Rep. William C. Wampler** (R-Va.) is the Ranking Minority Member of the House Agriculture Committee. He has been a member of the House of Representatives from the 9th District of Virginia since 1966. Consumer groups rate his voting record on consumer issues a very poor one.

# List of Major Food Corporations

*The following list of major food corporations is taken from* **Everybody's Money Complaint Directory for Consumers.** *The entire book is available for $2.50 from Everybody's Money, Credit Union National Association, Inc., Box 431B, Madison, WI 53701.*

**A & W International, Inc.**
922 Broadway
Santa Monica, CA 90406
Tele: (213) 395-3261
Francis X. Dwyer, President

**American Home Foods**
685 Third Ave.
New York, NY 10017
Tele: (212) 878-6300
David P. Jaicks, President

**Arby's Inc.**
One Piedmont Center
Atlanta, GA 30305
Tele: (404) 262-2729
Jefferson McMahon, President

**Archway Cookies, Inc.**
5451 W. Dickman
Battle Creek, MI 49015
Tele: (616) 962-6205
G.J. Markham, President

**Armour and Company**
Greyhound Tower
Phoenix, AZ 85077
Tele: (602) 248-4000
John W. Teets, President

**Arthur Treacher's Fish & Chips, Inc.**
5501 Tabor Rd.
Philadelphia, PA 19120
Tele: (215) 537-0410
J.M. Gleason, President

**Banquet Foods Corporation**
100 N. Broadway
St. Louis, MO 63102
Tele: (314) 425-0200
F.F. Smiley, President

**Beatrice Foods Company**
2 North La Salle
Chicago, IL 60602
Tele: (312) 782-3820
Donald Eckrich, President

**Bonanza International, Inc.**
1000 Campbell Centre
8350 North Central Exwy.
Dallas, TX 75206
Tele: (214) 987-6400
John Boylan, President

**Borden's, Inc.**
180 E. Broad
Columbus, OH 43215
Tele: (212) 573-4000
Eugene J. Sullivan,
Chairman of Board

**Burger Chef Systems, Inc.**
Box 927
Indianapolis, IN 46206
Tele: (317) 875-8400
T.A. Collins, President

**Burger King Corporation**
Box 520783
Miami, FL 33152
Tele: (305) 596-7011
Louis P. Neeb, Chairman of Board

**CPC International, Inc.**
International Plaza
Englewood Cliffs, NJ 07632
Tele: (201) 894-4000
James R. Eiszner, President

**California Canners and Growers**
3100 Ferry Building
San Francisco, CA 94106
Tele: (415) 981-0101
Robert L. Gibson, President

**Campbell Soup Company**
Box 176
Rancocas, NJ 08073
Tele: (609) 871-6550
Raymond E. Johnston, President

**Carnation Company**
5045 Wilshire Blvd.
Los Angeles, CA 90036
Tele: (213) 931-1911
Dwight L. Stuart, President

**Castle & Cooke, Inc.**
Box 2990
Honolulu, HI 96802
Tele: (808) 548-6611
Donald J. Kirchhoff, President

**Coca Cola Company Foods Div.**
Box 2079
Houston, TX 77001
Tele: (713) 868-8100
Ira C. Herbert, President

**Consolidated Foods Corporation**
135 South La Salle
Chicago, IL 60603
Tele: (312) 726-6414
John J. Cardwell, President

**Cudahy Foods Company**
100 West Clarendon
Phoenix, AZ 85013
Tele: (602) 264-7272
T.T. Day, President

**Dean Foods Company**
3600 N. River Rd.
Franklin Park, IL 60131
Tele: (312) 625-6200
H.M. Dean, Jr., President

**Del Monte Corporation**
Box 3575
San Francisco, CA 94105
Tele: (415) 442-4000
Richard G. Landis, President

**First National Supermarkets, Inc.**
17000 Rockside Rd.
Maple Heights, OH 44137
Tele: (216) 587-7100
Richard J. Bogomolny, President

**Frito-Lay, Inc.**
Frito Lay Tower
Dallas, TX 75235
Tele: (214) 351-7000
Wayne Calloway, President

**General Foods Corporation**
250 North
White Plains, NY 10625
Tele: (914) 683-2500
R. Barzelay, President

**General Host Corporation**
22 Gate House Rd.
Stamford, CT 06902
Tele: (203) 357-9900
Harris J. Ashton, President

**General Mills, Inc.**
9200 Wayzata Blvd.
Minneapolis, MN 55426
Tele: (612) 540-2311
H.B. Atwater, Jr., President

**Gerber Products Company**
445 State St.
Fremont, MI 49412
Tele: (616) 928-2000
Carl G. Smith, President

**Green Giant Company**
608 Second Ave.
Minneapolis, MN 55402
Tele: (612) 330-4966
Winston Wallin, President

**H.J. Heinz Company**
1062 Progress
Pittsburgh, PA 15212
Tele: (412) 237-5757
A.J.F. O'Reilly, President

**George A. Hormel and Company**
501 16th Ave., N.E.
Austin, MN 55912
Tele: (507) 437-5611
Richard Knowlton, President

**Hardee's Food Systems, Inc.**
Box 1619
Rocky Mount, NC 27801
Tele: (919) 977-2000
Jack Laughery, President

**Hunt-Wesson Foods, Inc.**
1645 W. Valencia Dr.
Fullerton, CA 92634
Tele: (714) 871-2100
Frederick B. Rentschler, President

**International Dairy Queen, Inc.**
Box 35286
Minneapolis, MN 55437
Tele: (612) 830-0200
Harris Cooper, President

**ITT Continental Baking Co., Inc.**
Box 7547
Charlottesville, VA 22906
Tele: (804) 971-9100
Robert W. Bracken, President

**Keebler Company**
1 Hollow Tree Ln.
Elmhurst, IL 60126
Tele: (312) 833-2900
T.T. Garvin, President

**Kellogg Company**
235 Porter St.
Battle Creek, MI 49016
Tele: (616) 966-2000
William E. LaMothe, President

**Kentucky Fried Chicken Corp.**
Box 32070
Louisville, KY 40232
Tele: (502) 459-8600
William A. Ready, President

**Kraft, Inc.**
Kraft Court
Glenview, IL 60025
Tele: (312) 998-2000
A.W. Woelfle, President

**The Kroger Company**
1014 Vine
Cincinnati, OH 45201
Tele: (513) 762-4000
Lyle Everingham, President

**Lawry's Foods, Inc.**
570 W. Ave. 26
Los Angeles, CA 90065
Tele: (213) 225-2491
Richard N. Frank, President

**Libby, McNeill & Libby**
200 South Michigan Ave.
Chicago, IL 60604
Tele: (312) 341-4111
I.W. Murray, President

**McDonald's Corporation**
McDonalds Plaza
Oak Brook, IL 60521
Tele: (312) 887-3200
Edward H. Schmitt, President

**Mr. Steak, Inc.**
5100 Race Court
Denver, CO 80216
Tele: (303) 292-3070
James C. Shearon, President

**Nabisco, Inc.**
East Hanover, NJ 07936
Tele: (201) 884-0500
Val B. Diehl, President

**National Tea Company**
9701 West Higgins Rd.
Rosemont, IL 60018
Tele: (312) 693-5100
V. Schulz, President

**The Nestle Company, Inc.**
100 Bloomingdale Rd.
White Plains, NY 10605
Tele: (914) 682-6000
David E. Guerrant,
Chairman of Board

**Oscar Mayer and Company**
Box 7188
Madison, WI 53707
Tele: (608) 241-3311
Jerry M. Hiegel, President

**Pet Incorporated**
Box 392
St. Louis, MO 63166
Tele: (314) 621-5400
Boyd F. Schenk, President

**Pillsbury Company**
608 Second Ave. South
Minneapolis, MN 55402
Tele: (612) 330-4966
Winston R. Wallin, President

**Pizza Hut, Inc.**
9111 E. Douglas
Wichita, KS 67207
Tele: (316) 681-9000
Arthur Gunther, President

**Ralston Purina Company**
Checkerboard Square
St. Louis, MO 63188
Tele: (314) 982-1000
William Stiritz, President

**Red Barn System**
550 Midtown Tower
Rochester, NY 14604
Tele: (615) 325-7032
Jim Lucas, President

**Sambo's Restaurants, Inc.**
6400 Cindy Lane
Carpinteria, CA 93013
Tele: (805) 687-1969
Daniel Shaughnessy, President

**Shakeys, Inc.**
3600 First Int'l. Bldg.
Dallas, TX 75270
Tele: (214) 741-5801
Louis A. Cappello, President

**Standard Brands, Inc.**
625 Madison Ave.
New York, NY 10022
Tele: (212) 759-4400
Martin Emmett, President

**Swift and Company**
115 West Jackson Blvd.
Chicago, IL 60604
Tele: (312) 431-2000
Joseph P. Sullivan, President

**Welch Foods, Inc.**
Westfield, NY 14787
Tele: (716) 326-3131
T.C. Whitney, President

**Winn Dixie Stores, Inc.**
Box B
Jacksonville, FL 32203
Tele: (904) 783-5000
B.L. Thomas, President

# ADDITIONAL MATERIALS YOU CAN OBTAIN

From Center for Science in the Public Interest, 1755 S Street, NW, Washington, D.C., 20009:
**Eater's Digest: The Consumer's Factbook of Food Additives,** by Michael F. Jacobson, published by Anchor Doubleday, 260 pages, $4.00.
**Jack Spratt's Legacy: The Science and Politics of Fat & Cholesterol,** by Patricia Houseman, Richard Marek Publishers, 288 pages, $6.95.
**Nutrition Scoreboard: Your Guide to Better Eating,** by Michael F. Jacobson, Avon Books, 213 pages, $3.00.
**Free Catalog** of additional books, posters, teaching aids, and T-shirts.

From Institute for Food and Development Policy, 2588 Mission Street, San Francisco, CA 94110. Please include 15% ($1.00 minimum) for shipping:
**Food First: Beyond the Myth of Scarcity,** by Francis Moore Lappe and Joseph Collins (with Cary Fowler), Ballantine Books, $3.95.
**What Can We Do? Food and Hunger: How You Can Make a Difference,** by William Valentine and Francis Moore Lappe, Ballantine Books, $2.95.
**Diet for a Small Planet,** by Frances Moore Lappe, Ballantine Books, $2.75.

From Sierra Club Books, P.O. Box 3886, Rincon Annex, San Francisco, CA 94119 (please include $1.75 for shipping):
**The Politics of Cancer,** by Samuel Epstein, MD, 1978, $12.50.
**Who's Poisoning America,** edited by Ralph Nader, Ronald Brownstein, and John Richard, 1981, $12.95.

Also available:
**Feeding the Few: Corporate Control of Food,** by Susan George, available from the Institute for Policy Studies, 1901 Q St., NW, Washington, DC 20009, $1.75 plus $1.00 for shipping.
**Healthy People: The Surgeon General's Report on Health Promotion and Disease Prevention,** #017-001-00416-2, 1979; available from Superintendent of Documents, U.S. Government Printing Office, Washington, D.C. 20402, $5.50.

# FREE INFORMATION SOURCES

From Office of Consumer Affairs, USDA, Room 16-85, 5600 Fishers Lane, Rockville, MD 20857:
**Caffeine and Pregnancy:** contains list of caffeine-containing foods and drugs and the amounts each contains.

From Publications Center, OGPA, USDA, Washington, DC 20250:
**How to Buy Economically.** 1981, 27 pp.
**How to Buy Canned and Frozen Vegetables.** (G-167) 1977. 24 pp.*
**How to Buy Cheese.** (G-167) 1977. 24 pp.*
**How to Buy Poultry.** (G-157) 1977. 8 pp. leaflet*
* *Also available in Spanish.*

From Consumer Information Center, Pueblo, CO 81009:
**Consumer Information Catalogue** (lists more than 250 federal government publications available).

# FOOD ACTION GROUPS

While not all of the groups listed below are concerned exclusively with food-related issues, they all are active in at least one of the areas of food safety discussed in this book.

**Center for Science in the Public Interest (CSPI),** 1755 S Street NW, Washington, DC 20005.
**Clean Water Action Project (CWAP),** 1341 G Street NW, Washington, DC 20005.
**Community Nutrition Institute (CNI),** 1146 19th St., NW, Washington, DC 20036.
**Environmental Defense Fund (EDF),** 1525 18th St. NW, Washington, DC 20036.
**Food Research and Action Center (FRAC),** 1319 F St., NW, Washington, DC 20004.
**Health Research Group (HRG),** 2000 P St. NW No. 700, Washington, DC 20036.
**Institute for Food and Development Policy,** 2588 Mission St., San Francisco, CA 94110.
**Natural Resources Defense Council,** 122 E. 42nd St., New York, NY 10168.
**Rodale Press,** 33 E. Minor St., Emmaus, PA 18049.
**Consumers Union,** 1535 Mission St., San Francisco, CA 94103.

## Publications from the Center for Study of Responsive Law

*These books, some of them already classics in their field, were written to inform, to involve, and to mobilize. They are designed to change one or more injustices and deficiencies through citizen action and awareness. A higher quality of daily democracy requires a higher quality of daily citizenship. These works can help guide the way toward that delightful commitment.*

—Ralph Nader

**For the People,** by Joanne Manning Anderson, with introduction by Ralph Nader (Addison-Wesley, 1977) — a consumer handbook for community action. 379 pp. ($5.95)

**The Madness Establishment,** by Franklin D. Chu & Sharland Trotter (Grossman Pub., NY, 1974) — an analysis of federal government programs on mental health centers. 232 pp. ($8.95 hard)

**Congress Project Profile Kit** — includes two profiles of past Members of Congress with 21-page bibliographic guideline for students to profile their current Member of Congresss. ($2.50)

**Company State,** by James Phelan and Robert Pozen, with introduction by Ralph Nader (Grossman Pub., NY, 1973) — a study group report on Dupont in Delaware, 464 pp. ($3.95)

**Hucksters in the Classroom,** by Sheila Harty (CSRL, 1979) — a review of industry propaganda in schools, 190 pp. with illustrations. ($10.00)

**Politics of Land,** by Robert C. Fellmeth, with introduction by Ralph Nader (Grossman Pub., NY, 1973) — a study group report on land use in California, 715 pp. ($5.95)

**How to Appraise and Improve Your Daily Newspaper,** by David Bollier, with introduction by Ralph Nader (Disability Rights Center, 1980) — a manual for readers. ($5.00)

**Banding Together: How Check-Offs Will Revolutionize the Consumer Movement,** by Andrew Sharpless and Sarah Gallup, 1981. ($5.00)

**Disposable Consumer Items: The Overlooked Mercury Pollution Problem,** by John Abbots, 1981. ($5.00)

**Energy Conservation: A Campus Guidebook,** by Kevin O'Brien and David Corn, 1981. ($5.00)

*Order from:* **Center For Study of Responsive Law, P.O. Box 19367, Washington, DC 20036.**